Proceedings of the International Conference on Linear Statistical Inference LINSTAT '93

Mathematics and Its Applications

Volume 306

Proceedings of the International Conference on Linear Statistical Inference LINSTAT '93

edited by

T. Caliński

and

R. Kala

Department of Mathematical and Statistical Methods,
Agricultural University of Poznań,
Poznań, Poland

Production Editor:
I. Siatkowski

SPRINGER SCIENCE+BUSINESS MEDIA, B.V.

A C.I.P. Catalogue record for this book is available from the Library of Congress

ISBN 978-0-7923-3136-0 ISBN 978-94-011-1004-4 (eBook)

DOI 10.1007/978-94-011-1004-4

Printed on acid-free paper

Originally published by Kluwer Academic Publishers in 1994

CONTENTS

Estimation of Variance Components

Nonlinear Generalizations

Design and Analysis of Experiments

Miscellanea

PREFACE

The International Conference on Linear Statistical Inference LINSTAT'93 was held in Poznań, Poland, from May 31 to June 4, 1993. The purpose of the conference was to enable scientists, from various countries, engaged in the diverse areas of statistical sciences and practice to meet together and exchange views and results related to the current research on linear statistical inference in its broadest sense. Thus, the conference programme included sessions on estimation, prediction and testing in linear models, on robustness of some relevant statistical methods, on estimation of variance components appearing in linear models, on certain generalizations to nonlinear models, on design and analysis of experiments, including optimality and comparison of linear experiments, and on some other topics related to linear statistical inference. Within the various sessions 22 invited papers and 37 contributed papers were presented, 12 of them as posters. The conference gathered 94 participants from eighteen countries of Europe, North America and Asia. There were 53 participants from abroad and 41 from Poland.

The conference was the second of this type, devoted to linear statistical inference. The first was held in Poznań in June, 4-8, 1984. Both belong to the series of conferences on mathematical statistics and probability theory organized under the auspices of the Committee of Mathematics of the Polish Academy of Sciences, due to the initiative and efforts of its Mathematical Statistics Section. In the years 1973-1993 there were held in Poland nineteen such conferences, some of them international. The last one, LINSTAT'93, was the seventh organized as international.

According to the opinion of many participants, expressed during and after the conference, it seems that the conference can be regarded as quite successful in bringing new ideas and results related to different areas linked with linear statistical inference. This was possible due to all participants who in various ways devoted their time and efforts to make the conference fruitfull and enjoyable. First and foremost, thanks should go to all the speakers and authors, the chairmen of sessions, and the discussants. Their excellent jobs were highly appreciated by the conference attendants. A smooth organizational run of the conference is to be attributed to the efforts of the local organizing committee, and particularly to its chairman Dr Idzi Siatkowski, who also took care of the technical preparation of the present proceedings. Without his hard and skilful work, and his stimulating enthusiasm, neither the conference nor the proceedings could be materialized.

The conference was financially supported by the Ministry of National Education of Poland, jointly with the Polish Scientific Research Council and the Polish Academy of Sciences, by both Universities and by others whose names are listed on page viii. All deserve our warmest thanks for their invaluable support of the conference and its rich social programme. We are particularly indebted to the Voivode of the Poznań Province for inviting all paricipants to the concert of the Poznań Chamber Choir given for them in the historical Town Hall.

The volume contains more than half of the 59 papers presented at the conference. This collection represents main features and results of LINSTAT'93. We are happy to express thanks to all authors who decided to prepare their papers for publication in our proceedings. The standard of the papers owes a great deal to the fruitful mutual co-operation between the authors, the referees and the editors. The help of the numerous referees (see the list at the end of the volume) has been very much appreciated not only by us but also by many authors. We would like to apologize to those who submitted their papers and, perhaps, found the referee criticisms too restrictive.

The editorial preparation of the volume would not be possible without the administrative and technical help of the Department of Mathematical and Statistical Methods, Agricultural University of Poznań. The generous support offered by its head and staff is highly appreciated. Last, but not least, let us also express thanks to the staff of the Kluwer Academic Publishers whose encouragement and advice have been very helpful.

Poznań,
March 1994

Tadeusz Caliński and Radosław Kala

ORGANIZERS

The conference was organized by

- The Committee of Mathematics of the Polish Academy of Sciences, Section of Mathematical Statistics
- Adam Mickiewicz University of Poznań, Institute of Mathematics
- Agricultural University of Poznań, Department of Mathematical and Statistical Methods

HONORARY COMMITTEE

B. Ceranka - Head of the Department of Mathematical and Statistical Methods of the Agricultural University of Poznań
Z. Ciesielski - Chairman of the Section of Mathematics of the Polish Scientific Research Council
J. Fedorowski - Rector of the Adam Mickiewicz University of Poznań
R. Ganowicz - Rector of the Agricultural University of Poznań
M. Krzyśko - Chairman of the Section of Mathematical Statistics of the Committee of Mathematics of the Polish Academy of Sciences
W. Łęcki - Voivode of the Poznań Province
J. Musielak - Director of the Institute of Mathematics of the Adam Mickiewicz University of Poznań

SCIENTIFIC PROGRAMME COMMITTEE

T. Caliński (chairman, Poznań)
S. Gnot (Wrocław)
R. Kala (Poznań)
C. Stępniak (Lublin)
R. Zmyślony (Wrocław)

LOCAL ORGANIZING COMMITTEE

B. Bogacka
H. Chudzik
J. Hauke
H. Kiełczewska
J. Krzyszkowska
K. Moliński
I. Siatkowski (chairman)

from the Department of Mathematical and Statistical Methods, Agricultural University of Poznań

The conference was supported by

The Ministry of National Education of Poland

The Polish Scientific Research Council

The Committee of Mathematics of the Polish Academy of Sciences

The Adam Mickiewicz University of Poznań

The Agricultural University of Poznań

The Voivode of the Poznań Province

Powszechny Bank Gospodarczy S.A. Łódź, I Oddział Poznań

The Polish Airlines LOT

SCIENTIFIC PROGRAM OF THE CONFERENCE

Opening Addresses:

R. Ganowicz - *Rector of the Agricultural University of Poznań*
W. Łęcki - *Voivode of the Poznań Province*
Z. Ciesielski - *Chairman of the Section of Mathematics of the Polish Scientific Research Council*

SESSION I.A
Chair: B. Ceranka (Poznań, Poland)

Invited opening lectures:

L. C. A. Corsten - *Agricultural University of Wageningen, The Netherlands*
Spatial interpolation by linear prediction
S. Audrain and R. Tomassone - *Institut de Recherches Internationales Servier and Institut National Agronomique Paris-Grignon, France*
Prediction domain in nonlinear models

SESSION I.B
Chair: C.W. Dunnett (Hamilton, Canada)

Invited papers:

G. J. S. Ross - *Rothamsted Experimental Station, U.K.*
The geometry of non-linear inference: how parameter transformations allow linear methods to be used with minimum distortion
A. Pázman - *Comenius University, Bratislava, Slovakia*
The geometry of nonlinear inference: accounting of prior and boundaries

Contributed paper:

C. M. Cuadras, C. Arenas, J. Fortiana and J. Llopis - *Universitat de Barcelona, Spain*
Predicitve dimensionality and other aspects on a distance based model for prediction with mixed variables

SESSION I.C
Chair: P. Westfall (Lubbock, U.S.A.)

Contributed papers:

J. Volaufová - *Slovak Academy of Sciences, Bratislava, Slovakia*
Estimation of parameters in a special type of random effects model
M. Lejeune - *ENSAE et ESCP, Paris, France*
A generic look at factor analyses

C. P. Kitsos - *Athens University of Business and Economics, Greece*
Multiple-multivariate-sequential T^2-comparisons
D. Szynal and A. Krajka - *UMCS, Lublin, Poland*
On Q-covariance and its applications
T. Müller-Gronbach - *Freie Universität Berlin, Germany*
Optimal designs for approximating the path of a stochastic process

SESSION II.A
Chair: G.J.S. Ross (Rothamsted, U.K.)

Invited papers:

R. A. Bailey - *Goldsmiths' College, University of London, U.K.*
General balance: artificial theory or practical relevance ?
J. M. Azaïs - *Université Paul Sabatier, Toulouse, France*
Design of experiments and neighbour models

Contributed paper:

C. Stępniak - *Agricultural University of Lublin, Poland*
Comparing linear experiments

SESSION II.B
Chair: Bimal K. Sinha (Baltimore, U.S.A.)

Invited papers:

Bikas K. Sinha - *Indian Statistical Institute, Calcutta, India*
Choosing between two experiments: a Bayesian risk analysis
S. Ghosh - *University of California, Riverside, U.S.A.*
Measuring the influence of runs in response surface designs

Contributed paper:

B. Heiligers - *RWTH Aachen, Germany*
E-optimal designs in weighted polynomial regression

SESSION II.C
Chair: A. Pázman (Bratislava, Slovakia)

Invited papers:

H. Drygas and H. Läuter - *Universität Gesamthochschule Kassel and Universität Potsdam, Germany*
On the representation of the linear minimax estimator in the convex linear model
G. Trenkler - *Universität Dortmund, Germany*
Characterizations of oblique and orthogonal projectors

Contributed paper:

R. Kala - *Agricultural University of Poznań, Poland*
On the perpendicularity

SESSION II.D
Chair: R.A. Bailey (London, U.K.)

Contributed papers:

S. Kageyama and A. Das - *Hiroshima University, Japan, and Indian Statistical Institute, Calcutta, India*
On generalized binary proper efficiency-balanced block designs
H. Monod and A. Kobilinsky - *INRA-Versailles, France*
Efficient juxtaposition of fractional designs using the complex linear model
B. Bogacka and S. Mejza - *Agricultural University of Poznań, Poland*
Optimality of generally balanced experimental block designs
J. Hauke and A. Markiewicz - *Agricultural University of Poznań, Poland*
Notes on Shaked and Tong theorem on comparison of experiments

SESSION III.A
Chair: R. Zieliński (Warsaw, Poland)

Invited papers:

C. W. Dunnett - *McMaster University, Hamilton, Canada*
Recent results in multiple testing between several treatments and a specified treatment
T. Mathew - *University of Maryland, U.S.A.*
Some hypotheses tests using generalized p-values in linear models

Contributed paper:

A. Kornacki - *Agricultural University of Lublin, Poland*
Validity of invariant linearly sufficient statistics in the general Gauss-Markov model

SESSION III.B
Chair: H. Nyquist (Umeå, Sweden)

Invited papers:

C. Domański - *University of Łódź, Poland*
Tests of the homoscedasticity for the linear trend
V. L. Girko - *Kiev University, Ukraine*
The linear models with interior noises

Contributed paper:

J. T. Mexia - *Universidade Nova de Lisboa, Portugal*
F tests and checking of assumptions in subnormal models

SESSION IV.A
Chair: H. Drygas (Kassel, Germany)

Invited papers:

K. Nordström - *University of Helsinki, Finland*
Model-robustness in linear models

R. Zieliński - *Institute of Mathematics, Polish Academy of Sciences, Warsaw, Poland*
Robust statistical procedures in linear models

Contributed paper:

Ch. Müller - *Freie Universität Berlin, Germany*
Optimal bias bounds for robust estimation in linear models

SESSION IV.B
Chair: G. Trenkler (Dortmund, Germany)
Contributed papers:

D. Rasch and A. Tuchscherer - *Agricultural University of Wageningen, The Netherlands, and FBN, Dummerstorf, Germany*
The robustness of statistical methods against nonnormality
M. L. Bougeard and C. Michelot - *Observatoire de Paris and Université de Paris, France*
Geometric characterization of Huber's M estimators
N. Benda - *Freie Universität Berlin, Germany*
Robust estimation and design in the linear model

SESSION IV.C
Chair: T. Mathew (Baltimore, U.S.A.)

Invited paper:

H. Nyquist - *University of Umeå, Sweden*
On diagnosing collinearity - influential points in linear regression

Contributed papers:

D. von Rosen - *Uppsala University, Sweden*
PLS, linear models and invariant spaces
H. Knautz, G. Trenkler and S. Gnot - *Universität Dortmund, Germany, and Agricultural University of Wrocław, Poland*
Using nonnegative minimum biased quadratic estimation for variable selection in the linear regression model

SESSION IV.D
Chair: Bikas K. Sinha (Calcutta, India)

Contributed papers:

M. Krzysztofiak - *University of Gdańsk, Poland*
Nonclassical linear regression
V. Slivinskas and V. Šimonytė - *Lithuanian Academy of Sciences, Vilnius, Lithuania*
Cramer-Rao bound for the estimates of frequencies and damping factors of superimposed quasipolynomials in noise
J. M. ten Vregelaar - *Agricultural University of Wageningen, The Netherlands*
Parameter estimation for a specific errors-in-variables model

General discussion:

R. A. Bailey, S. Ghosh, L. R. LaMotte, T. Mathew, Bimal K. Sinha, P. Westfall - main contributors

SESSION V.A
Chair: D. von Rosen (Uppsala, Sweden)

Invited papers:

P. Westfall - *Texas Tech University, U.S.A.*
Efficiency and nonnormality in mixed ANOVA models
T. Mathew, A. Niyogi and Bimal K. Sinha - *University of Maryland, U.S.A.*
Improved nonegative estimation of variance components in balanced multivariate mixed models

Contributed paper:

S. Gnot, D. Stemann and G. Trenkler - *Agricultural University of Wrocław, Poland, and Universität Dortmund, Germany*
Nonnegative admissible invariant quadratic estimation in linear models with two variance components

SESSION V.B
Chair: M. Krzyśko (Poznań, Poland)

Invited paper:

T. Bednarski - *Institute of Mathematics, Polish Academy of Sciences, Wrocław, Poland*
On robust estimation of variance components

Contributed paper:

S. Zontek - *Institute of Mathematics, Polish Academy of Sciences, Wrocław, Poland*
On asymptotic normality of some invariant quadratic admissible estimators of variance components

Invited closing lecture:

L. R. LaMotte - *Louisiana State University, U.S.A.*
Geometrical relations among variance component estimators

POSTER SESSION

A. Boratyńska and R. Zieliński - *University of Warsaw and Institute of Mathematics, Polish Academy of Sciences, Warsaw, Poland*
Bayes robustness via the Kolmogorov metric
B. Ceranka and K. Katulska - *Agricultural University of Poznań and Adam Mickiewicz University of Poznań, Poland*
Relations between optimum biased spring balance weighing designs and optimum chemical balance weighing designs with non-homogeneity of the variances of errors
V. Guiard - *FBN, Dummerstorf, Germany*
About the multimodality of the likelihood function when estimating the variance components in a one-way classification by means of the ML or REML method

T. Jelenkowska - *Agricultural University of Lublin, Poland*
Bayesian estimation of the intraclass correlation coefficients in standard random model
K. Kłaczyński, A. Molińska and K. Moliński - *Agricultural University of Poznań, Poland*
A new view on estimating parameters in a mixed linear model
M. Krzyśko and W. Wołyński - *Adam Mickiewicz University of Poznań, Poland*
Statistical group classification rules for the multivariate Student's t-distribution
G. Nürnberg and V. Guiard - *FBN, Dummerstorf, Germany*
MSE-comparison of ANOVA- and ML-estimators in dependence on skewness and kurtosis of the underlying distributions
Z. Piasta - *Kielce University of Technology, Poland*
Methods of determining compromise solutions in multicriterion optimization problems with regression objective functions
I. Siatkowski - *Agricultural University of Poznań, Poland*
The efficiency factor of balanced two-way elimination of heterogeneity designs
A. Tuchscherer and G. Herrendörfer - *FBN, Dummerstorf, Germany*
Mean estimation in the random one-way model and its robustness
A. Zieliński - *Warsaw Agricultural University, Poland*
Box-Cox transformation in some mixed linear models
R. Zmyślony and S. Zontek - *Institute of Mathematics, Polish Academy of Sciences, Wrocław, Poland*
Comparison of experiments for robust estimation of parameters in block designs under mixed model

INCREMENTS FOR (CO)KRIGING WITH TREND AND PSEUDO-COVARIANCE ESTIMATION

L.C.A. CORSTEN
Department of Mathematics
Agricultural University
Dreijenlaan 4
6703 HA Wageningen
The Netherlands

Abstract. This paper presents an integrated account of the interpolation techniques of universal kriging and cokriging by way of the solution of a prediction problem in a situation akin to linear regression. Polynomial expectations of degree k in the coordinates of a region are eliminated by consideration of increments only, followed by the introduction of a covariance structure by polynomial pseudo-covariance functions. Then the best linear unbiased predictor is derived as well as the variance of the ensuing prediction error, both interpretable by some analogy to the linear regression situation, but essentially different because of the non-uniqueness of the pseudo-covariance functions and the absence of unconditional positivity of the pseudo-covariance matrices involved. Estimation of essential coefficients of the pseudo-covariance functions is accomplished by the restricted maximum likelihood method which may be helpful in deciding about the value of k as well.

Key words: Spatial interpolation, Kriging, Cokriging, Predictor, Prediction error, Linear regression, Increments, Pseudo-covariance function, Maximum likelihood method.

1. Introduction

Consider first a spatial prediction problem concerning a main or target variable only. Let $y(z)$ be a locally varying property of practical interest, like the soil content of a specific mineral, rain intensity in a certain period, depth of ground water table, surface altitude, density of a specific plant species in a vegetation, soil fertility measured by yield in a uniformity trial, amount of pollution in the air or in soil at a certain time, etc., to be measured at a location z. On the basis of measurements $y_1, \ldots, y_n$ at n locations $z_1, \ldots, z_n$, respectively, it is required to guess or to predict the value y_0 of y at any unvisited point z_0. With the availability of such a prediction procedure, based on assumptions to be specified, one may repeatedly apply it, e.g. at the nodes of a fine-meshed grid. Next the resulting predictions may be visualized by different grey tones or colours, by contour lines or by a three-dimensional picture for a two-dimensional phenomenon, in short, by a map.

2. Specification and Assumptions

Assume that at each location z the value of y is the realization of a random variable $\underline{y}$ (the symbol for a random variable or a random vector will be underlined). Each $\underline{y}$ has provisionally expectation $\mu(z)$ which is a polynomial of degree k in the (orthonormal)

T. Caliński and R. Kala (eds.),
Proceedings of the International Conference on Linear Statistical Inference LINSTAT '93, 1–11.

coordinates of z. E.g., with $k = 1$ in $\mathbb{R}^2$ there are three terms or regressors: $1, \xi_1, \xi_2$, with corresponding coefficient vector $\beta = (\beta_0, \beta_1, \beta_2)'$. Similarly, with $k = 2$ in $\mathbb{R}^2$ there are six regressors $1, \xi_1, \xi_2, \xi_1^2, \xi_2^2, \xi_1\xi_2$, with regression coefficient vector $\beta = (\beta_0, \beta_1, \beta_2, \beta_{11}, \beta_{22}, \beta_{12})'$. The vector β will be unknown.

We aim at linear prediction of $\underline{y_0}$ by $\underline{t} = \lambda_1 \underline{y_1} + \ldots + \lambda_n \underline{y_n}$, the inner product $\lambda' \underline{y}$ of vectors λ and $\underline{y}$ such that it is unbiased, i.e. having prediction error $\underline{t} - \underline{y_0}$ with expectation zero, and among the unbiased linear predictors the best one, that is with minimum variance of prediction error, called Best Linear Unbiased Predictor (BLUP). From here we proceed to assumptions concerning the covariance structure of $\underline{y}$. This procedure will be motivated by the requirement of unbiasedness of $\underline{t}$. Further, we recognize the fact that estimation of the covariance structure on the basis of observation residuals with respect to estimated expectations, possibly followed by some iterations, may be considerably biased, even for the simple case of one common but unknown expectation for all possible observations. See, e.g., Matheron (1971, p. 196, Exercises 1 and 2) and Cressie (1987, pp. 427–428).

The previous assumption about expectations can be expressed as $E\underline{y} = X\beta$, where $\underline{y}$ is now a column of n observations (n-vector), β is the p-vector of regression coefficients, and X is an $n \times p$ matrix of full column rank whose i-th row ($i = 1, \ldots, n$) consists of the p appropriate monomials, including mixed ones, of the coordinates of location z_i (at least the one of power zero). Let the monomials corresponding to the unvisited point z_0 be given in the p-vector x_0. Then unbiasedness of $\underline{t} = \lambda' \underline{y}$ requires that $E(\underline{t} - \underline{y_0}) = E(\lambda' \underline{y} - \underline{y_0}) = \lambda' X \beta - x_0' \beta$ vanishes for all β, which is equivalent to

$$-x_0' + \lambda' X = 0.$$

In other words, the coefficient vector $(-1, \lambda')$ of the $n+1$ random variables $\underline{y_0}, \underline{y_1}, \ldots, \underline{y_n}$ in the prediction error should be orthogonal to each column of the $(n+1) \times p$ matrix consisting of X extended with an additional first row x_0'.

Each linear combination of possible observations y_1 up to y_n whose coefficient vector λ is orthogonal to the space spanned by the p columns of the matrix X of monomials up to degree k will be called an increment of order $k + 1$. In particular, the coefficient vector of first order increments will be orthogonal to the vector 1_n (consisting of n ones) only, and therefore these are the same as contrasts among y_1 up to y_n. An essential property of all increments of order $k+1$ is that their expectation vanishes. By restricting attention to such increments only, the knowledge about regression coefficients will become immaterial: it implies the formation of equivalence classes among vectors y to the extent that all vectors which are equal except for a polynomial of degree k or less belong to the same equivalence class. In other words, by considering only increments of order $k+1$ any expectation of degree k or less will have been filtered out. Thus the complications of estimating regression coefficients β can be avoided. In the sequel, assumptions concerning expectations and covariance structure will only involve increments of a certain order. It should be noted that the prediction error we are aiming at must be as well an increment of order $k + 1$, now for $y_0, y_1, y_2, \ldots, y_n$, X being extended with an additional first row x_0', and with specific coefficient -1 for y_0.

In our second assumption any increment $\lambda' \underline{y}$ of order $k + 1$ with expectation zero has variance $\text{Var}(\lambda' \underline{y}) = E(\lambda' \underline{y})^2$ which will be $\lambda' G \lambda$, where $G = (g_{ij})$, and g_{ij}

is called the generalized covariance or pseudo-covariance between $\underline{y_i}$ and $\underline{y_j}$. The variance $\lambda' G \lambda$ must be positive for all permissible λ (except the null vector). There is sign indifference with respect to λ other than the permissible ones; e.g., $\lambda' G \lambda$ for $\lambda = (1, 0, \ldots, 0)$ may be negative, and Var($\underline{y_i}$) may even be non-existent as we shall see in a moment. Further, G will not be unique, although $\lambda' G \lambda$ will be unique for permissible λ. These facts explain the use of the terms pseudo- or generalized covariance. Similarly, the covariance between permissible increments $\lambda_1' \underline{y}$ and $\lambda_2' \underline{y}$ will be equal to $\lambda_1' G \lambda_2$.

The third assumption is that the pseudo-covariance between $\underline{y_i}$ and $\underline{y_j}$ is dependent on $h = z_i - z_j$, and that h affects g_{ij} only by its length $|h| = r$, so-called isotropy. The pseudo-covariance will thus be a function of r to be denoted as $g(r)$. We recall that any ordinary covariance function $c(r)$ under isotropy has to satisfy the condition of positive-definiteness, i.e., $\lambda' G \lambda > 0$ for all λ, except $\lambda = 0$, and that this condition can be checked by Fourier transformation of $c(r)$ to a spectral density. Similarly, the positive definiteness of pseudo-covariance functions for permissible λ's, belonging to a subspace only, can be checked by Fourier transformation due to an extension of Bochner's theorem, within the framework of generalized function theory. See Gelfand and Shilov (1964), and Gelfand and Vilenkin (1964) for details.

Two examples for $g(r)$ if $k = 0$ are: (1) $A[1 - \exp(-r/a)]$ with $A < 0$ and $a > 0$, and (2) Ar^d with $A < 0$ and $0 < d < 2$. Compare Christakos (1984). The simplest contrast is $\underline{y_i} - \underline{y_j}$ with variance $g_{ii} - 2g_{ij} + g_{jj} = 2[g(0) - g(r)]$. Due to the observation that $\lambda' G \lambda$ is invariant against addition of a common constant α to all g_{ij}, i.e. $\lambda' G \lambda$ is equivalent to $\lambda'(G + \alpha 1_n 1_n')\lambda$, one may freely choose $g(0)$ in Var($\underline{y_i} - \underline{y_j}$). With the particular choice $g(0) = 0$ one obtains Var($\underline{y_i} - \underline{y_j}$)$/2 = -g(r)$, the so-called semivariogram, by definition zero for $r = 0$, and useful for exploration purposes concerning the covariance structure if $k = 0$, i.e. under the absence of trend. If for $k = 0$ the variance for individual observations $c(0)$ exists, then this $-g(r)$ will be equal to $c(0) - c(r)$, i.e., the covariance function and the semivariogram sum to variance $c(0)$ for all n. In the second example with $A = -1$ and $d = 1$ we have $g(r) = -r$, and due to $c(\infty) = 0$ the variance at a point, which should be $-g(\infty)$, does not exist.

For $k \geq 0$, polynomials to degree $2k + 1$ are sufficiently rich to describe $g(r)$ according to Matheron (1973), and have the attractive property to be linear in the parameters. The dependence of the degree of the polynomials $g(r)$ on k is partially explainable by the fact that the increasingly imposed restrictions on increments by increasing k are counterbalanced by increasing flexibility for pseudo-covariance functions. On the other hand, the degree of $g(r)$ can be at most $2k + 1$ in order that increments of order $k + 1$ be without trend, a property shown by Matheron to be equivalent to $g(r)/r^{2k+2}$ tending to zero as r tends to infinity. The polynomials of degree $2k + 1$ are easily shown (Corsten, 1989) to be invariant against addition of any term of even degree $\leq 2k$. So we have for $\mathbb{R}^2$ and $k = 2$ in the simplest form $g(r) = \alpha_1 r + \alpha_2 r^3 + \alpha_3 r^5$, however only with one or two lowest degree terms for $k = 0$ and $k = 1$, respectively, together with restrictions supplied by Fourier transformation: $\alpha_1 \leq 0, \alpha_3 \leq 0, \alpha_2 \geq -10(\alpha_1 \alpha_3)^{1/2}/3$. For $\mathbb{R}^1$ and $\mathbb{R}^3$, conditions are somewhat different, but not of interest for the sequel, as we restrict attention to spatial phenomena in $\mathbb{R}^2$.

To each $g(r)$ of the previous examples with zero limit from the right one may add a discontinuity by the term $\alpha_0\delta$, where $\alpha_0 \geq 0$ while $\delta = 1$ at $r = 0$, and $\delta = 0$ elsewhere. This term introduces an additional component at zero distance only, representing measurement error variance, sometimes called a nugget effect, that is approximately the same as relatively large variance of point pair differences at a very small distance.

3. Best Linear Unbiased Prediction

Returning to the subject of BLUP, we remember that it is required to find a coefficient vector ϕ for an increment of order $k+1$ pertaining to $\underline{y} = (\underline{y_0}, \underline{y_1}, \ldots, \underline{y_n})'$ of the form $(-1, \lambda')'$ such that this increment will have minimum variance. Consider any increment coefficient vector δ of the form $(0, \delta_1, \ldots, \delta_n)'$, again orthogonal to the columns of $(x_0, X')'$; the expectation of such an increment involving only $\underline{y_1}, \ldots, \underline{y_n}$ will vanish and therefore is said to induce a predictor of zero. The set of all such δ is a vector subspace M of dimension $n - p$. The set of all vectors inducing an unbiased prediction error is the shifted subspace $M + \phi$, where ϕ is any particular vector of that set. The difference between any pair in this collection belongs to M. In order to find the element ϵ in $M + \phi$ which induces minimum variance, we use the equivalence with vanishing covariance between $\epsilon'\underline{y}$ and all $\delta'\underline{y}$ with $\delta \in M$. See the Appendix for a proof of this equivalence.

So we combine the orthogonality of $\epsilon = (-1, \lambda')'$ to the columns of $(x_0, X')'$, i.e.

$$-x_0' + \lambda' X = 0 \tag{1}$$

with

$$\epsilon' G_{n+1} \delta = 0, \tag{2}$$

where

$$G_{n+1} = \begin{pmatrix} g_{00} & g_0' \\ g_0 & G \end{pmatrix},$$

is the pseudo-covariance matrix of $\underline{y}$. Since the first element of δ vanishes, the first column of G_{n+1} may be omitted in (2) and we are left with an $(n+1) \times n$ matrix of which the lower $n \times n$ submatrix is the pseudocovariance matrix G of the observations and the upper row g_0' contains the pseudocovariances of $\underline{y_0}$ with the observations. So (2) is equivalent to $\epsilon'(g_0, G)'\delta_{(0)} = 0$, where $\delta_{(0)}$ is the n-vector remaining after the removal of the first element 0 from δ. This in turn is equivalent to the orthogonality (with respect to ordinary metric) of $(g_0, G)\epsilon$ to all $\delta_{(0)}$ orthogonal to the columns of X. Hence, $(g_0, G)\epsilon$ must be a vector in the space spanned by the columns of X, and therefore equals $-X\tau$ for some τ. So (2) is equivalent to

$$G\lambda + X\tau = g_0. \tag{3}$$

By premultiplication of (3) with $X'G^{-1}$ and elimination of λ by inserting (1), we obtain

$$\tau = (X'G^{-1}X)^{-1}(X'G^{-1}g_0 - x_0), \tag{4}$$

and (3) yields

$$\lambda = G^{-1}g_0 + G^{-1}X(X'G^{-1}X)^{-1}(x_0 - X'G^{-1}g_0). \tag{5}$$

The required predictor $\underline{t} = \lambda'\underline{y}$ for $\underline{y_0}$ will be

$$\underline{t} = g_0'G^{-1}\underline{y} + x_0'(X'G^{-1}X)^{-1}X'G^{-1}\underline{y} - g_0'G^{-1}X(X'G^{-1}X)^{-1}X'G^{-1}\underline{y}.$$

Although we dismissed the idea of estimating β, by considering increments only, let a pseudo-generalized least squares (pseudo-GLS) estimator of the non-identifiable expectation $x_0'\beta$ be denoted purely formally as $x_0'\underline{\hat{\beta}}$ with

$$\underline{\hat{\beta}} = (X'G^{-1}X)^{-1}X'G^{-1}\underline{y}. \tag{6}$$

Then $\underline{t} = \lambda'\underline{y}$ can be written as

$$\underline{t} = x_0'\underline{\hat{\beta}} + g_0'G^{-1}(\underline{y} - X\underline{\hat{\beta}}). \tag{7}$$

There is a striking resemblance between this expression and the BLUP of $\underline{y_0}$ if the matrix G_{n+1} would have been the unique positive definite covariance matrix of the vector $\underline{y}$ in the classical linear prediction problem for $\underline{y_0}$. The first term looks like an estimated expectation in z_0, but is not unique due to the non-uniqueness of G and thus of $\underline{\hat{\beta}}$; similarly, the second term looks like the best linear approximation of the disturbance in z_0 on the basis of n estimated disturbances by residuals contained in the vector $\underline{y} - X\underline{\hat{\beta}}$, but again is not unique, although the sum of the two terms in (7) is unique. It should be stressed that in general $\underline{\hat{\beta}}$ is not the best linear unbiased estimator of β, since G_{n+1} will not be an unconditionally positive definite covariance matrix. The indeterminateness of regression coefficient estimators, residuals and covariance matrices renders the present procedure essentially different from the classical one which almost belongs to statistical folklore, in particular among animal breeders. Compare, e.g., Henderson (1963) and more recently Kackar and Harville (1984). Although expectations (and their estimates) as well as residuals are not uniquely defined by increments, this interpretation helps to understand and to remember the structure of the derived predictor.

If $z_0 = z_i$, an observation point, then g_0 in (7) is to be replaced with the i-th column of G and $g_0'G^{-1}(\underline{y} - X\underline{\hat{\beta}})$ is the 'residual' at z_i; so the predictor regenerates the observation in z_i, and $\underline{t}$ is called an exact interpolator. If $g(r)$ is continuous in r, also in $r = 0$, the interpolation surface will be continuous and passing through the observation points. If there is a nugget effect, observations will appear as discrete deviations from a surface, smooth everywhere else, but obviously different from a surface that fits expectations optimally in any sense.

The prediction error variance $\mathrm{Var}(\underline{t} - \underline{y_0})$ will be $\epsilon' G_{n+1}\epsilon$ with $\epsilon' = (-1, \lambda')$, or $g_{00} - g_0'\lambda - \lambda' g_0 + \lambda' G\lambda$, where g_{00} is the pseudo-variance at z_0. Inserting (3) for g_0 into the third term yields $g_{00} - g_0'\lambda - \lambda' X\tau$, and replacing λ with $G^{-1}g_0 - G^{-1}X\tau$, again according to (3), leads to $g_{00} - g_0'G^{-1}g_0 + \tau' X' G^{-1}X\tau$. Inserting (4) for τ gives finally

$$\mathrm{Var}(\underline{t} - \underline{y_0}) = (g_{00} - g_0'G^{-1}g_0) + x_a'(X'G^{-1}X)^{-1}x_a, \tag{8}$$

where $x_a' = x_0' - g_0' G^{-1} X$.

The first part of (8) in brackets resembles the residual variance of the (here not well-defined) disturbance at z_0 with respect to the best linear approximation by those at $z_1, \ldots, z_n$, while the second part looks like the variance of an expectation estimator uncorrelated with the former disturbance residual, however not at the point z_0 with regressor row x_0', but at a point with an adjusted regressor row x_a'; the adjustment of x_0' to x_a' is a linear combination of the rows of X with the vector of approximation coefficients for the realised 'residuals' in predictor (7) as the coefficient vector. This is again a mnemonic device for seeming comprehension of (8). Though the two terms are dependent on the infinitely many choices for G_{n+1}, their sum remains unique. The actual choice of G_{n+1} by the free coefficients of even powers of r in $g(r)$ will often be motivated by computational considerations.

In addition to a map of the predicted or interpolated values one may produce as well a map of the associated standard deviations of prediction error, revealing the locations where additional observations could improve the accuracy of the map.

This concludes an exposition on the subject of universal kriging or optimal prediction, as invented by D. Krige (1951), and given a mathematical basis by G. Matheron (1971) in an admirable but not very accessible fashion. More explanations are in Delfiner (1976), Journel and Huijbregts (1978), and Ripley (1981). The book by Cressie (1991) is a reference work for specialists in spatial analysis. The present exposition is the completion of considerations aimed at embedding the subject in a linear regression type of context, as started in Corsten (1989) and Stein and Corsten (1991).

4. Estimation of the Pseudo-Covariance Function

Since $g(r)$ in its simplest form without even powers of r in most practical cases is unknown, it must be estimated, that is the essential coefficients of the polynomial form for $g(r)$ must be estimated before the kriging procedure can be applied. For this purpose we have recourse to the maximum likelihood method under the additional assumption that the possible observations, or rather the possible increments of order $k+1$, have a Gaussian distribution. Using only increments, independent of regression coefficients, once more in order to avoid serious bias, leads to the restricted maximum likelihood method, due to Patterson and Thompson (1975). This procedure is also one of the three methods of analysis compared in Corsten and Stein (1994).

A basis for all observable increments is obtained by applying the orthogonal projection $I - P_D$ to $\underline{y}$, where P_D represents the orthogonal projection $X(X'X)^{-1}X'$ on the p-dimensional space D spanned by the columns of X, and next omitting in a sensible manner p rows from $(I - P_D)\underline{y}$ resulting into $\underline{u} = C\underline{y}$, where C is an $(n-p) \times n$ matrix of rank $n - p$. Obviously $E\underline{u} = 0$ and $\text{Cov}(\underline{u}) = W = CGC'$ of order and rank $n - p$. Since $g(r) = \alpha_0\delta(r) + \alpha_1 r + \alpha_2 r^3 + \alpha_3 r^5$, G is of the form

$$\sum_i \alpha_i G_i = G_\alpha,$$

each element of $G_i (i \geq 1)$ being a power of the distance between the two observation points concerned, and G_0 being equal to I_{n-p}; further, W equals $W_\alpha = \sum \alpha_i W_i$, where $W_i = CG_iC'$ for $i = 0, 1, 2, 3$.

Minimization of $-\ln L = \frac{1}{2}\ln\det W_\alpha + \frac{1}{2}u'W_\alpha^{-1}u$ with respect to the elements $\alpha_0, \alpha_1, \alpha_2, \alpha_3$ of the coefficient vector α leads to the likelihood equations $g(\alpha) = 0$, where

$$[g(\alpha)]_i = \frac{1}{2}\mathrm{tr}(W_\alpha^{-1}W_i) - \frac{1}{2}u'W_\alpha^{-1}W_iW_\alpha^{-1}u \quad (i = 0, 1, 2, 3).$$

Solution of these equations is equivalent to equating each $u'W_\alpha^{-1}W_iW_\alpha^{-1}u$ to its expectation, $\mathrm{tr}(W_\alpha^{-1}W_i)$.

One may start the solution of α with a preliminary estimate $\alpha_{(0)}$ of α, obtained, e.g., by ordinary least squares regression of the $n - p$ squared elements of u on the (at most) four corresponding regressors provided by diagonal elements of W_0, W_1, W_2, W_3. If such a preliminary estimate does not satisfy the conditions stated in Section 2, one chooses the appropriate boundary instead and recalculates the other elements of $\alpha_{(0)}$; the same strategy will be followed in the sequel of the minimization process.

The Gauss–Newton procedure, extended with a line search method and consisting of the iteration steps

$$\alpha_{(j+1)} = \alpha_{(j)} - \rho_j \left[\frac{\partial g(\alpha)}{\partial \alpha}\right]_{\alpha=\alpha_{(j)}} g(\alpha_{(j)}) \quad (j = 0, 1, 2, \ldots),$$

where $0 \leq \rho_j \leq 1$ is chosen as the first non-negative integer power of $1/2$ such that $-\ln L(\alpha_{(j+1)}) < -\ln L(\alpha_{(j)})$, can be simplified. To this end the matrix $\partial g(\alpha)/\partial\alpha$ will (according to Fisher's scoring method) be replaced by its expectation with typical element $E[g(\alpha)]_i[g(\alpha)]_{i'}$. The present information matrix (if α is the maximum likelihood solution) has been proved by Kitanidis (1983) to be equal to $\frac{1}{2}F(\alpha)$ with typical element

$$\frac{1}{2}\mathrm{tr}(W_\alpha^{-1}W_iW_\alpha^{-1}W_{i'}).$$

Thus, we arrive at the iteration procedure

$$\alpha_{(j+1)} = \alpha_{(j)} - \rho_j \left[\frac{1}{2}F(\alpha_{(j)})\right]^{-1} g(\alpha_{(j)}). \tag{9}$$

This usually rapidly convergent process will generate the required solution $\underline{\alpha}$ as well as its asymptotically valid covariance matrix $[\frac{1}{2}F(\alpha)]^{-1}$. The most appropriate value of k according to Akaike's information criterion minimizes

$$\min_\alpha[-2\ln L(\alpha)] + 2(k + 2),$$

$k + 2$ being the number of estimated parameters.

5. Use of an Auxiliary Variable

Consider the situation that, in addition to the target variable y_1, a second, auxiliary variable y_2 has been observed, with most of those observations at different locations but some of which may coincide (but not necessarily) with the locations of the previous measurements. Now the value y_0 of the first variable at an unvisited point z_0 should be predicted as a linear unbiased minimum variance function of main and auxiliary variable observations together. Use of a relatively large amount of relatively cheap measurements of the auxiliary variable may hopefully lead to a smaller prediction variance by cokriging than from kriging the target variable only. See, e.g., the three papers by Stein et al. (1991). Cokriging will be presented as a rather simple extension of the kriging procedure described in the preceding sections.

Let $\underline{y}$ consist of the vector $(\underline{y}_1', \underline{y}_2')'$ where $\underline{y_1}$ contains n_1 possible observations of the primary variable, and $\underline{y_2}$ contains n_2 possible observations of the secondary variable. Let $n_1 + n_2 = n$. Under similar assumptions concerning polynomial expectations of degree k as before, we have $E\underline{y} = X\beta$, where

$$X = \begin{pmatrix} X_1 & 0 \\ 0 & X_2 \end{pmatrix}$$

is a partitioned matrix with n rows and $2p$ columns, X_1 and X_2 being similar to X in the previous situation, and β being equal to $(\beta_1', \beta_2')'$ where β_1 contains the 'regression coefficients' of the target variable on the location monomials, and β_2 those of the auxiliary variable on the location monomials. The unvisited point z_0, where the prediction of the target variable is required, will appear as a regressor row starting with the p-vector x_{10}' followed by p zeroes, and will be denoted as x_0'. Hence the expectation of $\underline{y_1}$ at z_0 will be

$$x_{10}'\beta_1 = x_0'\beta. \tag{10}$$

Again we introduce increments $\lambda' y$ of order $k+1$, with vanishing expectations due to the orthogonality of λ to the columns of X, possibly extended with x_0' as first row, while their variance is governed by a pseudo-covariance matrix G_{n+1} concerning $(y_0, y')'$. This has the partitioned pattern

$$\begin{pmatrix} g_{00} & g_{01}' & g_{02}' \\ g_{01} & G_{11} & G_{12} \\ g_{02} & G_{21} & G_{22} \end{pmatrix} \text{ or, in abridged form, } \begin{pmatrix} g_{00} & g_0' \\ g_0 & G \end{pmatrix}.$$

Here g_{00} is the pseudo-variance of the target variable $\underline{y_{10}}$ at z_0, G_{11} the pseudo-covariance matrix of the elements of $\underline{y_1}$ in $\underline{y}$, G_{22} that of the elements of $\underline{y_2}$ in $\underline{y}$, G_{12} the pseudo-covariance matrix of the elements of $\underline{y_1}$ with those of $\underline{y_2}$, each of which is a so-called pseudo-cross-covariance, G_{21} is the transpose of G_{12}, g_{01} a vector of pseudo-covariances of $\underline{y_{10}}$ with the elements of $\underline{y_1}$, and g_{02} a vector of pseudo-cross-covariances of $\underline{y_{10}}$ with the elements of $\underline{y_2}$. Due to isotropy there is a pseudo-covariance function $g_u(r)$ for the target variable, determining g_{00}, g_{01} and G_{11}, a pseudo-covariance function $g_v(r)$ for the auxiliary variable, determining G_{22}, and a pseudo-cross-covariance function $g_{uv}(r)$ determining g_{02} and G_{12}. All three of them are polynomials of degree at most $2k+1$, possibly extended with a nugget effect. In order that both $g_u(r)$ and $g_v(r)$ are positive-definite for $\mathbb{R}^2$, the two sets of coefficients, $\alpha_{u0}, \alpha_{u1}, \alpha_{u2}, \alpha_{u3}$ and $\alpha_{v0}, \alpha_{v1}, \alpha_{v2}, \alpha_{v3}$, respectively, must obey similar restrictions as before, that is at most 8 restrictions together for $k = 2$, and less for lower k. Necessary and sufficient conditions for G or G_{n+1} to be positive-definite for $\mathbb{R}^2$ are not yet known, but additional sufficient conditions, to be obeyed by the coefficients $\alpha_{uv0}, \alpha_{uv1}, \alpha_{uv2}, \alpha_{uv3}$ for $g_{uv}(r)$, wherein α_{uv0} belongs to nugget effect occurring only at points where both target and auxiliary variable will be observed or considered, are the following six for $k = 2$ (three of which are for $k = 1$, and two for $k = 0$):

$$\alpha_{u0}\alpha_{v0} - \alpha_{uv0}^2 \geq 0, \quad \alpha_{ui}\alpha_{vi} - \alpha_{uvi}^2 \geq 0 \quad (i = 1, 3),$$

$$\alpha_{u1}\alpha_{v2} - 2\alpha_{uv1}\alpha_{uv2} + \alpha_{u2}\alpha_{v1} \leq 0, \quad \alpha_{u2}\alpha_{v3} - 2\alpha_{uv2}\alpha_{uv3} + \alpha_{u3}\alpha_{v2} \leq 0,$$

$$9(\alpha_{u2}\alpha_{v2} - \alpha_{uv2}^2) + 25(\alpha_{u1}\alpha_{v3} - 2\alpha_{uv1}\alpha_{uv3} + \alpha_{u3}\alpha_{v1}) \geq 0.$$

These have been derived from the required positivity of the determinant of the Fourier transform $f(\omega)$ of the 2×2 matrix $(g_{uv}(r))$ at page 710 of the paper by Stein, van Eijnsbergen and Barendregt (1991); they are different from the non-sufficient conditions (10) at page 711 of the same paper.

The predictor of the target variable at z_0 whose prediction error has vanishing expectation and minimal variance turns out to be in similarity with (7),

$$\underline{t} = x_0'\underline{\hat{\beta}} + g_0'G^{-1}(\underline{y} - X\underline{\hat{\beta}}),$$

where now $\underline{\hat{\beta}} = (\hat{\beta}_1', \hat{\beta}_2')' = (X'G^{-1}X)^{-1}X'G^{-1}\underline{y}$, the 'GLS estimator' of the joint 'regression' coefficient vector $(\beta_1', \beta_2')'$. Although $\hat{\beta}_2$ is irrelevant for the first term of $\underline{t}$ due to (10), it does occur in the second term as an indispensable ingredient for the 'residuals' of both target and auxiliary variable observations. Of course, $\hat{\beta}_1$ will be different in general from $\hat{\beta}$ obtained by kriging the target variable only. Similarly, the prediction error variance will be as in (8), where the last p elements of $x_a' = x_0' - g_0'G^{-1}X$ will not vanish in general.

It can be shown that cokriging has no effect compared to kriging if the locations where the target variable and the auxiliary variable have been measured coincide, and the coefficient vectors of the three pseudo-covariance functions $g_u(r), g_v(r)$ and $g_{uv}(r)$ are proportional.

Estimation of the coefficients of $g_u(r), g_v(r)$ and $g_{uv}(r)$ will follow similar lines as before under the normality assumption of increments. The matrix G is a linear combination of at most four known matrices of order n if $k \leq 2$. Now $u = Cy$ is obtained by subtraction of the orthogonal projection $P_D y$ from y, and omitting twice p elements, one set of p connected with the perpendicular on X_1-space and one with that on X_2-space. Again, a preliminary estimate of the coefficients of $g_u(r)$ and $g_v(r)$ can be obtained by ordinary least squares regression of the $n - 2p$ squared elements of u on the corresponding diagonal elements of the $W_i = CG_iC'$ matrices. For a preliminary estimate of the coefficients of $g_{uv}(r)$ one may use in a similar manner least squares regression. Each of the $\min(n_1 - p, n_2 - p)$ regressands will be the product of an observed increment of the target variable from u with an increment of the auxiliary variable, each of these increments being used at most once. The regressors are provided by the appropriate elements of the W_i matrices. The extended Gauss–Newton procedure (9) will yield the required corrections.

The generalization to more than one auxiliary variable will be analogous.

6. Epilogue

In this paper an integrated account of kriging and of cokriging has been reached by embedding them into regression-like situations, but with special dependence structure of observations and predictands. Particular features are in the first place that variances of individual observations may be non-existent, but that only variances of increments with vanishing expectations do exist, rendering the estimation of observation expectations immaterial. Secondly, no use has been made of Lagrange multipliers in minimizing prediction error variance, but rather elementary and straightforward procedures, guaranteeing not only a stationary value, but a minimum indeed. Thirdly, Lagrange multipliers do not appear in the predictors, nor in the better understandable prediction error variances in contrast to the writings of Matheron and his followers. Finally, the estimation of pseudo-covariance function coefficients, independently from estimation of unknown regression coefficients for observation expectations, involving quadratic forms of observed increments to be equated to their expectations, seems to be quite natural, even without the assumption about increments to be Gaussian.

The subject at hand is somewhat related to stochastic processes and to time series analysis. Points of agreement with time series are the appearance of covariance functions or their generalizations in the form of pseudo-covariance functions, as well as the prediction problem. The agreement with stochastic processes lies in the implicit assumption of second order stationarity at least in the absence of trend, i.e. one common expectation for all possible observations.

An essential difference with time series is that there is no specific time direction. Even if the prediction problem is posed in $\mathbb{R}^1$, with possible measurements along a transect, there is no distinction here between moving forward or backward. For $\mathbb{R}^1$ as well as phenomena in $\mathbb{R}^2$ or $\mathbb{R}^3$ one loses second order stationarity as soon as expectations are not constant in the area considered. This is remedied by stationarity of increments of a certain order in kriging as well as in cokriging. In addition, the assumption of isotropy appeared in the two subjects at hand, leading to the possibility of introducing (pseudo)-covariance functions in an extension of shift invariance in stochastic processes. The whole treatment of the subjects seems to be more down-to-earth than that of time series and stochastic processes. On the other hand, an assumption like that of isotropy requires some kind of verification on the basis of data. Such techniques of graphical type are available; see Journel and Huybregts (1978).

Appendix

In connection with the step from the first paragraph of Section 3 to the second one, it is proved that ϵ in $M+\phi$ inducing minimum variance is equivalent to vanishing covariance between $\epsilon'\underline{y}$ and all $\delta'\underline{y}$ with $\delta \in M$.

1. Let $\epsilon \in M+\phi$ and let $\text{Cov}(\epsilon'\underline{y},\ \delta'\underline{y}) = 0$ for all $\delta \in M$. Then with $\phi_0 \in M+\phi$ we have $\text{Var}(\phi_0'\underline{y}) = \text{Var}[(\phi_0-\epsilon)'\underline{y} + \epsilon'\underline{y}] = \text{Var}[(\phi_0-\epsilon)'\underline{y}] + 2\text{Cov}[(\phi_0-\epsilon)'\underline{y}, \epsilon'\underline{y}] + \text{Var}(\epsilon'\underline{y})$, where the second term vanishes since $\phi_0 - \epsilon \in M$. Hence $\text{Var}(\phi_0'\underline{y}) \geq \text{Var}(\epsilon'\underline{y})$ for all $\phi_0 \in M+\phi$. Thus ϵ induces minimum variance.

2. Let ϵ and ϕ_0 both belong to $M+\phi$, while $\text{Var}(\phi_0'\underline{y}) \geq \text{Var}(\epsilon'\underline{y})$ for all ϕ_0.

Choose an arbitrary element $\delta \in M$ and consider the perpendicular η (with respect to pseudo-covariance metric) from ϵ on the space spanned by δ, i.e. $\eta = \epsilon - [\mathrm{Cov}(\epsilon'\underline{y}, \delta'\underline{y})/\mathrm{Var}(\delta'\underline{y})]\delta$. Since $\eta \in M+\phi$ (as well as ϵ) we have $\mathrm{Var}(\eta'\underline{y}) \geq \mathrm{Var}(\epsilon'\underline{y})$. By simple reduction it follows that $\mathrm{Var}(\eta'\underline{y}) = \mathrm{Var}(\epsilon'\underline{y}) - \mathrm{Cov}^2(\epsilon'\underline{y}, \delta'\underline{y})/\mathrm{Var}(\delta'\underline{y})$. Hence $\mathrm{Cov}(\epsilon'\underline{y}, \delta'\underline{y}) = 0$, which holds for all $\delta \in M$.

References

Christakos, G. (1984). On the problem of permissible covariance and variogram models. *Water Resources Research* **20**, 251–265.

Corsten, L.C.A. (1989). Interpolation and optimal linear prediction. *Statistica Neerlandica* **43**, 69–84.

Corsten, L.C.A. and Stein, A. (1994). Nested sampling for estimating spatial variograms compared to other designs. *Applied Stochastic Models and Data Analysis*. To appear.

Cressie, N. (1987). A nonparametric view of generalized covariances for kriging. *Journal of the International Association for Mathematical Geology* **19**, 425–453.

Cressie, N. (1991). *Statistics for Spatial Data*. Wiley, New York.

Delfiner, P. (1976). Linear estimation of nonstationary spatial phenomena. In: M. Guarascio et al., Eds., *Advanced Geostatistics in the Mining Industry*. Reidel, Dordrecht, The Netherlands, 49–68.

Gelfand, L.M. and Shilov, G.E. (1964). *Generalized Functions Vol. 1*. Academic Press, New York.

Gelfand, L.M. and Vilenkin, N. (1964). *Generalized Functions Vol. 4*. Academic Press, New York.

Henderson, C.R. (1963). Selection index and expected genetic advance. In: W.D. Hanson and H.F. Robinson, Eds., *Statistical genetics and Plant Breeding*. National Research Council Publication 982, National Academy of Sciences, Washington, D.C., 141–163.

Journel, A.G. and Huijbregts, C.J. (1978). *Mining Geostatistics*. Academic Press, New York.

Kackar, R.N. and Harville, D.A. (1984). Approximations for standard errors of estimators of fixed and random effects in mixed linear models. *Journal of the American Statistical Association* **79**, 853–862.

Kitanidis, P.K. (1983). Statistical estimation of polynomial generalized covariance functions and hydrologic applications. *Water Resources Research* **19**, 901–921.

Krige, D.G. (1951). A statistical approach to some mine valuation problems on the Witwatersrand. *Journal of the Chemical Metallurgical and Mining Society of South Africa* **52**, 119–138.

Matheron, G. (1971). *The Theory of Regionalized Variables and its Applications*. Les Cahiers du Centre de Morphologie Mathématique de Fontainebleau, No. 5, Ecole Nationale Supérieure des Mines de Paris.

Matheron, G. (1973). The intrinsic random functions and their applications. *Advances in Applied Probability* **5**, 439–468.

Patterson, H.D. and Thompson, R. (1975). Maximum likelihood estimation of components of variance. In: L.C.A. Corsten and T. Postelnicu, Eds., *Proceedings of the 8th International Biometric Conference*. Editura Academiei Republicii Socialiste România, Bucureşti, 197–207.

Ripley, B.D. (1981). *Spatial Statistics*. Wiley, New York.

Stein, A. and Corsten, L.C.A. (1991). Universal kriging and cokriging as a regression procedure. *Biometrics* **47**, 575–587.

Stein, A., van Eijnsbergen, A.C. and Barendregt, L.G. (1991). Cokriging nonstationary data. *Mathematical Geology* **23**, 703–719.

Stein, A., Staritsky, I.G., Bouma, J., van Eijnsbergen, A.C. and Bregt, A.K. (1991). Simulation of moisture deficits and areal interpolation by universal cokriging. *Water Resources Research* **27**, 1963–1973.

ON THE REPRESENTATION OF THE MINIMAX LINEAR ESTIMATOR IN THE CONVEX LINEAR MODEL

HILMAR DRYGAS
Universität Gesamthochschule Kassel
Fachbereich 17, Mathematik-Informatik
D-34109 Kassel
Germany

and

HENNING LÄUTER
Universität Potsdam
Fachbereich Mathematik, WIP
Am Neuen Palais 10, D-14469 Potsdam
Germany

Abstract. The linear model under ellipsoidal constraints on the expectation parameter is considered and the problem of incorporating the additional information into the process of estimating the parameters is reexamined. After reviewing the spectral and the Bayesian approaches to minimax linear estimation, the Läuter–Hoffman representation of the minimax linear estimator of the parameters is investigated. Some complementary results on the topic are also given.

Key words: Linear model under ellipsoidal constraints, Spectral minimax linear estimation, Bayesian minimax linear estimation, Läuter–Hoffman minimax linear estimator.

1. Introduction

We consider the linear model (written in matrix notation with the relevant dimensions indicated underneath)

$$\underset{n\times 1}{y} = \underset{n\times k}{X}\ \underset{k\times 1}{\beta} + \underset{n\times 1}{\varepsilon}, \quad E\varepsilon = 0, \quad E\varepsilon\varepsilon' = \sigma^2 \underset{n\times n, p.d.}{V} \tag{1}$$

under the ellipsoidal constraints

$$\beta \in \mathcal{B} = \{\beta : (\beta - \beta_0)'T(\beta - \beta_0) \leq a\}, \tag{2}$$

where $a > 0$ and T is a positive definite (p.d.) $k \times k$ matrix. By transformation and reparameterization ($y \to \sigma^{-1}V^{-\frac{1}{2}}(y - X\beta_0)$, $X \to \sigma^{-1}V^{-\frac{1}{2}}X(a^{+\frac{1}{2}}T^{-\frac{1}{2}})$, $\beta \to a^{-\frac{1}{2}}T^{\frac{1}{2}}(\beta-\beta_0)$) the problem reduces to the case $\sigma^2 V = I$, $a = 1$, $\beta_0 = 0$ and $T = I$, i.e., the unit ball. In the sequel we will assume this simplification for our further analysis.

The problem is to incorporate the additional information on β given by (2) into the process of estimating the parameter β. Kuks (1972), Olman (1983) as well as Kuks and Olman (1971, 1972) have proposed to apply the minimax-principle in its linear form in order to estimate β, i.e., according to them, one should minimize

T. Caliński and R. Kala (eds.),
Proceedings of the International Conference on Linear Statistical Inference LINSTAT '93, 13–26.

$$\psi(C,d) = \sup_{\beta'\beta\leq 1} E(\| B(Cy+d-\beta) \|^2), \tag{3}$$

subject to $C \in \mathbb{R}_{k\times n}$ and $d \in \mathbb{R}^k$. B denotes a given $m \times k$ matrix of rank m. The matrix $B'B$ is called the loss-matrix. A complete solution for this problem in the case $m = 1$ was already given by Kuks and Olman. In the case $a = 1$, it can be expressed by the ridge estimator

$$B\overset{\wedge}{\beta} = B[(I+X'X)^{-1}X'(y - X\beta_0) + \beta_0], \tag{4}$$

i.e., using $C = (I+X'X)^{-1}X'$ and $d = (I - CX)\beta_0$. For $m \geq 2$ the problem is more difficult and many efforts have been made to cope with this problem. Among them are Läuter (1975), Hoffman (1979), Bunke (1975), Pilz (1986, 1991), Gaffke and Heiligers (1989), Gaffke and Mathar (1990) and others. Many references can be found in the recent book by Pilz (1991).

A new direction in the analysis of the linear minimax-problem was introduced by Girko (1988, p. 72–74). The function to be minimized is

$$\begin{aligned}\psi_0(C) &= \lambda_{\max}\{(CX-I)'B'B(CX-I)\} + \mathrm{tr}(BB'CC') \\ &= \lambda_{\max}\{B(CX-I)(CX-I)'B'\} + \mathrm{tr}(BB'CC'),\end{aligned} \tag{5}$$

since from a theorem of Pilz (1986) it follows easily that $d = 0$ due to the symmetry of $\mathcal{B}$. We will assume that $B'B$ is regular or $k = m$ (see Drygas, 1993). The more general case $m \leq k$ will be dealt with in a subsequent paper. In this case it follows that $\psi_0(C)$ is a strictly convex function and it therefore possesses a unique minimum. Unfortunately, $\psi_0(C)$ in general is not differentiable. $\psi_0(C)$ is differentiable if and only if the maximal eigenvalue $\lambda_{\max}\{B(CX-I)(CX-I)'B'\}$ of $B(CX-I)(CX-I)B'$ is a simple one. In this case we get as a necessary and sufficient optimality condition the Girko equation

$$u_1u_1'B(CX-I)X' + BC = 0, \tag{6}$$

see Drygas (1993), where u_1 is an eigenvector of length one of the bias-matrix $B(CX-I)(CX-I)'B'$ corresponding to the largest eigenvalue of this matrix. It can be proved that then u_1 should be an eigenvector corresponding to an eigenvalue ρ of the matrix

$$B(I+X'X)^{-1}B', \tag{7}$$

obeying the inequality $\rho u_1'(BB')^{-1}u_1 \geq \lambda_{\max}\ (I - u_1u_1')BB'(I - u_1u_1')$. In the general case when the maximal eigenspace of the bias-matrix (to be understood as the eigenspace corresponding to the maximal eigenvalue of $B(CX-I)(CX-I)'B'$ is j-dimensional and possesses an orthonormal basis $\{w_1,\ldots,w_j\}$, then (6) has to be replaced by a continuum of inequalities, namely

$$\begin{aligned}\mathrm{tr}(-B'BC\Theta') \leq{}& \frac{1}{2}\lambda_{\max}\,(P'[B(CX-I)X'\Theta'B' + \\ & B\Theta X(CX-I)'B']P) \quad \text{for any } \Theta \in \mathbb{R}_{k\times n},\end{aligned} \tag{8}$$

where $P = (w_1,\ldots,w_j)$. See Drygas (1991) and Kiefer (1974).

2. The Spectral and the Bayesian Approach to Minimax Linear Estimation

Girko (1990, 1993) proved the following theorem.

Theorem 1. *Necessary for* (8) *to hold is that there is an orthonormal basis* $\{u_1, \ldots, u_j\}$ *of the maximal eigenspace of the bias-matrix and numbers* $p_1, \ldots, p_j$, $p_i \geq 0$, $\sum_{i=1}^{j} p_i = 1$, *such that*

$$(S_1^u) \qquad \left(\sum_{i=1}^{j} p_i u_i u_i'\right) B(CX - I)X' + B\,C = 0\,. \tag{9}$$

From (S_1^u) we can simply derive an alternative equivalent Spectral Equation (S_1^v) as follows. Let $B(CX - I)(CX - I)'B' = \lambda_1^2 \sum_{i=1}^{j} u_i u_i' + \sum_{i=j+1}^{r} \lambda_i^2 u_i u_i'$ be the spectral decomposition of the bias-matrix and let, for $\lambda_i > 0$,

$$v_i = \lambda_i^{-1}(B(CX - I))'u_i\,, \quad \lambda_i = \lambda_1\,, \quad i = 1, \ldots, j\,. \tag{10}$$

Then $v_t' v_s = \delta_{ts}$ (Kronecker delta) and

$$B(CX - I) = \lambda_1 \sum_{i=1}^{j} u_i v_i' + \sum_{i=j+1}^{r} \lambda_i u_i v_i',$$

where $\lambda_1 > \lambda_{j+1} \geq \lambda_{j+2} \geq \ldots \geq \lambda_r > 0$, $\lambda_{r+1} = \ldots = \lambda_k = 0$, is a singular value decomposition of $B(CX - I)$. The spectral equation (S_1^u) is then equivalent to

$$\begin{aligned} 0 &= \lambda_1 \sum_{i=1}^{j} p_i u_i v_i' X' + BC \\ &= \lambda_1 \left(\sum_{i=1}^{j} u_i v_i'\right)\left(\sum_{i=1}^{j} p_i v_i v_i'\right) X' + BC \\ &= B(CX - I)\left(\sum_{i=1}^{j} p_i v_i v_i'\right) X' + BC. \end{aligned} \tag{11}$$

Thus we get (see also Drygas, 1993; Drygas and Pilz, 1993)

Theorem 2. *The spectral equation* (S_1^u) *is equivalent to the spectral equation*

$$(S_1^v) \qquad B(CX - I)\left(\sum_{i=1}^{j} p_i v_i v_i'\right) X' + BC = 0,$$

where $\{v_1, \ldots, v_j\}$ *is an orthonormal basis of the maximal eigenspace of* $(CX - I)'B'B(CX - I)$.

Theorem 3. *The spectral equation* (S_1^u) *is sufficient for* Cy *to be the Minimax Linear Estimator* (MILE).

Proof. We prove that the relation (8) is valid. By assumption (9),

$$-B'BC = B'\left(\sum_{i=1}^{j} p_i u_i u_i'\right) B(CX - I)X' \tag{12}$$

for numbers $p_i \geq 0$, such that $\sum_{i=1}^{j} p_i = 1$, and an orthonormal basis $\{u_1, \ldots, u_j\}$ of the maximal eigenspace of $B(CX - I)(CX - I)'B'$. Thus $u_i = Pz_i$, $||z_i|| = 1$, $z_i \in \mathbb{R}^j$, and

$$\operatorname{tr}(-B'BC\Theta') = \tfrac{1}{2}\sum_{i=1}^{j} p_i z_i' P'[B(CX - I)X'\Theta'B' + B\Theta X(CX - I)'B']Pz_i. \tag{13}$$

Since $||z_i|| = 1$, it follows that

$$\begin{aligned} &z_i'\{P'[B(CX - I)X'\Theta'B' + B\Theta X(CX - I)'B']P\}z_i \\ &\leq \lambda_{\max}\{P'[B(CX - I)X'\Theta'B' + B\Theta X(CX - I)'B']P\}. \end{aligned} \tag{14}$$

By multiplying (14) with $p_i \geq 0$ and summing up over i we get that

$$\operatorname{tr}(-B'BC\Theta') \leq \frac{1}{2}\lambda_{\max}\{P'[B(CX - I)X'\Theta'B' + B\Theta X(CX - I)'B']P\}, \tag{15}$$

i.e., (8). □

An alternative proof of the spectral equation, using convex analysis, was given in Stahlecker and Drygas (1993) and Christopeit and Helmes (1991).

Remark 1. The proof of Theorem 3 shows that the theorem would also be valid if $\sum_{i=1}^{j} p_i \leq 1$. If X is not of full rank then it can happen that $(Xv_i)' = v_i'X' = 0$. Therefore, the term $p_i u_i v_i' X'$ can be omitted and $\{u_i, \ldots, u_j\}$ is only an orthonormal basis of a subspace of the maximal eigenspace. (The same happens if you omit terms for which $p_i = 0$.) Moreover, if $p_i > 0$, then $\sum_{i=1}^{j} p_i < 1$.

Theorem 4. *Let $X'X$ be non-singular. Then the Minimax-problem consists in finding*

(i) *a number $j, 1 \leq j \leq k$, numbers $p_i \geq 0$, $i = 1, \ldots, j$, such that $\sum_{i=1}^{j} p_i = 1$,*
(ii) *an orthonormal basis $\{u_1, \ldots, u_j\}$,*
(iii) *an orthonormal basis $\{v_1, \ldots, v_j\}$ such that*

$$v_i = -\frac{1}{\lambda_1}(I + p_i X'X)^{-1}B'u_i$$

for some number $\lambda_1 > 0$ such that

(iv) $B(I + p_i X'X)^{-1}B'u_i \in \operatorname{span}\{u_1, \ldots, u_j\}$,
(v) $\lambda_1^2 > \lambda_{\max}\left(I - \sum_{i=1}^{j} u_i u_i'\right) BB' \left(I - \sum_{i=1}^{j} u_i u_i'\right)$.

Then the MILE *is given by*

$$
\begin{aligned}
Cy = \hat{\beta} &= B^{-1}\left(\lambda_1 \sum_{i=1}^{j} u_i v_i' + \sum_{i=1}^{j} u_i u_i' B\right) X^+ y \\
&= \lambda_1^2 (B'B)^{-1}\left(X'X \sum_{i=1}^{j} p_i v_i v_i' + I\right)\left(\sum_{i=1}^{j} p_i v_i v_i'\right) X'y . \qquad (16)
\end{aligned}
$$

Proof. We investigate the first expression of the estimator (16). Under the assumption $\mathrm{im}(BC) \subseteq \mathrm{span}\{u_1, \ldots, u_j\}$, we obtain from the singular value decomposition of $B(CX - I)$, in the case $X^+X = I$, that:

$$BCXX' - BX' = \lambda_1 \sum_{i=1}^{j} u_i v_i' X' - \left(I - \sum_{i=1}^{j} u_i u_i'\right) BX', \qquad (17)$$

$$-BC = -\left(\lambda_1 \sum_{i=1}^{j} u_i v_i' + \sum_{i=1}^{j} u_i u_i' B\right) X^+, \qquad (18)$$

and

$$\left(\sum_{i=1}^{j} p_i u_i u_i'\right) B(CX - I)X' = \lambda_1 \sum_{i=1}^{j} p_i u_i v_i' X'. \qquad (19)$$

Since $\mathrm{im}\,(X^+) \subseteq \mathrm{im}\, X'$, it follows, from (11), that (19) equals (18) if and only if

$$\lambda_1 \sum_{i=1}^{j} p_i u_i v_i' X'X = -\lambda_1 \sum_{i=1}^{j} u_i v_i' X^+X - \sum_{i=1}^{j} u_i u_i' BX^+X . \qquad (20)$$

Mulitplying this equation from the left with u_i' we get that

$$\lambda_1((I + p_i X'X)v_i)' X^+X = -u_i' BX^+X \qquad (21)$$

or

$$\lambda_1 X^+X(I + p_i X'X)\left(\frac{1}{\lambda_1}(I + p_i X'X)^{-1} B'u_i + v_i\right) = 0 . \qquad (22)$$

If $X'X$ is regular then $X^+X = I$ and thus

$$v_i = -\frac{1}{\lambda_1}(I + p_i X'X)^{-1} B'u_i, \quad i = 1, \ldots, j . \qquad (23)$$

If the v_i are orthogonal and normed then [assuming that $\mathrm{im}\,(BC) \subseteq \mathrm{span}\{u_1, \ldots, u_j\}$]

$$
\begin{aligned}
F_0 &= B(CX - I)(CX - I)'B' \\
&= \left(\lambda_1 \sum_{i=1}^{j} u_i v_i' - \left(I - \sum_{i=1}^{j} u_i u_i'\right) B\right)\left(\lambda_1 \sum_{i=1}^{j} u_i v_i' - \left(I - \sum_{i=1}^{j} u_i u_i'\right) B\right)' \\
&= \lambda_1^2 \sum_{i=1}^{j} u_i u_i' + \left(I - \sum_{i=1}^{j} u_i u_i'\right) BB' \left(I - \sum_{i=1}^{j} u_i u_i'\right) \\
&\quad -\lambda_1 \left(I - \sum_{i=1}^{j} u_i u_i'\right) B \sum_{i=1}^{j} v_i u_i' - \lambda_1 \sum_{i=1}^{j} u_i v_i' B' \left(I - \sum_{i=1}^{j} u_i u_i'\right) . \qquad (24)
\end{aligned}
$$

In order that $F_0 u_i = \lambda_1^2 u_i$, it is necessary and sufficient that

$$-\lambda_1 \left(I - \sum_{i=1}^{j} u_i u_i' \right) B v_i = 0, \quad i = 1, \ldots, j, \tag{25}$$

i.e., $Bv_i \in \operatorname{span}\{u_1, \ldots, u_j\}$. Moreover, λ_1^2 should be the largest eigenvalue of F_0. Let, therefore, $u_{j+1}, \ldots, u_k$ be the eigenvectors of F_0, which are orthogonal to $u_1, \ldots, u_j$. Then, if $F_0 u_r = \lambda_r^2 u_r, r = j+1, \ldots, k$,

$$\begin{aligned} \lambda_r^2 u_r &= \left(I - \sum_{i=1}^{j} u_i u_i' \right) BB' \left(I - \sum_{i=1}^{j} u_i u_j' \right) u_r - \lambda_1 \sum_{i=1}^{j} u_i v_i' B' u_r \\ &= \left(I - \sum_{i=1}^{j} u_i u_i' \right) BB' \left(I - \sum_{i=1}^{j} u_i u_i' \right) u_r, \end{aligned} \tag{26}$$

in view of $Bv_i \in \operatorname{span}\{u_1, \ldots, u_j\}$, $i = 1, \ldots, j$. This implies that u_r are eigenvectors of the symmetric n.n.d. matrix $(I - \sum_{i=1}^{j} u_i u_i') BB' (I - \sum_{i=1}^{j} u_i u_i') = R_0$. R_0 has the j-fold eigenvalue zero with eigenvectors $u_1, \ldots, u_j$. Indeed from $u_r \perp u_1, \ldots, u_j$ and $R_0 u_r = 0$ it follows that $BB' \left(I - \sum_{i=1}^{j} u_i u_i' \right) u_r = BB' u_r = 0$. But BB' is regular and therefore $u_r = 0$. Thus u_r cannot be an eigenvector of R_0 corresponding to the zero eigenvalue.

From the formula (23) of v_i it follows that

$$u_i = -\lambda_1 B'^{-1} (I + p_i X'X) v_i . \tag{27}$$

Introduce the matrix $U_1 = (u_1, \ldots, u_j)$ and $V_1 = (v_1, \ldots, v_j)$. Then $U_1' U_1 = V_1' V_1 = I_j$ and

$$U_1 = -\lambda_1 B'^{-1} (X'XV_1 P + V_1), \tag{28}$$

where $P = \operatorname{diag}(p_1, \ldots, p_j) = (p_i \delta_{ij})$. Moreover,

$$\begin{aligned} Cy &= B^{-1}(\lambda_1 U_1 V_1' + U_1 U_1' B) X^+ y \\ &= B^{-1}(U_1(\lambda_1 V_1' + U_1' B)) X^+ y \\ &= B^{-1}(U_1(\lambda_1 V_1' - \lambda_1 V_1' - \lambda_1 P V_1' X'X) X^+ y \\ &= \lambda_1^2 (B'B)^{-1} (X'XV_1 P + V_1) P V_1' (X'X) X^+ y \\ &= \lambda_1^2 (B'B)^{-1} (X'XV_1 P V_1' + I) V_1 P V_1' X' y. \end{aligned} \tag{29}$$

Since $V_1 P V_1' = \sum_{i=1}^{j} p_i v_i v_i'$, the last assertion of the theorem follows. □

Theorem 5. *Let $V_1 P V_1' = M_0 = CC'$, $C \in \mathbb{R}_{k \times n}$. Then C obeys the non-linear eigenvalue equation*

$$\lambda_1^2 (CC'X'X + I)(B'B)^{-1}(X'XCC' + I) C = C. \tag{30}$$

Moreover, λ_1^2 and C have to be determined in such a way that

$$\lambda_1^2 (1 + \operatorname{tr}(M_0 X'X M_0)) \tag{31}$$

is minimized subject to $\mathrm{tr}(M_0) = 1$.

Proof. In addition to the spectral theory of minimax estimation, we use the Bayesian theory of experimental design, as devolopped in Pilz (1991). The MILE can be characterized as a Bayesian estimator with the most unfavourable moment-matrix $E(\beta\beta') = M_0$, and yields the formula

$$\begin{aligned} Cy &= M_0(M_0 + (X'X)^{-1})^{-1}(X'X)^{-1}X'y \\ &= (X'X)^{-1}(M_0 + (X'X)^{-1})^{-1}M_0X'y \\ &= (M_0X'X + I)^{-1}M_0X'y. \end{aligned} \tag{32}$$

In Drygas and Pilz (1993) it has been shown that $M_0 = V_1PV_1'$. Since X is of full rank the two representations coincide if and only if

$$\lambda_1^2(B'B)^{-1}(X'XM_0 + I)M_0 = (M_0X'X + I)^{-1}M_0 \tag{33}$$

or, equivalently,

$$\lambda_1^2(M_0X'X + I)(B'B)^{-1}(X'XM_0 + I)M_0 = M_0\,. \tag{34}$$

Since $CC'X' = 0$ if and only if $C'X' = 0$, (34) is indeed equivalent to (30). Pilz concludes from the Minimax theorem and the theory of optimal experimental design that M_0 has to be determined in such a way that

$$f(M_0) = \mathrm{tr}[(X'X)^{-1}B'B] - \mathrm{tr}[(X'X)^{-1}B'B(X'X)^{-1}(M_0 + (X'X)^{-1})^{-1}] \tag{35}$$

has to be maximized. Since $\mathrm{tr}(M_0) = \mathrm{tr}(V_1PV_1') = \mathrm{tr}(P) = \sum_{i=1}^{j} p_i = 1$, the side condition on M_0 is clear. The expression (35) can be rewritten as follows:

$$\begin{aligned} f(M_0) &= \mathrm{tr}[(X'X)^{-1}B'B(I - (X'X)^{-1}(M_0 + (X'X)^{-1})^{-1}] \\ &= \mathrm{tr}[(X'X)^{-1}B'BM_0(M_0 + (X'X)^{-1})^{-1}] \\ &= \mathrm{tr}[(M_0 + (X'X)^{-1})^{-1}(X'X)^{-1}B'BM_0] \\ &= \mathrm{tr}[(X'XM_0 + I)^{-1}B'BM_0] \\ &= \mathrm{tr}[M_0B'B(M_0X'X + I)^{-1}] \\ &= \mathrm{tr}[B'B(M_0X'X + I)^{-1}M_0]. \end{aligned} \tag{36}$$

From equation (33) it follows that

$$B'B(M_0X'X + I)^{-1}M_0 = \lambda_1^2(X'XM_0 + I)M_0 \tag{37}$$

and thus

$$\begin{aligned} f(M_0) &= \lambda_1^2(\mathrm{tr}(M_0) + \mathrm{tr}(M_0X'XM_0)) \\ &= \lambda_1^2(1 + \mathrm{tr}(M_0X'XM_0)). \quad \square \end{aligned} \tag{38}$$

3. The Läuter-Hoffmann Representation of the Minimax Linear Estimator

We now consider the representation of the MILE with

$$M_0 = a_0^{\frac{1}{2}}(F+A)^{\frac{1}{2}} - S^{-1}, \tag{39}$$

$a_0 > 0$, $a_0 \in \mathbb{R}$, $A \in \mathbb{R}_{k\times k}$, $A \geq 0$, $S = X'X$, $F = S^{-1}B'BS^{-1}$. This approach has been followed by Läuter (1975) and Hoffmann (1979). Of course, a_0 and A have to be determined in such a way that M_0 is n.n.d. Let $M_0 = V_1PV_1'$, $P > 0$. Then $a_0^{\frac{1}{2}}(F+A)^{\frac{1}{2}} = V_1PV_1' + S^{-1}$ and

$$\begin{aligned} B(CX-I) &= -BS^{-1}(V_1PV_1' + S^{-1})^{-1} \\ &= -a_0^{-\frac{1}{2}}BS^{-1}(F+A)^{-\frac{1}{2}}, \end{aligned} \tag{40}$$

$$\begin{aligned} R_0 &= (B(CX-I))'B(CX-I) = a_0^{-1}(F+A)^{-\frac{1}{2}}F(F+A)^{-\frac{1}{2}} \\ &= a_0^{-1}I - a_0^{-1}(F+A)^{-\frac{1}{2}}A(F+A)^{-\frac{1}{2}}. \end{aligned} \tag{41}$$

Since $A \geq 0$, it follows that $a_0^{-1} = \lambda_1^2$ is the largest eigenvalue of R_0. Equation (34) can be written in the equivalent form

$$\lambda_1^2(M_0 + S^{-1})F^{-1}(M_0 + S^{-1})M_0 = M_0. \tag{42}$$

This yields for the setup according to (39) that

$$\lambda_1^2 a_0(F+A)^{\frac{1}{2}}F^{-1}(F+A)^{\frac{1}{2}}M_0 = M_0 \tag{43}$$

or

$$(F+A)^{-\frac{1}{2}}F(F+A)^{-\frac{1}{2}}M_0 = M_0 \tag{44}$$

or

$$(F+A)^{-\frac{1}{2}}A(F+A)^{-\frac{1}{2}}M_0 = 0. \tag{45}$$

Since A is n.n.d. this equation is equivalent to

$$A(F+A)^{-\frac{1}{2}}M_0 = a_0^{\frac{1}{2}}A - A(F+A)^{-\frac{1}{2}}S^{-1} = 0 \tag{46}$$

or, equivalently,

$$SA = a_0^{-\frac{1}{2}}(F+A)^{-\frac{1}{2}}A. \tag{47}$$

Another equivalent condition is $PV_1'SA = 0$ or $V_1'SA = 0$ if $P > 0$. The equation (47) has already been derived in Läuter (1975) and Hoffmann (1979). However, still the question arises whether M_0 can be represented in the form (39). This equation means that $(V_1PV_1' + S^{-1})^2 = \lambda_1^{-2}(F+A)$ or

$$\begin{aligned} A &= \lambda_1^2(V_1PV_1' + S^{-1})^2 - F \\ &= \lambda_1^2(S^{-1}(SV_1PV_1' + I)(V_1PV_1'S + I)S^{-1}) - F. \end{aligned} \tag{48}$$

Now, from (28), $-\lambda_1 B'^{-1}SV_1P = U_1 - \lambda_1 B'^{-1}V_1$. Inserting this relation into (48) yields

$$\begin{aligned} A &= S^{-1}B'(U_1V_1' + \lambda_1 B'^{-1}(I - V_1V_1'))(V_1U_1' + \lambda_1(I - V_1V_1')B^{-1})BS^{-1} \\ &- S^{-1}B'BS^{-1}. \end{aligned} \tag{49}$$

Since BS^{-1} is regular, A is n.n.d. if and only if

$$(U_1V_1' + \lambda_1 B'^{-1}(I - V_1V_1'))\,(V_1U_1' + \lambda_1(I - V_1V_1')B^{-1}) \geq I \tag{50}$$

in the sense of the Loewner ordering. We get

$$\begin{aligned} &(U_1V_1' + \lambda_1 B'^{-1}(I - V_1V_1'))\,(V_1U_1' + \lambda_1(I - V_1V_1')B^{-1}) \\ &= U_1U_1' + \lambda_1^2 B'^{-1}(I - V_1V_1')B^{-1} \geq I = U_1U_1' + U_2U_2' \end{aligned} \tag{51}$$

if and only if

$$\lambda_1^2 B'^{-1}V_2V_2'B^{-1} \geq U_2U_2'. \tag{52}$$

From the singular value decomposition $B(CX - I) = \lambda_1 U_1V_1' + U_2\Lambda_2V_2'$, giving $BC = (\lambda_1 U_1V_1' + U_2\Lambda_2V_2' + B)X^+$, and from the spectral equation (9), it follows for $X^+X = I$, i.e. for regular $X'X$, that $\lambda_1 U_1PV_1'X'X + \lambda_1 U_1V_1' + U_2\Lambda_2V_2' + B = 0$. Premultiplying this with U_2' yields $\Lambda_2V_2' = -U_2'B$ or $V_2' = -\Lambda_2^{-1}U_2'B$, $V_2 = -B'U_2\Lambda_2^{-1}$. Thus (52) becomes equivalent to

$$\lambda_1^2 U_2\Lambda_2^{-2}U_2' \geq U_2U_2'. \tag{53}$$

From $\Lambda_2^2 \leq \lambda_1^2 I$ it follows that $\Lambda_2^{-2} \geq \lambda_1^{-2}I$ and hence $\lambda_1^2U_2\Lambda_2^{-2}U_2' \geq U_2U_2'$. Thus $A \geq 0$ is proved. We get the following theorem.

Theorem 6. *The least unfavourable moment matrix* $M_0 = V_1PV_1'$ *of the* MILE *can be represented in the form*

$$M_0 = a_0^{\frac{1}{2}}(F + A)^{\frac{1}{2}} - S^{-1}, \tag{54}$$

where $S = X'X$, $F = S^{-1}B'BS^{-1}$, $a_0^{\frac{1}{2}} = \lambda_1^{-1}$, *and where*

$$A = S^{-1}B'U_2\Lambda_2^{-1}(\lambda_1^2 I - \Lambda_2^2)\Lambda_2^{-1}U_2'BS^{-1} = S^{-1}V_2(\lambda_1^2 I - \Lambda_2^2)V_2'S^{-1} \geq 0$$

if $B(CX - I)(CX - I)B' = \lambda_1^2U_1U_1' + U_1\Lambda_2^2U_1'$, $(CX - I)'B'B(CX - I) = \lambda_1^2V_1V_1' + V_2\Lambda_2^2V_2'$ *and* $\lambda_1^2 I \geq \Lambda_2^2$. *Moreover,* A *obeys the equation*

$$SA = \lambda_1(F + A)^{-\frac{1}{2}}A = \frac{\operatorname{tr}((F + A)^{\frac{1}{2}})}{1 + \operatorname{tr}(S^{-1})}(F + A)^{-\frac{1}{2}}A, \tag{55}$$

which is also equivalent to $V_1'SA = 0$ *if* $M_0 = V_1PV_1'$, $P > 0$, $\operatorname{tr}(P) = 1$.

The last relation in (55) follows from $\operatorname{tr}(M_0) = 1$ and $a_0^{\frac{1}{2}} = \lambda_1^{-1}$. This non-linear equation can perhaps be solved iteratively by $A_{n+1} = \lambda_1 S^{-1}(F + A_n)^{-\frac{1}{2}}A_n$ or by the generalized Newton algorithm.

4. Some Complementary Results

We give now some other relevant results on the topic discussed in this paper.

Theorem 7. *$Cy = \lambda_1^2(B'B)^{-1}(X'XM_0+I)M_0X'y$ is the* MILE *of β if and only if $M_0 = CC' = V_1PV_1', V_1'V_1 = I, \mathrm{tr}(P) = 1, P > 0$ and the following hold:*
(i) $\lambda_1^2(CC'X'X+I)(B'B)^{-1}(X'XCC'+I)C = C$,
(ii) $U_1 = -\lambda_1 B'^{-1}(X'XV_1P+V_1)$,
(iii) $\lambda_1^2 I \geq (I-U_1U_1')BB'(I-U_1U_1')$,
(iv) $(I-U_1U_1')BV_1 = 0$.

Proof. We first show that $U_1'U_1 = I$. We get

$$U_1'U_1 = \lambda_1^2 V_1'(M_0X'X+I)(B'B)^{-1}(X'XM_0+I)V_1 .$$

Multiplying from the right with PV_1' yields, by (34),

$$U_1'U_1PV_1' = \lambda_1^2 V_1'(M_0X'X+I)(B'B)^{-1}(X'XM_0+I)M_0 = V_1'M_0 = PV_1' .$$

Multiplying this equation from the right with V_1P^{-1} yields indeed $U_1'U_1 = I$. In a next step we show that

$$\begin{aligned}\lambda_1^2(B'B)^{-1}(X'XM_0+I)M_0X' &= B^{-1}(\lambda_1U_1V_1'+U_1U_1'B)X^+\\ &= B^{-1}(\lambda_1U_1V_1'+U_1U_1'B)(X'X)^{-1}X' .\end{aligned}$$

We have

$$\begin{aligned}(X'XM_0+I)M_0 &= (X'XV_1P+V_1)PV_1'\\ &= (X'XV_1P+V_1)PV_1'X'X(X'X)^{-1} .\end{aligned}$$

Now

$$\begin{aligned}X'XV_1P+V_1 &= -\tfrac{1}{\lambda_1}B'U_1 ,\\ X'XV_1P &= -\tfrac{1}{\lambda_1}B'U_1 - V_1 ,\\ (X'XV_1P)' = PV_1'X'X &= -\left(\tfrac{1}{\lambda_1}U_1'B+V_1'\right) .\end{aligned}$$

Thus

$$\begin{aligned}(X'XM_0+I)M_0 &= +\tfrac{1}{\lambda_1}B'U_1\left(\tfrac{1}{\lambda_1}U_1'B+V_1'\right)(X'X)^{-1}\\ &= \left(\tfrac{1}{\lambda_1^2}B'U_1U_1'B+\tfrac{1}{\lambda_1}B'U_1V_1'\right)(X'X)^{-1}\end{aligned}$$

and, therefore,

$$\begin{aligned}\lambda_1^2(B'B)^{-1}(X'XM_0+I)M_0X' &= B^{-1}(\lambda_1U_1V_1'+U_1U_1'B)(X'X)^{-1}X'\\ &= B^{-1}(\lambda_1U_1V_1'+U_1U_1'B)X^+ .\end{aligned}$$

From $U_1 = -\lambda_1B'^{-1}(X'XV_1P+V_1)$ it follows that $u_i = -\lambda_1B'^{-1}(I+p_iX'X)v_i$, consequently $v_i = -\frac{1}{\lambda_1}(I+p_iX'X)^{-1}B'u_i$. The two other assumptions of the theorem ensure that the conditions (iv) and (v) of Theorem 4 are met. Thus the theorem is proved. □

Remark 2. The representation of the MILE is

$$Cy = B^{-1}\left(\lambda_1 \sum_{i=1}^{j} u_i v_i' + \sum_{i=1}^{j} u_i u_i' B\right) X^+ y.$$

If $p_i = 0$, then $v_i = -\frac{1}{\lambda_1} B' u_i$. Therefore $\lambda_1 u_i v_i' = -u_i u_i' B$, and this term also appears with opposite sign in the second term of the representation. Thus the representation of Cy can be shortened as follows:

$$\begin{aligned} Cy &= B^{-1}\left(\lambda_1 \sum_{i:p_i \neq 0} u_i v_i' + \sum_{i:p_i \neq 0} u_i u_i' B\right) X^+ y \\ &= B^{-1}(\lambda_1 \tilde{U}_1 \tilde{V}_1' + \tilde{U}_1 \tilde{U}_1' B) X^+ y. \end{aligned}$$

If we now consider the eigenvalue equation

$$\left(I - \sum_{i:p_i \neq 0} u_i u_i'\right) BB' \left(I - \sum_{i:p_i \neq 0} u_i u_i'\right) u_r = \lambda_r^2 u_r \tag{56}$$

it may happen that some of the λ_i^2 equal λ_1^2. Thus the optimality condition becomes

$$\begin{aligned} \lambda_1^2 &\geq \lambda_{\max}(I - \tilde{U}_1 \tilde{U}_1') BB' (I - \tilde{U}_1 \tilde{U}_1') \\ &= \lambda_{\max}(B'(I - \tilde{U}_1 \tilde{U}_1') B) \end{aligned}$$

Remark 3. We have shown that

$$a_0^{-1}(V_1 P V_1' + S^{-1})^2 = F + A, \quad A = S^{-1} B' U_2 \Lambda_2^{-1} (\lambda_1^2 I - \Lambda_2^2) \Lambda_2^{-1} U_2' B S^{-1}.$$

Does this already imply that

$$(F + A)^{\frac{1}{2}} = a_0^{-\frac{1}{2}} (V_1 P V_1' + S^{-1})?$$

Yes, it does. In general $A^2 = B^2$ does not imply $A = B$. But if A and B are both n.n.d. then it does (uniqueness of the root). Since $(F+A)^{\frac{1}{2}}$ and $a_0^{-\frac{1}{2}}(V_1 P V_1' + S^{-1})$ are both n.n.d., this situation is given in our case.

Lemma. *$V_1 = B^{-1} U_1 \Lambda$ and $(\Lambda P)' = \Lambda P$ for some Λ.*

Proof. The MILE is given by

$$Cy = -\lambda_1 B^{-1}(U_1 P V_1') X' y \ \text{(Spectral approach)} \tag{57}$$

or by

$$Cy = M_0 (X' X M_0 + I)^{-1} X' y \ \text{(Bayesian approach)}, \tag{58}$$

where $M_0 = V_1 P V_1'$. Uniqueness implies

$$-\lambda_1 B^{-1}(U_1 P V_1') = M_0 (X' X M_0 + I)^{-1}$$

or

$$\begin{aligned} BM_0 &= BV_1PV_1' = -\lambda_1 U_1 PV_1'(XX'V_1PV_1' + I) \\ &= -\lambda_1 U_1 P(V_1'X'XV_1PV_1' + V_1'). \end{aligned}$$

Postmultiplying with V_1 shows that this is equivalent to

$$BV_1P = -\lambda_1 U_1 PV_1'(X'XV_1P + V_1) = U_1PV_1'B'U_1 = U_1P(BV_1)'U_1$$

in view of (ii) in Theorem 7. From $(I - U_1U_1')BV_1 = 0$ it follows that $BV_1 = U_1\Lambda$ for some $\Lambda \in \mathbb{R}_{j\times j}$. Thus $U_1\Lambda P = U_1P\Lambda'U_1'U_1 = U_1P\Lambda'$. Premultiplying with U_1' yields that this is equivalent to $\Lambda P = P\Lambda' = (\Lambda P)'$. □

Corollary 1. *$C(X^+)' = -\lambda_1 B^{-1}U_1PV_1'$ is a symmetric matrix.*

Proof. $V_1 = B^{-1}U_1\Lambda$ implies $C(X^+)' = -\lambda_1 B^{-1}U_1P\Lambda'U_1'B'^{-1}$. Since $P\Lambda' = (\Lambda P)' = \Lambda P$, the corollary is proved. □

Corollary 2. *The eigenvalues of $C(X^+)' = -\lambda_1 B^{-1}U_1PV_1'$ lie between zero and one.*

Proof. We use the representation

$$\begin{aligned} C(X^+)' &= M_0(X'XM_0 + I)^{-1} \\ &= M_0(M_0 + (X'X)^{-1})^{-1}(X'X)^{-1} \\ &= (X'X)^{-1}(M_0 + (X'X)^{-1})^{-1}M_0 \\ &= M_0 - M_0(M_0 + (X'X)^{-1})^{-1}M_0. \end{aligned} \tag{59}$$

Since $M_0 + (X'X)^{-1} \geq 0$, it follows that $C(X^+)' \leq M_0$ in the sense of Loewner ordering. Since $M_0 \geq 0$ and $\operatorname{tr}(M_0) = 1$, we also get $C(X^+)' \leq M_0 \leq I$. This implies at first that the eigenvalues of $C(X^+)'$ are smaller than one. Moreover, we have to show that

$$M_0 \geq M_0(M_0 + (X'X)^{-1})^{-1}M_0. \tag{60}$$

This relation is clearly correct if M_0 is regular since $M_0^{-1} \geq (M_0 + (X'X)^{-1})^{-1}$. Now $M_0(M_0 + (X'X)^{-1})^{-1}M_0$ is a continuous function of M_0. The argument that the regular matrices are dense in the set of all matrices finishes the proof of the Corollary. □

Corollary 2 is the starting point of the paper by Alson (1988). His setup is that $Cy = A_0X'y$, where A_0 is a symmetric matrix with eigenvalues between zero and one. This implies that

$$A_0 = \sum_{i=1}^{k} \delta_i \beta_i \beta_i'$$

with $0 \leq \delta_i \leq 1$. So one could try to minimize first with respect to the δ_i for given β_i and then with respect to the β_i. Alson pursued this way, but some of his results are not correct (Gaffke and Heiligers, 1989).

Remark 4. The Läuter-Hoffmann representation can also be proved just by verification. In the sequel we give the proof. Let

$$B(C - I) = \lambda_1 U_1 V_1' + U_2 \Lambda_2 V_2' \tag{61}$$

be the singular value decomposition of $B(CX - I)$ corresponding to the spectral equation. Let, moreover,

$$\begin{aligned} A &= S^{-1} B' U_2 \Lambda_2^{-1} (\lambda_1^2 I_{k-j} - \Lambda_2^2) \Lambda_2^{-1} U_2' B S^{-1} \\ &= S^{-1} B' U_2 (\Delta - I) U_2' B S^{-1} \geq 0, \end{aligned} \tag{62}$$

where we define Δ by $\Delta - I = \Lambda_2^{-1}(\lambda_1^2 I_{k-j} - \Lambda_2^2)\Lambda_2^{-1}$. We remark that $\Lambda_2 \Delta \Lambda_2 = \lambda_1^2 I_{k-j}$. We again define F by $S^{-1}B'BS^{-1}$ and we then get

$$\begin{aligned} &(CX - I)'S(F + A)S(CX - I) \\ &= (B(CX - I))' B'^{-1} S(F + A) S B^{-1} B(CX - I) \\ &= (\lambda_1 V_1 U_1' + V_2 \Lambda_2 U_2')(I + U_2(\Delta - I)U_2')(\lambda_1 U_1 V_1' + U_2 \Lambda_2 V_2') \\ &= (\lambda_1 V_1 U_1' + V_2 \Lambda_2 U_2')(U_1 U_1' + U_2 U_2' + U_2(\Delta - I)U_2')(\lambda_1 U_1 V_1' + U_2 \Lambda_2 V_2') \\ &= (\lambda_1 V_1 U_1' + V_2 \Lambda_2 U_2')(U_1 U_1' + U_2 \Delta U_2')(\lambda_1 U_1 V_1' + U_2 \Lambda_2 V_2') \\ &= \lambda_1^2 V_1 V_1' + V_2 \Lambda_2 \Delta \Lambda_2 V_2' \\ &= \lambda_1^2 (V_1 V_1' + V_2 V_2') \\ &= \lambda_1^2 I_k \end{aligned} \tag{63}$$

in view of $\Lambda_2 \Delta \Lambda_2 = I$. From $(CX - I)'S(F + A)S(CX - I) = \lambda_1^2 I_k$ we easily conclude that

$$\lambda_1^{-2}(F + A) = S^{-1} B' \big(B(CX - I)(CX - I)'B'\big)^{-1} B S^{-1}. \tag{64}$$

From the Bayesian approach we get that $B(CX - I) = -B(M_0 X'X + I)^{-1}$. Thus (64) implies that

$$\lambda_1^{-2}(F + A) = (M_0 + S^{-1})(M_0 + S^{-1}). \tag{65}$$

Since $M_0 + S^{-1}$ is n.n.d. it follows that

$$\lambda_1^{-1}(F + A)^{\frac{1}{2}} = M_0 + S^{-1}, \tag{66}$$

i.e., $M_0 = \lambda_1^{-1}(F + A)^{\frac{1}{2}} - S^{-1}$.

Acknowledgement

The first author thanks the organizers of the LINSTAT conference for their invitation (Dziękuję bardzo za wasze zaproszenie!). Moreover, we are indebted to Prof.

Caliński and an anonymous referee. Their suggestions have led to a considerable improvement of the presentation. Finally the first author submits his thanks to "Stiftung Volkswagenwerk" in Hannover whose financial support has made this research possible.

References

Alson, P. (1988). Minimax proporties for linear estimators of a location parameter of a linear model. *Statistics* **19**, 153–171.

Bunke, O. (1975). Minimax linear, ridge and shrunken estimators for linear parameters. *Mathematische Operationsforschung und Statistik, Series Statistics* **6**, 697–701.

Christopeit, N. and Helmes, K. (1991). On Minimax Estimation in Linear Regression Models with Ellipsoidal Constraints. Discussion paper, Sonderforschungsbereich 303, Econometrics Unit, University of Bonn.

Drygas, H. (1991). On an extension of the Girko inequality in linear minimax estimation. In: A. Pázman and J. Volaufová, Eds., *Proceedings of the Probastat '91 Conference.* Bratislava, 3–10.

Drygas, H. (1993). Spectral methods in linear minimax estimation. Preprint 4/93, Kasseler Mathematische Schriften. In: *Proceedings of the Oldenburg Minimax Workshop.* To appear.

Drygas, H. and Pilz, J. (1993). On the equivalence of spectral theory and Bayesian analysis in minimax linear estimation. Preprint 13/93, Kasseler Mathematische Schriften. Submitted for publication.

Gaffke, N. and Heiligers, B. (1989). Bayes, admissible and minimax linear estimators in linear models with restricted parameter space. *Statistics* **20**, 478–508.

Gaffke, N. and Mathar, R. (1990). Linear minimax estimation and related Bayes L-optimal design. In: B. Fuchssteiner, B. Lengenauer, H. J. Skala, Eds., *Methods of Operations Research 60, Proceedings of the XIII Symposium on Operations Research*, 617–628.

Gaffke, N. and Heiligers, B. (1991). Note on a paper by P. Alson. *Statistics* **22**, 3–8.

Girko, V. L. (1988). *Multidimensional Statistical Analysis.* Nauka, Kiew (in Russian). An English translation in preparation.

Girko, V.L. (1990). S-estimators. *Vycislitel'naja i Prikladnaja Matematika.* Kievskii Gosudarstvennyi Univ., Kiev, 90–97 (in Russian).

Girko, V. L. (1993). Spectral theory of estimation. In: *Proceedings of the Oldenburg Minimax Workshop.* To appear.

Hofmann, K. (1979). Characterization of minimax linear estimators in linear regression. *Mathematische Operationsforschung und Statistik, Series Statistics* **10**, 19–26.

Kiefer, J. (1974). General equivalence theory for optimum design (Approximate theory). *The Annals of Statistics* **2**, 849–879.

Kuks, J. (1972). Minimax estimation of regression coefficients. *Izvestiya Akademii Nauk Estonskoy SSR* **21**, 73–78 (in Russian).

Kuks, J. and Olman, V. (1971). Minimax linear estimation of regression coefficients. *Izvestija Akademii Nauk Estonskoi SSR* **20**, 480–482 (in Russian).

Kuks, J. and Olman, V. (1972). Minimax linear estimation of regression coefficients. *Izvestija Akademii Nauk Estonskoi SSR* **21**, 66–72 (in Russian).

Läuter, H. (1973). A minimax linear estimator for linear parameters under restrictions in the form of inequalities. *Mathematische Operationsforschung und Statistik, Series Statistics* **6**, 689–695.

Olman, V. (1983). Estimation of linear regression coefficients in an antagonistic game. *Izvestiya Akademii Nauk Estonskoy SSR* **32**, 241–245 (in Russian).

Pilz, J. (1986). Minimax linear regression estimation with symmetric parameter restrictions. *Journal of Statistical Planning and Inference* **13**, 297–318.

Pilz, J. (1991). *Bayesian Estimation and Experimental Design in Linear Regression Models*, 2nd ed. Wiley, Chichester.

Stahlecker, P. and Drygas, H. (1992). Representation theorems in Linear Minimax Estimation. Report No. V-85-92, University of Oldenburg.

ESTIMATION OF PARAMETERS IN A SPECIAL TYPE OF RANDOM EFFECTS MODEL

JÚLIA VOLAUFOVÁ
Institute of Measurement Science
Slovak Academy of Sciences
Dúbravská 9
84219 Bratislava
Slovakia

Abstract. Consider a linear model, where the random effect is composed of two independent parts one of which has an unknown common covariance matrix and the covariance matrix of the second part is known to depend on the index of individuals. The aim is to find an estimator for the expectation and the unknown part of the covariance matrix. Besides a maximum likelihood a so-called natural estimator is suggested.

Key words: Random effects model, Variance-covariance components model, MINQE, Maximum likelihood estimation, Natural estimator.

1. Introduction

Consider a linear model given by

$$Y_i = \mu + \eta_i, \quad i = 1, \ldots, n, \tag{1}$$

where Y_i's are independent p dimensional vectors as the sum of an unknown expectation vector μ and η_i's error vectors (or random effect vectors) specified as

$$\eta_i = \xi_i + \varepsilon_i, \quad i = 1, \ldots, n. \tag{2}$$

The two parts of the vectors η_i, viz. ξ_i and ε_i, are independent and it is assumed that

$$E(\xi_i) = 0, \quad E(\varepsilon_i) = 0, \quad i = 1, \ldots, n,$$
$$E(\xi_i \xi_i') = \Sigma, \; E(\varepsilon_i \varepsilon_i') = \Sigma_i, \; i = 1, \ldots, n.$$

The matrices Σ_i, depending on the index i, are supposed to be known (or calculated from some previous independent experiment) and positive definite, while the covariance matrix Σ is unknown (belonging to the cone of nonnegative definite matrices). There are no specific assumptions on the distribution of the random vectors entering the model, it is merely assumed that the 3rd and 4th moments are finite.

The model (1) together with (2) can be expressed as a combined model in the form

$$\underline{Y} = (\underline{1} \otimes I)\mu + \underline{\xi} + \underline{\varepsilon}, \tag{3}$$

T. Caliński and R. Kala (eds.),
Proceedings of the International Conference on Linear Statistical Inference LINSTAT '93, 27–34.

where $\underline{1}$ is an n dimensional vector of one's, $\underline{Y} = (Y_1', \ldots, Y_n')'$, and, analogously, $\underline{\xi} = (\xi_1', \ldots, \xi_n')'$ and $\underline{\varepsilon} = (\varepsilon_1', \ldots, \varepsilon_n')'$, while the symbol '$\otimes$' denotes the Kronecker product of matrices. The covariance matrix of $\underline{Y}$ takes, in consequence, the form

$$\text{Cov}\,(\underline{Y}) = \Gamma = (I \otimes \Sigma) + \text{Diag}\,(\Sigma_i),$$

where Diag (Σ_i) is used to denote the block-diagonal matrix with Σ_i's on the diagonal,

$$\text{Diag}\,(\Sigma_i) = \begin{pmatrix} \Sigma_1 & 0 & \cdots & 0 \\ 0 & \Sigma_2 & \cdots & 0 \\ \cdots & \cdots & \cdots & \cdots \\ 0 & 0 & \cdots & \Sigma_n \end{pmatrix}.$$

The main purpose of this note is to estimate the vector μ and the matrix Σ.

The model (3) is in fact a special case of a more general linear model commonly written in the form

$$Y = X\beta + \varepsilon, \tag{4}$$

together with

$$E(\varepsilon) = 0, \quad E(\varepsilon\varepsilon') = \sum_{i=1}^{p} \vartheta_i V_i + V_{p+1} = V(\vartheta), \tag{5}$$

where $\vartheta = (\vartheta_1, \ldots, \vartheta_p)' \in \Theta \subseteq \mathbb{R}^p$, while the matrices V_i, $i = 1, \ldots, p+1$, are all known and symmetric. The parametric space Θ is assumed to contain an open subset.

In our case $V(\vartheta)$ can be expressed as

$$V(\vartheta) = \Gamma = \sum_{i \le j} \sigma_{ij}(1 + \delta_{ij})^{-1} \left(I \otimes (e_i e_j' + e_j e_i')\right) + \text{Diag}\,(\Sigma_i), \tag{6}$$

where σ_{ij} is the i,j-th entry of the unknown matrix Σ, δ_{ij} is the Kronecker δ, and the vector $e_i = (0, \ldots, 0, 1, 0, \ldots, 0)'$ with one on the i-th place. Using now the notation

$$V_{ij} = (1 + \delta_{ij})^{-1} \left(I \otimes (e_i e_j' + e_j e_i')\right), \quad V_{p,p+1} = \text{Diag}\,(\Sigma_i), \tag{7}$$

we observe that Γ is of the form (5).

There are several possible approaches to solve the problem of estimating μ and Σ under model (3). Without any additional assumptions on the distribution of η_i's from (2) one can use e.g. the MINQE(U,I) (Minimum Norm Quadratic Unbiased Invariant Estimator) for estimation Σ and then can plug it in the best linear estimator of μ. If the normality conditions for both ξ_i's and η_i's are met, then it is possible to get the maximum likelihood equations which leads to an iterative procedure. Finally, it is possible to express a so-called natural estimator for μ and then use an iterative procedure for μ and Σ simultaneously. The basic ideas for the last mentioned method are given, e.g., in Štulajter (1991), and in Volaufová and Witkovský (1991). The brief review of these methods is given in the next Section.

2. The MINQE(U,I) and the Plug-in Estimator

2.1. Estimation of Σ

The originally introduced MINQE (Minimum Norm Quadratic Estimator) of the linear function of the vector of variance components, (see C. R. Rao, 1971a, b) was considered as a quadratic form, say $Y'AY$, with a symmetric matrix A, and with some additional properties, as e.g. unbiasedness, invariance with respect to the translation in the mean, and with A minimizing the proper Euclidian norm. The commonly chosen norm is tr AV_0AV_0, where V_0 is the covariance matrix at a preassigned point in the parameter space.

In the case of model (1) or (3), the class of estimators should differ because of the special form of the covariance matrix, which contains the additive term Diag (Σ_i). Hence the original definition of the MINQE should also be modified. For details in a general case see Volaufová and Witkovský (1992), or Volaufová (1992).

Fix the matrix Σ_0 and let $\Gamma_0 = I \otimes \Sigma_0 + \text{Diag}\,(\Sigma_i)$. The choice of the matrix Σ_0 depends on the approximate knowledge of the covariance matrix, however, in general, the reader is referred to C. R. Rao (1971a).

Definition 1. Consider the parametric function $f(\sigma) = \sum_{i \leq j} f_{ij}\sigma_{ij}$. The statistic $T(\underline{Y})$ is MINQE(U,I) of $f(\sigma)$ if

(i) $T(\underline{Y}) \in \mathcal{A} = \{\underline{Y}'A\underline{Y} + b'\underline{Y} + c,\ A = A',\ b \in \mathbb{R}^{np},\ c \in \mathbb{R}^1\}$,

(ii) $T(\underline{Y})$ is unbiased for $f(\sigma)$, invariant with respect to the group of translations of the form $\underline{Y} \to \underline{Y} + (\underline{1} \otimes I)\mu$,

(iii) under normality of $\underline{Y}$ minimizes the variance at Γ_0, with a preassigned matrix Σ_0.

It is easy to show, that under the model (3) the class of estimators $\mathcal{A}$ reduces to $\mathcal{A}_* = \{\underline{Y}'A\underline{Y} + c, A = A', c \in \mathbb{R}^1\}$ with additional constrains on A, $A(\underline{1} \otimes I) = 0$, tr $AV_{ij} = f_{ij}$, for all $i \leq j$, and the constant c is given by $c = -\text{tr}\ AV_{p,p+1}$. Moreover, it is easy to show that each σ_{ij} is MINQE(U,I) estimable.

Theorem 1. *The MINQE(U,I) of σ_{kl} in the model* (3) *is given by*

$$\hat{\sigma}_{kl} = \sum_{i \leq j} \lambda_{ij}^{(kl)} q_{ij} + \sum_{i \leq j} \lambda_{ij}^{(kl)} w_{ij},$$

where

$$q_{ij} = \underline{Y}'(M\Gamma_0 M)^+ V_{ij} (M\Gamma_0 M)^+ \underline{Y},$$
$$w_{ij} = -\text{tr}\ (M\Gamma_0 M)^+ V_{ij} (M\Gamma_0 M)^+ \text{Diag}\,(\Sigma_i),$$

with $M = (I - 1/n\underline{1}\underline{1}') \otimes I$, and the vector λ is a solution of the system $H\lambda = e_{kl}$, in which entries of the matrix H are given by the formulae

$$\{H\}_{ij,kl} = \text{tr}\ (M\Gamma_0 M)^+ V_{ij} (M\Gamma_0 M)^+ V_{kl},$$

with the matrices V_{ij} defined as in (7), *while e_{kl} is the $p(p+1)/2$ dimentional vector of zeros except the ones on the k-th and l-th place.*

Proof. The brief scatch of the proof goes along the following lines. According to Definition 1, the estimator $T(\underline{Y}) = \underline{Y}'A\underline{Y} + c$ is MINQE(U,I) of σ_{ij} if the matrix A minimizes the variance of $T(\underline{Y})$ under the normality of $\underline{Y}$ at Γ_0 among all matrices fulfilling the conditions following from the unbiasedness and invariance requirement.

The minimization of the variance turns to minimization of the form tr $A\Gamma_0 A\Gamma_0$ under the conditions $A(\underline{1} \otimes I) = 0$, tr $AV_{ij} = 1$, and tr $AV_{kl} = 0$ for all k, l for which $k \neq i$ and $l \neq j$ what results to a well known formulas for the MINQE(U,I). □

2.2. Estimation of μ

Denote the MINQE(U,I) of the matrix Σ by $\hat{\Sigma}$. The locally best linear estimator of the vector μ is the generalized least squares estimator depending, of course, on the covariance matrix Γ. The natural procedure is to plug-in its estimate, say $\hat{\Gamma} = (I \otimes \hat{\Sigma}) + \text{Diag}\,(\Sigma_i)$, instead of the matrix Γ. In result the plug-in estimator $\hat{\mu}$ takes the form

$$\hat{\mu} = \left[\sum_{i=1}^{n} (\hat{\Sigma} + \Sigma_i)^{-1}\right]^{-1} \sum_{i=1}^{n} (\hat{\Sigma} + \Sigma_i)^{-1} Y_i.$$

Referring to Seely and Hogg (1982), the unbiasedness of the estimator $\hat{\mu}$ for μ is a simple consequence of the symmetry of the distribution of $\underline{Y}$.

3. The ML Estimator

Assume that the vector $\underline{Y}$ is normally distributed, i.e. $\underline{Y} \sim N_{np}((\underline{1} \otimes I)\mu, \Gamma)$. In that case the log-likelihood function is of the form

$$l(\mu, \Sigma) = -\frac{np}{2} \log 2\pi - \frac{1}{2} \sum_{i=1}^{n} \log |\Sigma + \Sigma_i| - \frac{1}{2} \sum_{i=1}^{n} (Y_i - \mu)'(\Sigma + \Sigma_i)^{-1}(Y_i - \mu).$$

Then the ML equations are

$$\sum_{i=1}^{n} (\Sigma + \Sigma_i)^{-1} \mu = \sum_{i=1}^{n} (\Sigma + \Sigma_i)^{-1} Y_i,$$

$$\sum_{k=1}^{n} e_i'(\Sigma + \Sigma_k)^{-1} e_j = \sum_{k=1}^{n} e_i'(\Sigma + \Sigma_k)^{-1}(Y_k - \mu) e_j'(\Sigma + \Sigma_k)^{-1}(Y_k - \mu), \quad i \leq j.$$

The solution for μ and Σ, say μ^* and Σ^*, respectively, could be calculated by iterative procedure, however, it is apparent that the system of equations is extremely complicated. The author is not familiar with numerical aspects of calculation. One possible suggestion which may work for $p = 2$ is to calculate from the first equation $\mu_{(0)}$ for some initial $\Sigma_{(0)}$ and substitute it for μ in the next three equations which turn to a nonlinear system of equations with respect to the entries of Σ.

4. Natural Estimator

The basic idea of this approach is to create an initial estimator of the whole covariance matrix Γ and then to project it onto the space of matrices having the desired structure (for more details see Volaufová and Komorník, 1992). We proceed as follows.

Without any knowledge about Σ the first step is to estimate the expectation vector μ utilizing the known matrices Σ_i. The commonly used generalized least squares estimator (GLSE) has the form

$$\tilde{\mu} = \left[\sum_{i=1}^{n} \Sigma_i^{-1}\right]^{-1} \sum_{i=1}^{n} \Sigma_i^{-1} Y_i .$$

As the natural estimator for Γ, the following matrix is used

$$\hat{\Gamma} = (\underline{Y} - (\underline{1} \otimes I)\tilde{\mu})\,(\underline{Y} - (\underline{1} \otimes I)\tilde{\mu})' .$$

Lemma 1. *The random matrix $\hat{\Gamma}$ is asymptotically unbiased for Γ with respect to increasing n.*

Proof. It suffices to notice that the expectation of the matrix $\hat{\Gamma}$ has the form

$$E(\hat{\Gamma}) = \Gamma - \frac{2}{n}(\underline{1}\underline{1}' \otimes I)\Gamma + \frac{1}{n^2}(\underline{1}\underline{1}' \otimes I)\Gamma(\underline{1}\underline{1}' \otimes I).\square$$

Let $\hat{\hat{\Gamma}} = \hat{\Gamma} - \text{Diag}\,(\Sigma_i)$ and let $\mathcal{D}$ be the space of all matrices of the form $I \otimes D$, where D is a $p \times p$ dimensional symmetric matrix, i.e. $\mathcal{D} = \{(I \otimes D), D = D'\}$, with the inner product $< A, B >= \text{tr}\ AB$. The space $\mathcal{D}$ can be equivalently expressed as

$$\mathcal{D} = \{V(d) = \sum_{i \leq j} d_{ij} V_{ij}\}.$$

The projection of the matrix $\hat{\hat{\Gamma}}$ onto $\mathcal{D}$ is given by $\sum_{ij} \hat{d}_{ij} V_{ij}$, where the $p(p+1)/2$ dimensional vector $\hat{d} = (\hat{d}_{11}, \hat{d}_{12}, \ldots, \hat{d}_{pp})'$ is the solution of the matrix equation $G\hat{d} = \lambda$, with $\{G\}_{ij,kl} =< V_{ij}, V_{kl} >$, and $\lambda_{ij} =< \hat{\hat{\Gamma}}, V_{ij} >$, $i \leq j$.

Lemma 2. *The matrix G is given by its entries*

$$\{G\}_{ij,kl} = \begin{cases} n, & \text{for } \ i = j = k = l, \\ 2n, & \text{for } \ i \neq j, i = k, j = l, \ \text{or } i = l, j = k, \\ 0, & \text{for } \ i \neq j \neq k \neq l \ \text{or } i \neq k, i = j, k = l. \end{cases}$$

Proof. To show the result it is enough to calculate the inner products $< V_{ij}, V_{kl} >$ for all ij, kl. Since

$$\begin{aligned} \text{tr}\ V_{ij}V_{kl} &= (1+\delta_{ij})^{-1}(1+\delta_{kl})^{-1}\text{tr}\ \left(I \otimes (e_i e_j' + e_j e_i')\right)\left(I \otimes (e_k e_l' + e_l e_k')\right) \\ &= \tfrac{n}{(1+\delta_{ij})(1+\delta_{kl})}\text{tr}\ (e_i e_j' e_k e_l' + e_i e_j' e_l e_k' + e_j e_i' e_k e_l' + e_j e_i' e_l e_k'), \end{aligned}$$

the proof is completed.$\square$

Lemma 3. *The entries λ_{ij} of the vector λ are given by the relation*

$$\lambda_{ij} = \begin{cases} \sum_{k=1}^{n} \left[(Y_{ki} - \tilde{\mu}_i)^2 - \{\Sigma_k\}_{ii}\right], & \text{if } i = j, \\ 2\sum_{k=1}^{n} \left[(Y_{ki} - \tilde{\mu}_i)(Y_{kj} - \tilde{\mu}_j) - \{\Sigma_k\}_{ij}\right], & \text{if } i \neq j, \end{cases} \tag{8}$$

where Y_{ki} denotes the i-th element of Y_k.

Proof. The proof goes along the same lines as of Lemma 2. It suffices to calculate the inner products $< \hat{\hat{\Gamma}}, V_{ij} >$ for all ij.□

As a direct consequence of previous lemmas the following theorem can be stated.

Theorem 2. *The natural estimator of Σ in the model (3) is the matrix $\hat{\Sigma}$ given by*

$$\hat{\Sigma} = \frac{1}{n} \sum_{i=1}^{n} ((Y_i - \tilde{\mu})(Y_i - \tilde{\mu})' - \Sigma_i).$$

The expectation of $\hat{\Sigma}$ is $\frac{n-1}{n}\Sigma$.

It should be pointed out that the estimator $\hat{\Sigma}$ need not be nonnegative definite. The reason is that the class $\mathcal{D}$ is larger then the convex cone formed by nonnegative definite matrices of the same structure. This problem can be solved in two steps. The first one is to create the estimator $\hat{\Sigma}$ and the second to find a nonnegative part of the $\hat{\Sigma}$ which risk is less than or equal to that of the original estimator. We proceed as follows. The symmetric matrix $\hat{\Sigma}$ can be decomposed into the sum $\hat{\Sigma} = \hat{\Sigma}_+ - \hat{\Sigma}_-$, by the use of the singular value decomposotion,

$$\hat{\Sigma} = PQP' = \sum_{i=1}^{p} \kappa_i P_i P_i',$$

where κ_i is the i-th eigenvalue (possibly zero) of the matrix $\hat{\Sigma}$ and the matrix P_i is formed from the eigenvectors corresponding to κ_i. The positive part of the matrix $\hat{\Sigma}$ is then $\hat{\Sigma}_+$, obtained by replacing the negative eigenvalues κ_i by zero. For the norm-distance defined by the relation $||A - B||^2 = \text{tr } \left[(A - B)^2\right]$ the inequality

$$||\hat{\Sigma} - \Sigma|| \geq ||\hat{\Sigma}_+ - \Sigma||$$

is valid, and hence the risk function defined by $R(\hat{\Sigma}, \Sigma) = \text{tr } E(\hat{\Sigma} - \Sigma)$ obeys

$$R(\hat{\Sigma}, \Sigma) \geq R(\hat{\Sigma}_+, \Sigma).$$

Let us concentrate now on the estimation of the vector μ. As we can notice the procedure of the construction of the natural estimator $\tilde{\mu}$ of μ can be proceeded iteratively. Denote $\hat{\Gamma}^{(k-1)} = (I \otimes \hat{\Sigma}^{(k-1)}) + \text{Diag } (\Sigma_i)$ and, consequently,

$$\tilde{\mu}^{(k)} = \left[\sum_{i=1}^{n} (\hat{\Sigma}^{(k-1)} + \Sigma_i)^{-1}\right]^{-1} \sum_{i=1}^{n} (\hat{\Sigma}^{(k-1)} + \Sigma_i)^{-1} Y_i.$$

Lemma 4. *The natural estimator $\tilde{\mu}^{(k)}$ is unbiased for μ provided the distribution of $\underline{Y}$ is symmetric with respect to its expectation.*

The proof of the lemma is omitted for its simplicity.

5. Conclusion

In most of experimental situations the main interest is focused on the estimation of the unknown vector μ. Three different types of estimators were presented in the paper: the plug-in estimator, the ML estimator, and the natural estimator.

The ML method leads to a complicated nonlinear system of equations. Although it is very convenient to have the asymptotic statistical properties of the estimators, even in a favourable case, when the system is solvable, the ML estimator, as resulting from the iterative procedure, is approximate in its nature. On the other hand, in a case when the number of individuals is not large enough, the theoretical statistical properties of estimators are disputable.

The plug-in estimator is simpler than the ML estimator. Moreover, if it is based on a quadratic invariant estimator of the covariance matrix Σ and the distribution of $\underline{Y}$ is symmetric, then it is unbiased. Nevertheless, it is hardly possible to tell more about its covariance matrix.

The greatest advantage of the natural estimator is its computational simplicity. Again, to get the unbiased natural estimator, it is enough to assume the symmetry of $\underline{Y}$. Unfortunately, as in the previous case, its covariance matrix is unknown.

Finally, the preference in use of any one of suggested estimators depends strongly on the conditions of the experiment. In consequence there is no general rule which could be used when choosing the estimator.

Acknowledgements

The research was supported by the grant No. 999366 of the Slovak Academy of Sciences. I wish to thank Professor B. K. Sinha whose perceptive question during the discussion led to some improvements in the contribution as well as to the two anonymous referees for their comments and suggestions which led to the final form of the paper.

References

Rao, C.R. (1971). Estimation of variance and covariance components — MINQUE theory. *Journal of Multivariate Analysis* **1**, 257–275.

Rao, C.R. (1971). Minimum variance quadratic unbiased estimation of variance components. *Journal of Multivariate Analysis* **1**, 445–456.

Seely, J. and Hogg, R.V. (1982). Symmetrically distributed and unbiased estimators in linear models. *Communications in Statistics–Theory and Methods* **11**, 721–729.

Štulajter, F. (1991). Consistency of linear and quadratic least squares estimators in regression models with covariance stationary errors. *Applications of Mathematics* **36**, 149–155.

Volaufová, J. (1992). A brief survey on the linear methods in variance-covariance components model. In: W. Müller, Ed., *Proceedings MODA-3*. St. Petersburg, Russia. Physica Verlag, Vienna. To appear.

Volaufová, J. and Komorník, J. (1992). Weighted multivariate regression estimates solved by random effects approach. Submitted for publication.

Volaufová, J. and Witkovský, V. (1991). Least squares and minimum MSE estimators of variance components in mixed linear models. *Biometrical Journal* **33**, 923–936.

Volaufová, J. and Witkovský, V. (1992). Estimation of variance components in mixed linear models. *Applications of Mathematics* **37**, 139–148.

RECENT RESULTS IN MULTIPLE TESTING: SEVERAL TREATMENTS VS. A SPECIFIED TREATMENT

CHARLES W. DUNNETT
Department of Mathematics and Statistics, and
Department of Clinical Epidemiology and Biostatistics
McMaster University
Hamilton, Ontario
Canada L8S 4K1

Abstract. We consider multiple hypothesis testing of $H_i : \theta_i \leq 0$ vs. $A_i : \theta_i > 0$ (or its two-sided analog) assuming the general linear model with normally distributed errors and constant variance. The case $\theta_i = \mu_i - \mu_0$ and $\hat{\theta}_i = \bar{X}_i - \bar{X}_0$ is an important example, the $\bar{X}$'s being means based on n_i observations in a one-way (parallel groups) design. Using t-statistics to test the H_i, we show how critical constants $c_1, \ldots, c_k$ can be determined, based on multivariate t, so that the familywise error rate is $\leq \alpha$, for both step-down and step-up testing. Instead of fixing α, p-values can be calculated. Power comparisons between the step-down and step-up approaches are shown. Comparisons with single-step methods are made. An application in medical testing is used to illustrate the methods. Other approaches are also described.

Key words: Familywise error rates, Multiple comparisons, Multiple testing, Multivariate t distribution, Adjusted p values, Simultaneous inference, Stepwise tests, K-ratio method, Order restricted inference.

1. Introduction

In the linear regression model $\boldsymbol{Y} = \boldsymbol{X\beta}+\boldsymbol{\varepsilon}$, where $\boldsymbol{Y}$ is an $N \times 1$ observation vector and $\boldsymbol{\varepsilon} \sim N(\boldsymbol{O}, \sigma^2\boldsymbol{I})$, we are usually interested in a set of parametric functions of the β's, $\theta_i = \boldsymbol{h}_i'\boldsymbol{\beta}$, for $i = 1, \ldots, k$. Let $\hat{\boldsymbol{\beta}}$ be any solution $(\boldsymbol{X}'\boldsymbol{X})^{-}\boldsymbol{X}'\boldsymbol{Y}$ to the normal equations; then the least squares estimate of θ_i is $\hat{\theta}_i = \boldsymbol{h}_i'\hat{\boldsymbol{\beta}}$ and the estimator $\hat{\boldsymbol{\theta}} = (\hat{\theta}_1, \ldots, \hat{\theta}_k)'$ of $\boldsymbol{\theta} = (\theta_1, \ldots, \theta_k)'$ is distributed as $N(\boldsymbol{\theta}, \sigma^2\boldsymbol{V})$ where $\boldsymbol{V} = \{v_{ij}\}, v_{ij} = \boldsymbol{h}_i'(\boldsymbol{X}'\boldsymbol{X})^{-}\boldsymbol{h}_j'$. Assume that an unbiased estimator S^2 of σ^2 is available, distributed as $\sigma^2\chi^2_\nu/\nu$. For instance, take S^2 as the mean square error (MSE), with $\nu = N - \text{rank}(\boldsymbol{X})$. Define $T_i = (\hat{\theta}_i - \theta)/(S\sqrt{v_{ii}})$, $i = 1, \ldots, k$, which is a Student's t variate. Then $\boldsymbol{T} = (T_1, \ldots, T_k)'$ is multivariate t with ν d.f. and correlation matrix $\Re = \{\rho_{ij}\}$, where $\rho_{ij} = v_{ij}/\sqrt{v_{ii}v_{jj}}$.

The problem is to make simultaneous inferences about the θ_i. For the particular application of comparing several treatments with a specified treatment in a one-way (parallel groups) design with n_i observations in the ith group ($i = 0, 1, \ldots, k$), we have $\theta_i = \mu_i - \mu_0$ denoting the difference between the means of the ith and specified treatment groups. For this problem, $\hat{\theta}_i = \bar{X}_i - \bar{X}_0$, the difference between the observed means, and

$$v_{ii} = 1/n_i + 1/n_0, \quad v_{ij} = 1/n_0 \quad (i \neq j),$$

T. Caliński and R. Kala (eds.),
Proceedings of the International Conference on Linear Statistical Inference LINSTAT '93, 35–46.

and hence

$$\rho_{ij} = \lambda_i \lambda_j, \tag{1}$$

where $\lambda_i = 1/\sqrt{1 + n_0/n_i}$. The structure (1) for ρ_{ij}, where $-1 < \lambda_i < 1$, is known as 'product-type' (Hochberg and Tamhane, 1987, p. 63). The computation of probability integrals of multivariate t random variables is simplified when their correlation structure has this form.

At the previous LINSTAT conference, the problem of obtaining simultaneous confidence interval estimates of the θ's was considered (Dunnett, 1985). The present paper is concerned with simultaneous hypothesis testing. Denote the hypotheses by $H_1, \ldots, H_k$, which may be either one-sided ($H_i : \theta_i \leq 0$ versus $A_i : \theta_i > 0$) or two-sided ($H_i : \theta = 0$ versus $A_i : \theta_i \neq 0$). We use the statistic $t_i = \hat{\theta}_i/(s\sqrt{v_{ii}})$, where s is the observed value of S, to test H_i. Our problem is to do this so that the familywise error rate (FWE), which is the probability of rejecting one or more true null hypotheses, is $\leq \alpha$ under any null configuration of the parameters. This is appropriate when the hypotheses in each family are part of the same research question and must be considered *jointly*. When there is only one family in the experiment, familywise and experimentwise error rates coincide. When each family consists of a single hypothesis, then familywise error rates become comparisonwise. Comparisonwise error rates are appropriate when the tests are separate from each other, being included in the same experiment only for reasons of efficiency.

Let $\boldsymbol{\theta}_m$ be any parameter configuration where the θ_i are null for $i = 1, \ldots, m$ and non-null for $i = m + 1, \ldots, k$. To satisfy FWE $\leq \alpha$, we require that

$$P_{\boldsymbol{\theta}_m} [\text{Accept } H_1, \ldots, H_m] \geq 1 - \alpha, \text{ for } m = 1, \ldots, k. \tag{2}$$

One way to achieve this is to reject any H_i for which $t_i \geq c = t^{\alpha}_{k,\nu,\Re_k}$, the α-point of k-variate Student's t with ν d.f. and the associated correlation matrix $\Re_k$. This uses the same critical constant as in the joint confidence interval estimates, and is called *single-step* (SS). In the next section, we describe some results for *stepwise* testing of the H_i, based on recent joint work with Professor A.C. Tamhane.

2. Stepwise Multiple Testing

For stepwise testing of the H_i, the statistics t_i are ordered from the least to the most significant, so that $t_1 \leq \ldots \leq t_k$, and the hypotheses H_i re-labelled to correspond to this observed ordering. (For two-sided testing, the $|\, t_i \,|$ are ordered.) Stepwise testing uses a set of critical constants $c_1, \ldots, c_k$ to determine which H_i are rejected.

In a step-down (SD) test, testing starts with H_k: if $t_k \geq c_k$ then H_k is rejected and we go to H_{k-1}. This continues as long as $t_m \geq c_m$ indicating that H_m is rejected, stopping the first time that $t_m < c_m$ in which case $H_1, \ldots, H_m$ are accepted. Thus, for H_m to be rejected in a SD test, it is necessary that $t_i \geq c_i$ for *all* $i = m, \ldots, k$.

In step-up (SU) testing, the hypotheses are tested in the reverse order, starting with H_1, then H_2 and so on. Testing continues as long as $t_m < c_m$, indicating that H_m is accepted, stopping the first time that $t_m \geq c_m$ is observed, in which case $H_m, \ldots, H_k$ are rejected. Thus, for H_m to be rejected in a SU test, it is necessary that $t_i \geq c_i$ for *some* $i = 1, \ldots, m$.

A step-down multiple testing procedure is sometimes described as 'sequentially rejective', after the terminology used by Holm (1979) in describing his method which uses Bonferroni critical constants. Hochberg's (1988) method, which uses the same critical constants as Holm's but in step-up fashion, along with a method by Hommel (1988, 1989), which uses a more complex algorithm for determining which hypotheses to reject, initiated the present interest in step-up procedures. See Dunnett and Tamhane (1993) for a comparison of the powers of some step-up testing procedures.

2.1. Critical Constants for the Step-Down (SD) Procedure

To satisfy Equation (2), the critical constants c_m for SD are determined so that

$$P[T_1 < c_m, \ldots, T_m < c_m] = 1 - \alpha, \quad \text{for } m = 1, \ldots, k. \tag{3}$$

Here, T_i denotes the random variable associated with t_i. $T_1, \ldots, T_m$ have a central m-variate t distribution with ν d.f. and $\Re_m$, the correlation matrix corresponding to the m smallest t statistics. For $m = 1$, c_1 is the α-point of univariate Student's t, t_ν^α. For $m > 1$, the solution is $c_m = t^\alpha_{m,\nu,\Re_m}$, the α-point of m-variate Student's t. Except for $m = 1$, the constants depend on the observed ordering of the t-statistics in general, as the correlation structure changes with m unless the ρ_{ij} are equal. The constants satisfy the monotonicity property $c_1 < \ldots < c_k$. Table I shows the values of the constants c_m for all distinct orderings of the t-statistics in a particular example of treatments vs. a specified treatment comparisons.

Two useful references for determining the values of the critical constants are (i) Bechhofer and Dunnett (1988) which provides an extensive set of tables for the equal-correlation case, covering $\rho = 0(.1).9$, $m = 2(1)16,18,20$, $\alpha = .01,.05,.10,.20$ (one-sided and two-sided), and (ii) Dunnett (1989) which gives a Fortran algorithm for computing multivariate normal and t probability integrals over rectangular regions for product-type correlation structure.

2.2. Critical Constants for the Step-Up (SU) Procedure

The constants c_m for SU are determined so that the following equation is satisfied,

$$P[(T_1, \ldots, T_m) : T_{(1)} < c_1, \ldots, T_{(m)} < c_m] = 1 - \alpha, \text{ for } m = 1, \ldots, k. \tag{4}$$

Here $T_{(1)} \leq \ldots \leq T_{(m)}$ denote the ordered values of the T_i defined in (3). Equation (4) is solved recursively for the c's, starting with $m = 1$ which gives $c_1 = t_\nu^\alpha$, the same as for SD. For $m = 2$, the left-side of (4) can be expanded to obtain:

$$P[(T_1, T_2) : T_{(1)} < c_1, T_{(2)} < c_2] = \\ P(T_1 < c_1,\ T_2 < c_2) + P(c_1 < T_1 < c_2,\ T_2 < c_1) = 1 - \alpha. \tag{5}$$

The two probabilities in the expansion are bivariate Student t probabilities with correlation coefficient ρ_{12} determined by the two smallest t-statistics. The equation can be solved for c_2, using the previously determined value of c_1. For $m > 2$, there is an analogous expression consisting of a sum of m! probabilities, each representing an m-variate t integral over a rectangular subregion of the region on the left-side of (4). A recursive formula is given in Dunnett and Tamhane (1994) that defines how this region is subdivided for any m. The probability over each subregion can be

TABLE I
Critical Constants for Single-Step, Step-Down and Step-Up Methods
Example: $k = 4$, $(n_0; n_i) = (10; 5, 5, 5, 15)$, $\nu = 35$, $\alpha = .05$

Method	Order	n_1	n_2	n_3	n_4	c_1	c_2	c_3	c_4
SS	1-4	-	-	-	-	2.271	2.271	2.271	2.271
SD	1	5	5	5	15	1.690	2.000	2.169	2.271
	2	5	5	15	5	1.690	2.000	2.155	2.271
	3	5	15	5	5	1.690	1.986	2.155	2.271
	4	15	5	5	5	1.690	1.986	2.155	2.271
SU	1	5	5	5	15	1.690	2.014	2.176	2.2756
	2	5	5	15	5	1.690	2.014	2.163	2.2761
	3	5	15	5	5	1.690	2.003	2.164	2.2758
	4	15	5	5	5	1.690	2.003	2.163	2.2761

evaluated for specified values of the c's, for the case of product-correlation structure, using the computer algorithm in Dunnett (1989). Equating the summation to $1-\alpha$ defines the solution for c_m.

The computations required to determine the c's for SU are more involved than for SD. In the equal-correlation case $\rho_{ij} = \rho \geq 0$, some simplification is possible and a computing algorithm has been given in Dunnett and Tamhane (1992a, Section 3.3). For unequal ρ_{ij}, a conservative approximation to the value of c_m for $m \geq 3$ can be obtained by using this algorithm with the $\binom{m}{2}$ correlation coefficients replaced by their arithmetic average $\bar{\rho}_m$. Another approximation is to estimate the c's by simulation. These are described in Dunnett and Tamhane (1994).

A set of tables giving the values of c_m for the equal-ρ case, covering $\rho = 0$, .1, .3, .5, .7, .9, $m = 1(1)8$, $\alpha = .01$, .05, .10 (one-sided and two-sided) is available from the author.

2.3. Adjusted p-values for SD and SU Procedures

The ***adjusted*** (for multiplicity) p-value for a hypothesis or t statistic is defined to be the smallest level α at which the hypothesis can be rejected using a stated multiple test procedure and the observed test statistics $t_1 \leq \ldots \leq t_k$; see Dunnett and Tamhane (1991) where it was called a *joint* p-value. To determine the ***adjusted*** p-value p_m corresponding to t_m, we first compute a set of multivariate t probabilities denoted by p'_i. Then p_m is defined in terms of these p'_i, as follows.

For SD, compute $p'_k, p'_{k-1}, \ldots, p'_m$, where

$$p'_m = 1 - P[T_1 < t_m, \ldots, T_m < t_m];$$

then define

$$p_m = \max(p'_m, \ldots, p'_k). \tag{6}$$

For SU, compute $p'_1, p'_2, \ldots, p'_m$, where for p'_m we define $c_m = t_m$ and compute constants $c_1, \ldots, c_{m-1}$ and p'_m to satisfy the following set of equations:

$$p'_m = 1 - P[(T_1, \ldots, T_i) : T_{(1)} < c_1, \ldots, T_{(i)} < c_i] \quad \text{for } i = 1, \ldots, m;$$

then define

$$p_m = \min(p'_1, \ldots, p'_m). \tag{7}$$

2.4. Power Comparisons

Power denotes the probability of detecting false hypotheses, expressed as a function of the parameter values. Denote by f the number of false H_i and $m = k - f$ the number which are true. Dunnett and Tamhane (1992a) defined power as the probability of rejecting at least r of the f false hypotheses. The values of r most likely to be of interest are $r = 1$ (reject *at least one*) and $r = f$ (reject *all*). Bristol (1989) and Hayter and Liu (1992) used the power of the SS test to determine the sample sizes needed for two-sided comparisons between treatments and the specified treatment, but limited their consideration to the values $m = 0$ and $r = 1$ in their definition of power.

Define a configuration $\boldsymbol{\theta}_m$ of $\boldsymbol{\theta}$ such that the first m of the θ's are zero ($H_1, \ldots, H_m$ are true) and the remaining θ's are positive. Then the power of any procedure R under this configuration is:

$$\Pi_{k,m,r}(\mathrm{R} \mid \boldsymbol{\theta}_m) = P_{\boldsymbol{\theta}_m}[\mathrm{R} \text{ rejects at least } r \text{ of } H_{m+1}, \ldots, H_k]. \tag{8}$$

For the treatments vs. specified treatment problem with sample size n_i in the ith group ($i = 0, 1, \ldots, k$), it is convenient to standardize the parameter values by defining $\gamma_i = \sqrt{n_0}\theta_i/\sigma$ (Dunnett and Tamhane, 1994). Then the power function can be expressed in terms of the γ's instead of the θ's which makes it independent of the sample sizes except through the ratios n_i/n_0, which determine the correlation structure as given in Equation (1). Dunnett and Tamhane (1992a) gave computable expressions for (8) applicable to the equal-ρ case and used them to make numerical comparisons between the powers of various procedures. When all k hypotheses are false and $r = k$, it was shown that the step-up (SU) procedure has uniformly higher power than the step-down (SD). When only one hypothesis is false and $t = 1$, it was conjectured that the reverse is true; although this was not proved, numerical results supported it. The single-step (SS) procedure is at a disadvantage compared with the stepwise procedures, except when $r = 1$.

For the unequal-ρ case, simulation may be used to determine power, as described in Dunnett and Tamhane (1994). In Table II, we show power comparisons of the SD, SU and SS (single-step) procedures for a particular example with $k = 3$ and $\nu = \infty$, obtained by simulation.

2.5. Testing Whether a Specified Number of Hypotheses are False

In some multiple testing problems, the false hypotheses must be at least a certain number, say h, for the experiment to be a success. Accordingly, consider the null hypothesis H: At least $k - h + 1$ H_i are true versus the alternative that at least h are false. Rejection of H means that the goal of the experiment has been achieved.

TABLE II
Power* Comparisons
Example: $k = 3,\ \rho_{ij} = 0.5,\ \nu = \infty,\ \alpha = .05$

Number of false H_i	r	γ_i	Single-Step (SS)	Step-Down (SD)	Step-Up (SU)
1	1	3.0	.524	.524	.521
		5.0	.930	.930	.928
2	1	3.0	.690	.690	.690
		5.0	.980	.980	.981
3	1	3.0	.771	.771	.778
		5.0	.991	.991	.992
2	2	3.0	.357	.416	.415
		5.0	.879	.907	.905
3	3	3.0	.272	.426	.448
		5.0	.839	.923	.927

*Power = Probability of rejecting at least r false hypotheses [see Equation (8)].

See Rüger (1978) who also adopted this formulation. Using the ordered t statistics $t_1 \leq \ldots \leq t_k$ for testing the individual hypotheses, a single-step test rejects H at level α if $t_{k-h+1} \geq t^{\alpha}_{k-h+1,\nu,\Re_{k-h+1}}$. Alternatively, either of the stepwise tests SD or SU can be used, rejecting H if at least h of the H_i are rejected.

The most useful cases are $h = 1$ and $h = k$. Suppose $h = 1$ (the experimenter hopes to show that at least one H_i is false) and denote the alternative hypothesis to H_i by A_i. In this case, H can be expressed as $\cap H_i$ and its alternative as $\cup A_i$. This is a classical simultaneous testing problem due to S.N. Roy and is called a union-intersection problem. The acceptance region for H is the intersection of the individual acceptance regions. Thus $H \equiv \cap H_i$ is accepted if $\max(t_i) = t_k < c$. To achieve a level α, we use $c = t^{\alpha}_{k,\nu,\Re_k}$, the α-point of k-variate Student's t. This is the single-step (SS) test described in Section 1. The step-down (SD) test uses the same critical constant for t_k but smaller constants for t_m $(m < k)$, chosen to satisfy (3). Thus, SD can be considered as a stepwise extension of SS.

Now suppose $h = k$ (the experimenter's goal is to show that all H_i are false). In this case, H can be expressed as $\cup H_i$ and its alternative as $\cap A_i$. Accordingly, we call it an intersection-union problem. The rejection region for H is the intersection of the individual rejection regions. Thus $H \equiv \cup H_i$ is rejected if $\min(t_i) = t_1 \geq c$. To achieve a level α for this problem, we choose $c = t^{\alpha}_{\nu}$, the α-point of univariate Student's t (see Berger 1982). This is the basis for Laska and Meisner's (1989) MIN test [so-called because it uses $\min(t_i)$ as its test statistic]. The step-up (SU) test uses the same critical constant for t_1 but larger constants chosen to satisfy (4) for testing t_m for $m > 1$. SU can be looked upon as a stepwise extension of the MIN test; see Dunnett and Tamhane (1992b).

The above concepts may be helpful in deciding whether to use SD or SU in a particular application. If the experimenter's goal is to find at least one false hypothesis, then $h = 1$ and he would use SD. On the other hand, if he hopes to show that all of them are false, then $h = k$ and he would use SU. (MIN test could also be used, if there is no interest in values $h < k$.) The example considered in the next section illustrates these choices.

2.6. An Application in Medical Testing

Consider a medical trial involving four treatments: a new drug, two reference standard drugs and a placebo. This example was proposed by D'Agostino and Heeren (1991) as an illustration of multiple treatment comparisons. Patients were assigned at random to one of the treatment groups and a response measured: the problem is to make inferences about certain differences between treatment means. Denote the true means for the four groups by μ_T, μ_1, μ_2 and μ_0, respectively. The differences of concern are $\mu_i - \mu_j$, where i, j denotes a particular treatment pair. Dunnett and Tamhane (1992b) proposed that the hypothesis tests of the treatment differences of interest be grouped into three separate families:

$$\begin{aligned} &1.\ H_1 : \mu_1 - \mu_0 \leq 0,\ H_2 : \mu_2 - \mu_0 \leq 0 \\ &2.\ H_3 : \mu_T - \mu_0 \leq 0 \\ &3.\ H_4 : \mu_T - \mu_1 \leq 0,\ H_5 : \mu_T - \mu_2 \leq 0 \end{aligned} \tag{9}$$

These were chosen because they correspond to the main research questions of interest in the study: 1. To establish the 'sensitivity' of the trial, which is its ability to detect activity in known active drugs (by rejecting H_1 and H_2); 2. To establish that the new drug is active (by rejecting H_3); and 3. To determine whether the new drug is better than either standard. (An additional test, $H_6 : \mu_1 - \mu_2 = 0$ vs $A_6 : \mu_1 - \mu_2 \neq 0$, could be added as another family, but was not because comparing the two known standards was not considered pertinent to the main purpose of the study.)

Since sensitivity requires both individual hypotheses H_1 and H_2 to be rejected, then $h = 2$ in the 1st family and it is tested as an intersection-union problem. Hence, either the MIN test or the SU test is appropriate, the SU test having the advantage in case it is not possible using MIN to show that both known actives differ from the placebo. The 2nd family, consisting of only a single hypothesis, is tested by Student's t. The 3rd family, where the research question is whether the new drug can be shown to be more effective than one or more of the standard drugs (which would strengthen the argument for approval by the regulatory agency), is an example where $h = 1$. Hence, SS is appropriate but, to achieve greater power in detecting superiority over both standard drugs, SD is preferred. (Of course, if it turned out that the test drug is superior to both standard drugs, SU would be even better. But it might not be possible to predict that this would be the case before performing the trial.)

The level α used in testing each family should be chosen according to the seriousness of making one or more Type I errors in that family. It is not necessary for the same value to be used for each. In this example, it is interesting to note that it is also possible to make a statement about the simultaneous error rate over all three families (i.e., the experimentwise error rate). Since the 3rd family of comparisons requires the previous two to be tested first and their respective hypotheses *rejected*, the experimentwise error rate cannot exceed the largest of the three individual α values used. For further discussion of this example, see Dunnett and Tamhane (1992b).

3. Other Approaches

3.1. Bayesian Approaches: Duncan's K-ratio Method

An alternative approach to multiple comparisons among treatment means which has long been advocated by Duncan and his followers is the 'K-ratio' method. The method derives its name from the fact that it focuses on the ratio (denoted by K) of the slopes of the linear loss functions associated with accepting A_i when H_i is true and accepting H_i when A_i is true, instead of considering error rates α. The method is Bayesian, assuming prior distributions for the unknown parameters and linear loss functions for the individual Type I and Type II errors, and the individual losses are assumed to be additive.

The K-ratio method has recently been extended by Brant, Duncan and Dixon (1992) to include the problem of comparing treatments with a specified treatment. The critical value c for testing any t-statistic is a function of the observed values of t_G, the t-statistic for testing the difference between the mean of all k test treatments and the specified treatment, and F_T, the F-statistic for comparing the mean square among test treatments (excluding the specified treatment) to the mean square for error. The critical value c is adaptive to the observed values of t_G and F_T, increasing when they are small indicating *a posteriori* that the treatment differences are small, and decreasing otherwise. For certain combinations of (t_G, F_T) values, the method leads to the value $c = 0$, indicating that *all* differences are to be declared significant regardless of the actual value of each $\bar{X}_i - \bar{X}_0$.

The feature of having a critical constant that is dependent upon the extent of heterogeneity observed in the data may be attractive in certain situations. Another feature is that the critical constant does not depend on the actual number of treatment comparisons being made, and in this respect the method behaves like a comparisonwise rather than a familywise error-controlling test. However, the assumptions made in the derivation of the method, such as linearity of the loss functions and a common normal prior distribution for the treatment means, may not be appropriate in many applications. Although it might be reasonable to assume a specific prior distribution in certain applications, such as screening for new drugs, the actual distribution is likely to be very non-normal. Hochberg and Tamhane (1987, p.333) have pointed out also that the assumption of additivity of the individual loss functions is a crucial one, and is inappropriate in an application where the simultaneous correctness of the separate hypotheses is required.

Tamhane and Gopal (1993) have also considered the Bayesian approach.

3.2. Order Restricted Methods: Multiple Contrast Tests

Recent results based on order restricted inference assume that the null hypothesis is $H : \mu_0 = \mu_1 = \ldots = \mu_k$ versus the alternative $A : \mu_0 \leq \mu_i$, with strict inequality for some i ($i = 1, \ldots, k$) (called a 'simple tree' alternative). This is similar to our one-sided joint hypothesis except that values $\mu_i < \mu_0$ are ruled out. Mukerjee, Robertson and Wright (1987) defined a class of tests, called multiple contrast tests, defined for a specified scalar r ($0 \leq r \leq 1$) by the following expression for the ith contrast $S_i(r)$, for the case of sample sizes $n_1 = \ldots = n_k = n$ and $n_0 = w_0 n$:

$$S_i(r) = rw_0 k(\tilde{X} - \bar{X}_0) + (1 - r)(w_0 + k - 1)(\bar{X}_i - \bar{X}_i^*),$$

where

$$\tilde{X} = (\bar{X}_1 + \ldots + \bar{X}_k)/k$$

and

$$\bar{X}_i^* = (w_0 \bar{X}_0 + \bar{X}_1 + \ldots + \bar{X}_{i-1} + \bar{X}_{i+1} + \ldots + \bar{X}_k)/(w_0 + k - 1). \tag{10}$$

$S_i(r)$ is a weighted average of two contrasts: $\tilde{X} - \bar{X}_0$, comparing the average treatment response with $\bar{X}_0$, and $\bar{X}_i - \bar{X}_i^*$, comparing the ith treatment with a weighted average of all the others including $\bar{X}_0$. A t-statistic t_i is defined for $S_i(r)$ and H is rejected if $\max(t_i) \geq c$ where c is based on multivariate t and is $c = t^{\alpha}_{k,\nu,\rho}$, where $\rho = \text{corr}(S_i, S_j)$ is the common correlation coefficient in the correlation matrix $\Re_k$.

The test SS already defined in this paper is a member of this class of tests, obtained by defining $r = 1/(w_0+1)$. The two extremes are given by $r = 0$ and $r = 1$. For $r = 0$, S_i is the contrast between $\bar{X}_i$ and all the others, including $\bar{X}_0$. The correlations between the contrasts are $\rho_{ij} = -1/(w_0 + k - 1)$ for this case. For $r = 1$, S_i is the single contrast comparing the average of all the treatments with $\bar{X}_0$. This corresponds to the test t_G used in the K-ratio method.

Clearly, there is no value of r that is optimum for all μ_i. The value $r = 0$ is best if there is exactly one $\mu_i - \mu_0 > 0$ and $r = 1$ is best if $\mu_i - \mu_0 = \delta > 0$ for all i. Mukerjee et al. (1987) recommend choosing r so that the contrasts are orthogonal ($\rho_{ij} = 0$) and provide an explicit formula for r to achieve this. The basis for this recommendation is that this test has approximately uniform power (using $\Pi_{k,0,1}$ in the notation of Section 2.4 for the definition of power) for constant values of $\sum_i(\mu_i - \mu_o)^2$, compared with other contrast tests which have greater power in particular directions. (In particular, the SS test has higher power in the direction corresponding to large values for all $\mu_i - \mu_0$, though not as high as that of the single contrast test.)

The likelihood ratio test of H under the assumed order restriction has been considered by Robertson and Wright (1985), Mukerjee et al. (1987) and Conaway, Pillars, Robertson and Sconing (1991). It is much harder to implement than the orthogonal contrast test but the latter has similar power characteristics, which is a point in the latter's favour. Cohen and Sackrowitz (1992) showed that all multiple contrast tests, including the orthogonal contrast test and SS, are 'inadmissible', apparently due to the fact that they use MSE for S^2 in estimating σ^2.

A disadvantage of the multiple contrast tests (except for SS) is the required order restriction $\mu_i \geq \mu_0$. This probably confines their use in drug trials to placebo comparisons, ruling out such important applications as comparing a new drug with a set of standards. Nevertheless, there is considerable potential for their application, which would be further enhanced if stepwise versions analogous to SD and SU could be developed.

4. Discussion

Although we have discussed stepwise testing in the context of treatments vs. specified treatment comparisons in a one-way model, it can be applied more generally. For instance, it applies immediately to any balanced design, such as BIB and BTIB designs (Bechhofer and Tamhane, 1985) where the balanced nature of the design ensures that the comparisons are equally correlated with correlation coefficient ρ dependent on the design. If accidental losses occur, causing an imbalance in the design, the general linear model theory can be invoked to determine the covariance matrix $V = \{v_{ij}\}$ associated with the estimates, from which the actual ρ_{ij} can be determined. Then the correlation coefficients for each $m \leq k$ may be averaged and the average ρ's, $\bar{\rho}_m$, used to determine either SD or SU critical constants based on equal-ρ multivariate t. This is more efficient (in terms of power) than using Šidák constants, as recommended by Fuchs and Sampson (1987) for simultaneous confidence intervals, since it takes the correlations into account. The use of these average-ρ's as an approximation usually results in conservative values for the constants (Dunnett and Tamhane, 1994).

Before adopting a stepwise testing procedure based on multivariate t, the assumptions of normality and homogeneous variances should be verified. The normality assumption is often not critical when means are involved in the estimates $\hat{\mu}$, since the central limit theorem can be invoked. But the homogeneous variance assumption is crucial. If a transformation cannot be found that equalizes the variances, it may be necessary to use separate variance estimates for each mean in the t statistics and approximate the critical constants. Another possibility is to use robust estimates of the means and variances as in Fung and Tam (1988). In a series of papers, Chakraborti and Desu (1988, 1990, 1991) have developed nonparametric methods, including methods to handle censored data, for treatments versus specified treatment comparisons.

Generalizations of the tests considered here are possible. For instance, Hoover (1991) has extended the problem to the case where there are several treatments and two or more controls and joint confidence intervals are required between each treatment and each control simultaneously. Cheung and Holland (1991, 1992) have extended the SS and SD methods to the case of more than one group of treatments, each group containing several treatments compared with a specified treatment, with the error rate covering all groups and treatment comparisons simultaneously.

The use of the SD (step-down) test in treatments vs. specified treatment problems has been known for a long time (see Dunnett and Tamhane, 1991, for the history). Nevertheless, it seems to be seldom used; references in the literature to the so-called "Dunnett's test" usually denoting the use of the critical constant $t^{\alpha}_{k,\nu,\Re_k}$ for all tests (viz., SS). If this test is to be called 'inadmissible', a better reason than the one given in Cohen and Sackrowitz (1992) is that the SD test has equal or greater power (with equality only if $r = 1$ in the definition of power in Section 2.4). Depending upon the particular problem, the step-up test SU may be preferable.

Acknowledgements

I owe a debt of gratitude to my friend and colleague, Professor Ajit Tamhane, whose collaboration over the past few years, carried out mostly by electronic-mail, has made the work enjoyable and this report possible. I acknowledge also the support of a research grant from the Natural Sciences and Engineering Research Council of Canada.

References

Bechhofer, R. E. and Dunnett, C. W. (1988). Tables of percentage points of multivariate t distributions. In: R.E. Odeh and J.M. Davenport, Eds., *Selected Tables in Mathematical Statistics, Vol. 11*. American Mathematical Society, Providence, Rhode Island, 1–371.

Bechhofer, R.E. and Tamhane, A.C. (1985). Tables of admissible and optimal BTIB designs for comparing treatments with a control. In: R.E. Odeh and J.M. Davenport, Eds., *Selected Tables in Mathematical Statistics, Vol. 8*. American Mathematical Society, Providence, Rhode Island, 41–139.

Berger, R.L. (1982). Multiparameter hypothesis testing and acceptance sampling. *Technometrics* **24**, 295–300.

Brant, L.J., Duncan, D.B. and Dixon, D.O. (1992). K-ratio t tests for multiple comparisons involving several treatments and a control. *Statistics in Medicine* **11**, 863–873.

Bristol, D.R. (1989). Designing clinical trials for two-sided multiple comparisons with a control. *Controlled Clinical Trials* **10**, 142–152.

Chakraborti, S. and Desu, M.M. (1988). Generalizations of Mathison's median test for comparing several treatments with a control. *Communications in Statistics - Simulation and Computation* **17**, 947–967.

Chakraborti, S. and Desu, M.M. (1990). Quantile tests for comparing several treatments with a control under unequal right-censoring. *Biometrical Journal* **32**, 697–706.

Chakraborti, S. and Desu, M.M. (1991). Linear rank tests for comparing treatments with a control when data are subject to unequal patterns of censorship. *Statistica Neerlandica* **45**, 227–254.

Cheung, S.H. and Holland, B. (1991). Extension of Dunnett's multiple comparison procedure to the case of several groups. *Biometrics* **47**, 21–32.

Cheung, S.H. and Holland, B. (1992). Extension of Dunnett's multiple comparison procedure with differing sample sizes to the case of several groups. *Computational Statistics & Data Analysis* **14**, 165–182.

Cohen, A. and Sackrowitz, B. (1992). Improved tests for comparing treatments against a control and other one-sided problems. *Journal of the American Statistical Association* **87**, 1137–1144.

Conaway, M., Pillers, C., Robertson, T. and Sconing, J. (1991). A circular-cone test for testing homogeneity against a simple tree order. *The Canadian Journal of Statistics* **19**, 283–296.

D'Agostino, R.B. and Heeren, T.C. (1991). Multiple comparisons in over-the-counter drug clinical trials with bost positive and placebo controls (with comments and rejoinder). *Statistics in Medicine* **10**, 1–31.

Dunnett, C.W. (1985). Multiple comparisons between several treatments and a specified treatment. In: T. Caliński and W. Klonecki, Eds., *Linear Statistical Inference. Lecture Notes in Statistics* **35**. Springer-Verlag, Berlin, 39–47.

Dunnett, C.W. (1989). Multivariate normal probability integrals with product correlation structure. Algorithm AS 251. *Applied Statistics* **38**, 564-579. Correction note. *Applied Statistics* **42**, 709.

Dunnett, C.W. and Tamhane, A.C. (1991). Step-down multiple tests for comparing treatments with a control in unbalanced one-way layouts. *Statistics in Medicine* **10**, 939–947.

Dunnett, C.W. and Tamhane, A.C. (1992a). A step-up multiple test procedure. *Journal of the American Statistical Association* **87**, 162–170.

Dunnett, C.W. and Tamhane, A.C. (1992b). Comparisons between a new drug and active and placebo controls in an efficacy clinical trial. *Statistics in Medicine* **11**, 1057–1063.

Dunnett, C.W. and Tamhane, A.C. (1993). Power comparisons of some step-up multiple test procedures. *Statistics & Probability Letters* **16**, 55–58.

Dunnett, C.W. and Tamhane, A.C. (1994). Step-up multiple testing of parameters with unequally correlated estimates. *Biometrics*. To appear.

Fuchs, C. and Sampson, A.R. (1987). Simultaneous confidence intervals for the general linear model. *Biometrics* **43**, 457–469.

Fung, K.Y. and Tam, H. (1988). Robust confidence intervals for comparing several treatment groups to a control group. *The Statistician* **37**, 387–399.

Hayter, A.J. and Liu, W. (1992). A method of power assessment for tests comparing several treatments with a control. *Communications in Statistics - Theory and Methods* **21**, 1871–1889.

Hochberg, Y. (1988). A sharper Bonferroni procedure for multiple tests of significance. *Biometrika* **75**, 800–802.

Hochberg, Y., and Tamhane, A.C. (1987). *Multiple Comparison Procedures.* Wiley, New York.

Holm, S. (1979). A simple sequentially rejective multiple test procedure. *Scandinavian Jounal of Statistics* **6**, 65–70.

Hommel, G. (1988). A stagewise rejective multiple test procedure based on a modified Bonferroni test. *Biometrika* **75**, 383–386.

Hommel, G. (1989). A comparison of two modified Bonferroni procedures. *Biometrika* **76**, 624–625.

Hoover, D.R. (1991). Simultaneous comparisons of multiple treatments to two (or more) controls. *Biometrical Journal* **33**, 913–921.

Laska, E.M. and Meisner, M.J. (1989). Testing whether an identified treatment is best. *Biometrics* **45**, 1139–1151.

Mukerjee, H., Robertson, T. and Wright, F.T. (1987). Comparison of several treatments with a control using multiple contrasts. *Journal of the American Statistical Association* **87**, 902–910.

Robertson, T. and Wright, F.T. (1985). One-sided comparisons for treatments with a control. *The Canadian Journal of Statistics* **13**, 109–122.

Rüger, B. (1978). Das maximale Signifikanzniveau des Tests: "Lehne H_0 ab, wenn k unter n gegebenen Tests zur Ablehnung führen". *Metrika* **25**, 171–178.

Tamhane, A.C. and Gopal, G.V.S. (1993). A Bayesian approach to comparing treatments with a control. In: F.M. Hoppe, Ed., *Multiple Comparisons, Selection, and Applications in Biometry.* Marcel Dekker, New York, 267–292.

MULTIPLE-MULTIVARIATE-SEQUENTIAL T^2-COMPARISONS

CHRISTOS P. KITSOS
Department of Statistics
Athens University of Business and Economics
Patision 76
Athens 10434
Greece

Abstract. In practice there are cases where multivariate T^2-comparisons are needed when the data set is augmented sequentially. Sometimes the entries are not indepentend, and the main assumption is violated. In the paper the multiple-multivariate-sequential T^2-comparisons are introduced through the likelihood function and Fisher's information matrix.

Key words: Multiple comparisons, Multivariate normal distribution, Sequential test, Hotelling's T^2-test, Fisher's information matrix.

1. Introduction

In clinical trials there are cases where multivariate T^2-comparisons are needed. A typical example is the data set obtained from many clinics, within or between hospitals. In such cases the data points are not always indepentent as the patients might re-enter the data set. Therefore, the problem is to create multiple-multivariate T^2-comparisons, for a sequentially augmented data set, where the entries might not be independent. The target of this paper is to apply a multivariate-sequential T^2-test which is obtained in Section 3, while in Section 2 the main points of a sequential test are discussed. The dependencies of the entries are discussed and overcome in Section 4. In Section 5 it is proved that the usual assumption of the equality of the means can also provide a sequential test. These results are finally used in the algorithm proposed in Section 6.

2. Background

Let $x_1, x_2, \ldots$ be random variables from the same distribution with pdf $f(.;\theta)$. We use the notation $f_n(x_n;\theta)$ to denote the pdf of $f(x_1,\ldots,x_n;\theta)$. Recall that Wald's sequential probability ratio test (SPRT) for testing

$$H_0 : \theta = \theta_0 \quad \text{vs.} \quad H_1 : \theta = \theta_1$$

stops sampling at stage (see the early work of Lai, 1981)

$$n_0 = \inf\{\, n \geq 1 \;:\; S_n \geq a \;\text{ or }\; S_n \leq b \,\},$$

T. Caliński and R. Kala (eds.),
Proceedings of the International Conference on Linear Statistical Inference LINSTAT '93, 47–51.

where a and b, $0 < b < 1 < a$, are two stopping bounds and

$$S_n = \sum_{i=1}^{n} Z_i \quad , \quad Z_i = \ln \frac{f_i(x_i, \theta_1)\, f_{i-1}(x_{i-1}, \theta_0)}{f_i(x_i, \theta_0)\, f_{i-1}(x_{i-1}, \theta_1)} ,$$

i.e. S_n is a modified expression of the likelihood ratio. We use the symbol $S(\beta, \alpha)$ to refer to SPRT with stopping bounds α, β.

The choice of the stopping bounds is directly related to type I, say $\alpha(\theta_0)$, and type II, say $\beta(\theta_1)$, errors as has been proved by Wald, i.e.,

$$\beta = \ln \frac{\beta(\theta_1)}{1 - \alpha(\theta_0)} \leq \min[0, b] , \quad \alpha = \ln \frac{1 - \beta(\theta_1)}{\alpha(\theta_0)} \geq \max[0, a]$$

with given $b, a \in$ R. The optimum property of SPRT is that $S(\beta, \alpha)$ is uniformly most efficient (UME) within the class of all sequential and fixed-sample tests. This actually means that $E_T(n; \theta_0) \geq E(n; \theta_0)$, $E_T(n; \theta_1) \geq E(n; \theta_1)$ for every test T from the class of all tests for H_0 with finite average sample number (ASN) and $\alpha_T(\theta_0) \leq \alpha(\theta_0)$, $\beta_T(\theta_1) \leq \beta(\theta_1)$. This remarkable property, due to Wald and Wolfowitz (1948), justifies the use of SPRT in practice.

3. Sequential T^2-test

Let $x_1, x_2, \ldots, x_n$ be i.i.d. from the k-variate normal distribution $N_k(\mu, \Sigma)$, with both μ and Σ unknown and $\det(\Sigma) > 0$. An invariant SPRT for testing

$$H_0 : \mu\Sigma^{-1}\mu' \leq \xi_0 \quad \text{vs.} \quad H_1 : \mu\Sigma^{-1}\mu' \geq \xi_1 > \xi_0$$

is constructing through the sample mean

$$\overline{x}(n) = (\overline{x}_1(n), \ldots, \overline{x}_k(n)), \text{ with } \overline{x}_j(n) = \frac{1}{n} \sum_{i=1}^{n} x_{ji}, \; j = 1, 2, \ldots, k ,$$

and the sample covariance matrix

$$S(n) = (S_{ij}(n)) \quad , \quad S_{ij}(n) = \frac{1}{n} \sum_{\ell=1}^{n} (x_{j\ell} - \overline{x}_j(n))(x_{i\ell} - \overline{x}_i(n)).$$

Then, for the (maximal invariant) statistic

$$V_n = \overline{x}(n)\, S^{-1}(n) \overline{x}'(n), \tag{1}$$

it can be proved, see Anderson (1984), that

$$F^* = \frac{(n-k)}{k} V_n \sim F_{k, n-k}(n\xi) , \quad \xi = \mu\Sigma^{-1}\mu' ,$$

i.e. V_n follows a noncentral F distribution, with the noncentrality parameter $\delta = n\xi$. However, the statistic F^* can be approximated by a statistic $\tilde{F}$ that has a central F distribution (see Patnaik, 1949), namely by

$$\tilde{F} = \frac{(n-k)}{k + n\xi} V_n \sim F_{m, n-k} \quad , \quad m = \frac{(k + n\xi)^2}{k + 2n\xi} .$$

The corresponding sequential probability ratio test is defined through a (nonsingular) transformation of V_n (Lai, 1981), $\Phi_n = V_n/(1+V_n)$, and is reduced to

$$Z_n = -\frac{n}{2}(\xi_1 - \xi_0) + \ln \frac{W(\frac{n}{2}, \frac{k}{2}; \frac{n\xi_1}{2}\Phi_n)}{W(\frac{n}{2}, \frac{k}{2}; \frac{n\xi_0}{2}\Phi_n)}, \tag{2}$$

with $W(\nu, \rho; \phi)$ defined through the gamma function $\Gamma(\cdot)$ as

$$W(\nu, \rho; \phi) = \sum_{j=0}^{\infty} \frac{\Gamma(\nu + j)\Gamma(\rho)\phi^j}{\Gamma(\rho + j)\Gamma(\nu)j!} .$$

Therefore, bounding V_n within $[L, U]$, Wald's approximated values, say α and β, are evaluated through $Z_n(L) = \beta$ and $Z_n(U) = \alpha$, respectively.

Because the random variable $(n-1)V_n$ is referred to as Hotelling's T^2-statistic with $n-1$ degrees of freedom, the above test is known as the multivariate sequential T^2-test.

4. Sequential Experimental Design

In the theory of experimental design the sequential principle has been faced under two different lines of thought. One line of thought approaches the experimental design problem through the SPRT discussed in Section 2 (see, among others, the early work of Ghosh, 1965, 1967) and the other approach is the optimization problem in experimental design (see Ford, Titterington and Wu, 1985; Ford, Kitsos and Titterington, 1989; Kitsos, 1989).

Both procedures admit the augmantation of the data set creating a sequence of points. Moreover, the optimization at each stage is using the estimates to get knew design points. In optimal experimental design theory it has been pointed out (see Ford, 1976, for linear models, and Kitsos, 1986, for nonlinear models) that the sequential nature of the design does not effect the likelihood function. Therefore the Fisher's information matrix $I(n)$ can be always evaluated and the sample covariance matrix $S(n)$ can be evaluated approximatelly through $I(n)$ as

$$S(n) \cong I^{-1}(n). \tag{3}$$

This result is the main point to perform multiple multivariate T^2-comparisons and we will use it in Section 6.

5. Transforming the Data

The sequential T^2-test mentioned in Section 4 cannot be applied directly for testing

$$H_0\colon \mu_1 = \mu_2 = \cdots = \mu_k \quad \text{vs.} \quad H_1 : H_0 \quad \text{not true.} \tag{4}$$

But, following Scheffé (1959, Section 8.1), a simple transformation of the data will provide a suitable statistic of the type V_n defined in (1). If X is the $n \times k$ data matrix of the i-th row as $(x_{1i}, \ldots, x_{ki})$, then the transformation is

$$Q : X \longmapsto X^* = XQ, \tag{5}$$

where the i-th row of the $n \times (k-1)$ matrix X^* is $(x_{1i} - x_{ki}, \ldots, x_{k-1i} - x_{ki})$, this being attained by the $k \times (k-1)$ matrix $Q = (q_{ij})$ defined as

$$q_{ij} = \begin{cases} 1, & i = j = 1, \ldots, k-1, \\ -1, & i = k, j = 1, \ldots, k-1, \\ 0, & \text{otherwise}. \end{cases} \tag{6}$$

Under the assumptions of Section 3, the rows of X^* are then again distributed independently, each according to $N(\mu^*, \Sigma^*)$, where $\mu^* = \mu Q$ and $\Sigma^* = Q'\Sigma Q$. In result one obtain the statistic

$$\begin{aligned} V_n^* &= (\overline{x}(n)Q)(Q'S(n)Q)^{-1}(\overline{x}(n)Q)' \\ &= \overline{x}^*(n)(S^*)^{-1}(\overline{x}^*(n))', \end{aligned} \tag{7}$$

where $\overline{x}^*(n) = \overline{x}(n)Q$ and $S^* = Q'S(n)Q$. Thus, V_n^* is of the same type as (1) with the corresponding distribution reduced now to

$$F^* = \frac{n-k+1}{k-1} V_n^* \sim F_{k-1,n-k+1}(n\xi^*), \quad \xi^* = \mu^* \Sigma^{*-1} \mu^{*\prime}$$

(cf. Scheffé, 1959, p. 272). It can also be approximated by the usual F, as

$$\tilde{F} = \frac{n-k+1}{k-1+n\xi^*} V_n^* \sim F_{m,n}\ , \quad m = \frac{(k-1+n\xi^*)^2}{k-1+2n\xi^*}.$$

Therefore, for the transformed data a multivariate T^2-test can be applied, to test the hypothesis (4), or more generally to test the hypothesis

$$H_0 \colon \mu^*(\Sigma^*)^{-1}(\mu^*)' \leq \xi_0^* \quad \text{vs.} \quad H_1 \colon \mu^*(\Sigma^*)^{-1}(\mu^*)' \geq \xi_1^* > \xi_0^*,$$

by the corresponding sequential probability ratio test, as described in Section 3.

It should also be noted that when replacing the $k \times (k-1)$ matrix Q in the above transformation by an $k \times 1$ vector $q = (q_1, \ldots, q_k)'$ such that $q_1 + \ldots + q_k = 0$, then the derived multivariate T^2-test can be applied to test a hypothesis concerning the comparison μq. If interest is in hypotheses concerning many comparisons of this type, a multiple-comparison method can be used, one of them being the S-method of Scheffé (1953, 1959, Section 3.1 and 8.1).

6. The Algorithm

There are different methods either graphical or analytical which lead to a minimum dimension subspace, say $E(x_1, x_2, \ldots, x_m)$, of the space spanned by the variables $x_1, x_2, \ldots, x_m$. A typical example is the principal component analysis.

The target of this paper is to provide a method for multiple-multivariate-sequential T^2-comparisons. The multivariate sequential T^2-test has been discussed already. The multiple comparisons problem either with the *best* (Piegorsh, 1981) or with the *worst* (Ruberg and Hsu, 1992), is well known, while different multi-comparison procedures are discussed by Westfall and Young (1993). Under the assumption that $x_1, x_2, \ldots, x_m$ are normal variables observed in ℓ different samples, the following algorithm can be proposed.

Step 1. Reduce dimensionality, i.e. obtain the k *common* variables for the ℓ samples due to $\max_\ell\{\min_m E(x_1, \ldots, x_m)\} = k$.

Step 2. Calculate for an initial sample size, say n_0, for the k variables Fisher's information matrix $I_j(n_0)$, $j = 1, \ldots, \ell$ for ℓ samples.

Step 3. For every two samples calculate the covariance matrix $S(n_0)$ as the inverse of the weighted average of Fisher's information matrices.

Step 4. Obtain $V^*_{n_0}$ as in (7) with $\overline{x}(n_0) = \overline{x}_i(n_0) - \overline{x}_j(n_0)$, $i \neq j = 1, \ldots, \ell$, and Q as in (6).

Step 5. Augmentation of data to certain level through Z_n as in (2) to obtain stopping rule.

Acknowledgements

I would like to thank the referees for their helpful comments.

References

Anderson, J.W. (1984). *An Introduction to Multivariate Statistical Analysis.* Wiley, New York.

Ford, I. (1976). Optimal static and sequential design. Ph.D thesis, University of Glasgow.

Ford I., Titterington D.M. and Wu F.J. (1985). Inference and sequential design. *Biometrika* **72**, 545–551.

Ford I., Kitsos C.P. and Titterington D.M. (1989). Recent advances in nonlinear experimental design. *Technometrics* **31**, 49–60.

Ghosh, B.K. (1965). Sequential range tests for components of variance. *Journal of the American Statistical Association* **60**, 826–836.

Ghosh, B.K. (1967). Sequential analysis of variance under random and mixed models. *Journal of the American Statistical Association* **62**, 1401–1417.

Kitsos, C.P. (1986). Design and inference in nonlinear problem. Ph.D. thesis, University of Glasgow.

Kitsos, C.P. (1989). Fully sequential procedures in nonlinear design problems. *Computational Statistics & Data Analysis* **8**, 13–19.

Lai, L.T. (1981). Asymptotic optimality of invariant sequential probability ratio tests. *The Annals of Mathematical Statistics* **8**, 318–333.

Patnaik, P.B. (1949). The non-central χ^2 and F-distributions and their application. *Biometrika* **37**, 78–87.

Piegorsch, W.W. (1991). Multiple comparisons for analysing dichotomous response. *Biometrics* **47**, 45–52.

Ruberg, S.J. and Hsu, J.C. (1992). Multiple comparison procedures for pooling batches in stability studies. *Technometrics* **34**, 465–472.

Scheffé, H. (1959). *The Analysis of Variance.* Wiley, New York.

Siegmund, D.(1985). *Sequential Analysis.* Springer-Verlag, New York.

Wald, A. and Wolfowitz, J. (1948). Optimum character of the sequential probability ratio test. *The Annals of Mathematical Statistics* **19**, 326–329.

Westfall, P.H. and Young, S.S (1993). *Resampling-Based Multiple Testing.* Wiley, New York.

ON DIAGNOSING COLLINEARITY-INFLUENTIAL POINTS IN LINEAR REGRESSION

HANS NYQUIST
Department of Biometry and Forest Management
Swedish University of Agricultural Sciences
S-901 83 Umeå
Sweden

Abstract. The influence function for the condition indexes of the data matrix in multiple linear regression is derived. Sample versions of the influence function are proposed to measure the influence of each observation on the condition indexes. In particular, it is found that a sample version of the influence function is equivalent to a proposal for detecting collinearity-influential observations. Other sample versions of the influence function yield alternative devices for detecting collinearity-influential points. A diagnostic based on the empirical influence function is found to be easy to compute and easy to interpret. This diagnostic is also related to the leverage component plots for detecting collinearity-influential points.

Key words: Collinearity-influential points, Condition indexes, High-leverage points, Influence function, Leverage components.

1. Introduction

Suppose X is an $n \times p$ full rank data matrix of n observations on p predictor variables and y is an $n \times 1$ vector of observations on a response variable related to X according to the multiple linear regression model

$$y = X\beta + \epsilon.$$

Here β is an unknown $p \times 1$ parameter vector to be estimated and ϵ is an $n \times 1$ error vector. A serious problem that can occur when analyzing multiple linear regression models is the presence of collinearities, i.e. approximate linear dependencies among the columns of X. An immediate consequence of collinearities is a highly unstable least squares estimation of the parameter vector β.

The purpose of diagnostic techniques is to point out the presence of potential problems in the analysis and to indicate a starting point for further investigation. Several diagnostics are suggested for detecting collinearity problems in multiple regression analysis (see, e.g., Mason, Gunst and Webster, 1975; Belsley, Kuh and Welsch, 1980; Belsley, 1991, for systematic treatment of the subject). While many of the proposed diagnostics fall short in providing information about all important aspects, Belsley (1991) suggests a more complete procedure based on condition indexes and variance-decomposition proportions. This procedure is presented in Section 2.

In order to increase the precision of an analysis when collinearity problems are present, it is necessary to identify the source of the problems and to select an appropriate inference technique (Mason et al., 1975). Recently it has been found that individual observations can create or hide a collinearity problem (Belsley et al., 1980; Hocking and Dunn, 1982; Mason and Gunst, 1985; Chatterjee and Hadi, 1988, p. 158). Observations that either create or hide a collinearity problem are referred to as collinearity-influential points. It has also been recognized that condition indexes are sensitive to collinearity-influential points. A careful analysis should therefore include an investigation whether a condition index is large because of the presence of observations that create collinearity problems or it is small because of the presence of observations that hide collinearity problems. In Sections 3 and 4 we propose an approach for assessing the influence of individual observations on condition indexes. Other diagnostics for detecting collinearity-influential points have been proposed by Dorsett (1982) and Walker (1989). A numerical illustration of the proposed approach is presented in the final section.

After diagnosing the data the next step of the analysis is to estimate and/or test the model parameters. One way to reduce or remove problems caused by observations that cause collinearity problems is to downweight or delete those observations. However, if the information provided by such observations is of vital interest, a deletion or downweighting would increase the problems rather than reduce them. It is therefore important to further investigate influential observations rather than automatically reduce their influence.

2. Condition Indexes and Variance-Decomposition Proportions

Let the singular value decomposition of the matrix X be $X = UDG^T$, where $U^TU = G^TG = I$ and D is the diagonal matrix with singular values in decreasing order, $m_1 > m_2 > \ldots > m_p$. Then the columns $g_1, \ldots, g_p$ of the orthonormal matrix G are the eigenvectors of X^TX corresponding to the eigenvalues $m_1^2, \ldots, m_p^2$, respectively.

The length of the vector Xg_h is m_h implying that for each near linear dependency there will be a 'small' singular value of X. For determining a basis for assessing smallness the singular values are compared with the largest singular value m_1, thus defining the set of condition indexes $k_h = m_1/m_h, \quad h = 1, \ldots, p$.

The covariance matrix of the least squares estimator $b = (X^TX)^{-1}X^Ty$ of β may now be written as

$$\mathrm{Var}(b) = \sigma^2(X^TX)^{-1} = \sigma^2 GD^{-2}G^T,$$

so that the variance of the j-th regression coefficient, b_j, is

$$\mathrm{Var}(b_j) = \sigma^2 \sum_{h=1}^{p} \frac{g_{jh}^2}{m_h^2} = \frac{\sigma^2}{m_1^2} \sum_{h=1}^{p} (k_h g_{jh})^2,$$

with $G = (g_{jh})$. This defines a decomposition of the variance of b_j into a sum in which each term is associated with exactly one singular value. If there is a near dependency in which a predictor variable is involved, then the corresponding numerator will be large and the denominator will be small, thus defining a term with a large contribution to the variance of the corresponding parameter estimator. This suggests that if large proportions of the variance of two or more parameter estimators appear for the same large condition index, then the corresponding variables are involved in a near linear dependency. It is therefore convenient to define the variance-decomposition proportions as

$$p_{hj} = \frac{g_{jh}^2}{m_h^2} / \sum_{r=1}^{p} \frac{g_{jr}^2}{m_r^2} = (k_h g_{jh}) / \sum_{r=1}^{p} (k_r g_{jr})^2, \qquad j, h = 1, \ldots, p.$$

A problem with condition indexes is that they are sensitive to column scaling of X. It is therefore recommended to use a standard scaling of the columns of X (i.e., letting the sum of squares of each column be equal to some constant) before computing the condition indexes (Belsley, 1984; Belsley, 1991, p. 65).

In summary, the diagnostic procedure consists of two steps. First identifying the number of near dependencies as the number of high condition indexes associated with high variance-decomposition proportions. Second, for each near dependency identifying the variables involved in it as those variables associated to high variance-decomposition proportions.

3. The Influence Function for the Condition Indexes

Let the rows of X be observations on a random vector $x = (x_1, \ldots, x_p)^T$ with cumulative distribution (cdf) F defined on $\mathbb{R}^p$ such that the second order moment matrix is $\Sigma = \int xx^T dF$. By allowing x to have a design measure we are able to describe problems with X fixed as well as random (cf., e.g., Hinkley, 1977). We are now interested in assessing changes in the singular values $m_1 > m_2 > \ldots > m_p$ and condition indexes $k_1 < k_2 < \ldots < k_p$ subject to a perturbation scheme. We therefore define functionals $\mu_1, \mu_2, \ldots, \mu_p, \kappa_1, \kappa_2, \ldots, \kappa_p$ mapping the set of all cdf's defined on $\mathbb{R}^p$ onto $\mathbb{R}$ such that if F_n is the sample distribution function, then

$$m_h = \mu_h(F_n), \qquad h = 1, \ldots, p,$$

$$k_h = \kappa_h(F_n) = \mu_1(F_n)/\mu_h(F_n), \quad h = 1, \ldots, p. \tag{1}$$

With these notations, $\mu_1 = \mu_1(F), \ldots, \mu_p = \mu_p(F)$ and $\kappa_1 = \kappa_1(F), \ldots, \kappa_p = \kappa_p(F)$ denote the population values, i.e. square roots of the eigenvalues and the condition indexes of Σ. Further, to the population singular value μ_h there corresponds the population eigenvector γ_h, and to the sample singular value m_h the sample eigenvector $g_h, h = 1, \ldots, p$.

The influence function (Hampel, 1974; Hampel, Ronchetti, Rousseeuw and Stahel, 1986) describes the behavior of the asymptotic value of an estimator when an infinitesimal contamination at a point x is introduced. Let δ_x be the distribution function with point mass 1 at $x = (x_1, \ldots, x_p)^T$ and let $\widetilde{F} = (1-s)F + s\delta_x$ be a contamination of F if $s \geq 0$ is sufficiently near zero. Then the influence function for a functional $\tau = \tau(F)$ is defined as

$$\mathrm{IF}(x; F, \tau) = \lim_{s \downarrow 0} \{\tau(\widetilde{F}) - \tau(F)\}/s$$

provided that the limit exists. If F is replaced by F_n and $s = (n+1)^{-1}$ the influence function can be interpreted as approximately $n+1$ times the change of $\kappa_h(F_n)$ caused by an additional observation at x. By substitution of (1) we obtain

$$\begin{aligned}\mathrm{IF}(x;F,\kappa_h) &= \lim_{s\downarrow 0}\left[\left\{\frac{\mu_1((1-s)F+s\delta_x)}{\mu_h((1-s)F+s\delta_x)}\right\}-\left\{\frac{\mu_1(F)}{\mu_h(F)}\right\}\right]/s \\ &= \{\mathrm{IF}(x;F,\mu_1)\mu_h - \mathrm{IF}(x;F,\mu_h)\mu_1\}/\mu_h^2,\end{aligned}$$

where $\mathrm{IF}(x;F,\mu_h)$ is the influence function for $\mu_h, h=1,\ldots,p$. Theorem 1 of Radhakrishnan and Kshirsagar (1981) states that the influence function for the eigenvalue λ_h satisfies the equality

$$\mathrm{IF}(x;F,\lambda_h) = \gamma_h^T \mathrm{IF}(x;F,\Sigma)\gamma_h,$$

where

$$\begin{aligned}\mathrm{IF}(x;F,\Sigma) &= \lim_{s\downarrow 0}\{\Sigma((1-s)F+s\delta_x)-\Sigma(F)\}/s \\ &= \lim_{s\downarrow 0}\{(1-s)\Sigma + sxx^T - \Sigma\}/s = xx^T - \Sigma,\end{aligned}$$

which in turn implies that

$$\mathrm{IF}(x;F,\lambda_h) = \gamma_h^T(xx^T-\Sigma)\gamma_h = (\gamma_h^T x)^2 - \lambda_h.$$

Since $\mu_h = \lambda_h^{1/2}$, we find that

$$\mathrm{IF}(x;F,\mu_h) = \lim_{s\downarrow 0}[\{\lambda_h((1-s)F+s\delta_x)\}^{1/2} - \{\lambda_h(F)\}^{1/2}]/s = \mathrm{IF}(x;F,\lambda_h)/(2\mu_h).$$

By substitution we obtain

$$\begin{aligned}\mathrm{IF}(x;F,\kappa_h) &= \left[\frac{1}{2\mu_1}\{(\gamma_1^T x)^2-\mu_1^2\}\mu_j - \frac{1}{2\mu_h}\{(\gamma_h^T x)^2 - \mu_h^2\}\mu_1\right]/\mu_h^2 \\ &= \frac{\kappa_h}{2}\left\{\frac{(\gamma_1^T x)^2}{\mu_1^2} - \frac{(\gamma_h^T x)^2}{\mu_h^2}\right\}. \qquad (2)\end{aligned}$$

Since $\gamma_h^T x$ is the value of the h-th principal component corresponding to an observation at x we find that the influence function of κ_h is proportional to the difference of squared standardized values on the first and h-th principal component, the constant of proportionality being $\kappa_h/2$.

By restricting x to be of length c, maximum and minimum of the influence function are obtained by equating the derivatives of the Lagrange function to zero,

$$\frac{\kappa_h}{2}\left\{\frac{(\gamma_1^T x)^2}{\mu_1^2} - \frac{(\gamma_h^T x)^2}{\mu_h^2}\right\} - a(x^T x - c^2) = 0,$$

where a is a Lagrange multiplier. Straightforward calculations yield the system of equations

$$\left\{\frac{\gamma_1\gamma_1^T}{\mu_1^2} - \frac{\gamma_h\gamma_h^T}{\mu_h^2}\right\}x = a^* x,$$

where $a^* = 2a/\kappa_h$. Hence, a^* is an eigenvalue and x is the corresponding eigenvector of $(\mu_1^{-2}\gamma_1\gamma_1^T - \mu_h^{-2}\gamma_h\gamma_h^T)$. The solution is therefore either $x = c\gamma_1$ (yielding $\mathrm{IF}(c\gamma_1; F, \kappa_h) = c^2\kappa_h/2\mu_1^2$) or $x = c\gamma_h$ (yielding $\mathrm{IF}(c\gamma_h; F, \kappa_h) = -c^2\kappa_h/2\mu_h^2$). Thus, the influence function for κ_h is increased when x is along γ_1 and is decreased when x is along γ_h. Its absolute value is maximized when x is in the direction of γ_h. The influence function is zero when x is on the hyperplane orthogonal to γ_1 and γ_h, or when x is on the line $c(\mu_1\gamma_1 + \mu_h\gamma_h)$ or $c(\mu_1\gamma_1 - \mu_h\gamma_h)$.

4. An Observations' Influence on the Condition Indexes

The influence function for κ_h can be interpreted as a measure of the influence on κ_h of adding one observation x to a very large sample. A measure of the relative change would then be

$$H_h(x) = \frac{\mathrm{IF}(x; F, \kappa_h)}{\kappa_h}. \tag{3}$$

However, the influence function is a large sample device so a finite sample approximation is required for assessing an observations influence on the condition indexes. There are several possibilities of constructing a finite sample version of an influence function. One is the sample influence function, SIC (Hampel et al., 1986, p. 92). It is obtained as the jackknife estimate of the influence function (Cook and Weisberg, 1982, p. 108)

$$\mathrm{SIC}(x_i; \kappa_h) = (n-1)(k_h - k_{h(i)}),$$

where $k_{h(i)}$ denotes the h-th condition index with the i-th observation removed. Substituting the sample influence function into (3) we get the single deletion measure

$$H_{hi}^S = (n-1)\frac{k_h - k_{h(i)}}{k_h}.$$

While the computational burden can be high for computing condition indexes with data points deleted, significant simplifications can be obtained by using approximations (see, e.g., Chatterjee and Hadi, 1988, p. 167; Wang and Nyquist, 1991; Hadi and Nyquist, 1993).

A second sample version of the influence function is the empirical influence function, EIC (Cook and Weisberg, 1982, p. 108) in which F is replaced by its sample counterpart F_n. Thus, $\mathrm{EIC}(x_i; \kappa_h)$ is obtained if population values of κ_h, μ_h and γ_h in (2) are replaced by sample values k_h, m_h and g_h, respectively. Hence,

$$\mathrm{EIC}(x_i; \kappa_h) = k_h(\omega_{i1}^2 - \omega_{ih}^2)/2,$$

where $\omega_{ih}^2 = (g_h^T x_i)^2/m_h^2, h = 1, \ldots, p$ are squared normalized principal components, also known as leverage components. By substituting the empirical influence function into (3) we get

$$H_{hi}^E = (\omega_{i1}^2 - \omega_{ih}^2)/2.$$

It is interesting here to compare this result with the leverage component (LC) plot suggested by Mason and Gunst (1985). LC plots are scatter plots of pairs of leverage components in which collinearity-influential points are supposed to be separated from the bulk of the other points, while the H_{ih}^E diagnostic utilizes the difference between the first and the h-th leverage component for detecting points with a high influence on the h-th condition index.

An immediate advantage of H_{hi}^E over H_{hi}^S as measures of an observations' influence on the condition indexes is that H_{hi}^E is much easier to compute. It should also be noted that H_{hi}^E admits a fairly simple interpretation. The definition of H_{hi}^E suggests that attention should be paid to observations along g_1 and g_h, the eigenvectors corresponding to the largest singular value and the h-th singular value, respectively. Points that are far out in these directions are particularly influential on the condition indexes and are therefore collinearity-influential. On the other hand, observations on a hyperplane orthogonal to g_1 and g_h, and observations on the lines $c(m_1 g_1 \pm m_h g_h)$ have no effect on H_{hi}^E. These observations are therefore classified as not influential on the h-th condition index.

High-leverage points are observations that are extreme relative to the bulk of the remaining observations on the predictor variables. In many applications there appears to be a close connection between collinearity-influential points and high-leverage points. An explanation of this is that high-leverage points tend to influence the eigenstructure of X and hence the condition number k_p of X. However, as has been illustrated in Chatterjee and Hadi (1988), not all high-leverage points have a large influence on k_p and not all points with a high influence on k_p are high-leverage points. One measure of the observations' leverage is the diagonal elements of the projection matrix, $p_{ii} = x_i^T(X^TX)^{-1}x_i$. Factoring X^TX as a product of the eigenvector and singular value matrices we find that

$$p_{ii} = \sum_{h=1}^{p} \frac{(g_h^T x_i)^2}{m_h^2} = \sum_{h=1}^{p} \omega_{ih}^2,$$

i.e., p_{ii} can be expressed as a sum of p leverage components while H_{hi}^E is half the difference between the first and the h-th term in the sum. Evidently, p_{ii} can be large when H_{hi}^E is small and vice versa.

5. A Numerical Illustration

In this section we illustrate the techniques for diagnosing collinearity-influential observations presented in the previous section. Data for the illustration are originally from Woods, Steinour and Starke (1932) and have been extensively analyzed by Daniel and Wood (1980) and Chatterjee and Hadi (1988). In this problem the heat generated during hardening of 14 samples of cement was related to the weights (measured as percentages of the total weight of each sample) of five clinker components in a linear regression model without intercept

$$y = \beta_1 X_1 + \beta_2 X_2 + \beta_3 X_3 + \beta_4 X_4 + \beta_5 X_5 + \epsilon. \tag{4}$$

Estimates of the regression parameters and their estimated standard errors are shown in Table I. Three parameters, β_2, β_3, and β_4 are significant by a t-test. R^2- and F-values are very high. After column scaling, condition indexes and variance-decomposition proportions are computed and reported in Table II. The two largest

TABLE I

Estimated parameters for the Cement Data Model (4). Estimated standard errors in parentheses. Statistics for full data: s=2.62, R^2=99.95%, F=3933 with 5 and 14 d.f. and with obs. 3 deleted: s=2.77, R^2=99.95%, F =3195 with 5 and 13 d.f.

	Parameter				
	β_1	β_2	β_3	β_4	β_5
Full data	0.327	2.025	1.297	0.558	0.354
	(0.17)	(0.21)	(0.06)	(0.05)	(1.06)
Obs. 3	0.307	1.971	1.251	0.512	2.150
deleted	(0.29)	(0.36)	(0.25)	(0.25)	(9.58)

TABLE II

Condition indexes and variance-decomposition proportions: Cement Data

Condition	Proportions				
index	var(b_1)	var(b_2)	var(b_3)	var(b_4)	var(b_5)
1.0	0.004	0.006	0.003	0.007	0.004
2.5	0.043	0.115	0.003	0.038	0.001
3.8	0.042	0.038	0.057	0.265	0.000
8.4	0.297	0.459	0.000	0.003	0.705
11.8	0.613	0.383	0.936	0.688	0.291

condition indexes are 11.8 and 8.4, values that usually are considered as acceptable. Scanning across the bottom row, we see that all variables, X_5 to a lesser extent, though, are involved in a near dependency. This is also reflected by some high pairwise correlations, -0.84 between X_1 and X_2, and -0.98 between X_3 and X_4. Large proportions of the variances of b_2 and b_5 are governed by the second largest condition index.

In order to investigate the influence of particular observations on the condition indexes we compute the H_{hi}^E statistics. These values are reported in Table III and for the two largest condition indexes graphically illustrated in Figures 1 and 2. Looking at H_{5i}^E it appears that observations 5 and 3 have the highest influences on the largest condition index. These two observations are also separated from the rest of the data in the LC plot of ω_{i5}^2 versus ω_{i1}^2 in Figure 3. However, their influence seems to be of limited importance. We conclude that the near dependency defined by the largest condition index is not generated by a few influential observations on it. On the other hand, examination of H_{4i}^E reveals that observation 3 has a large influence on the second largest condition index and observation 10 has a smaller influence. Further computations indicate that a removal of observation 3 would increase the condition index k_4 from 8.4 to 70.9, an increase of 744%. The effect of the removal of observation 3 on the parameter estimates and their standard errors are reported in Table I. We see that the precision with which parameters are estimated has been reduced, particularly for the estimator of β_5. Only two parameters, β_2 and

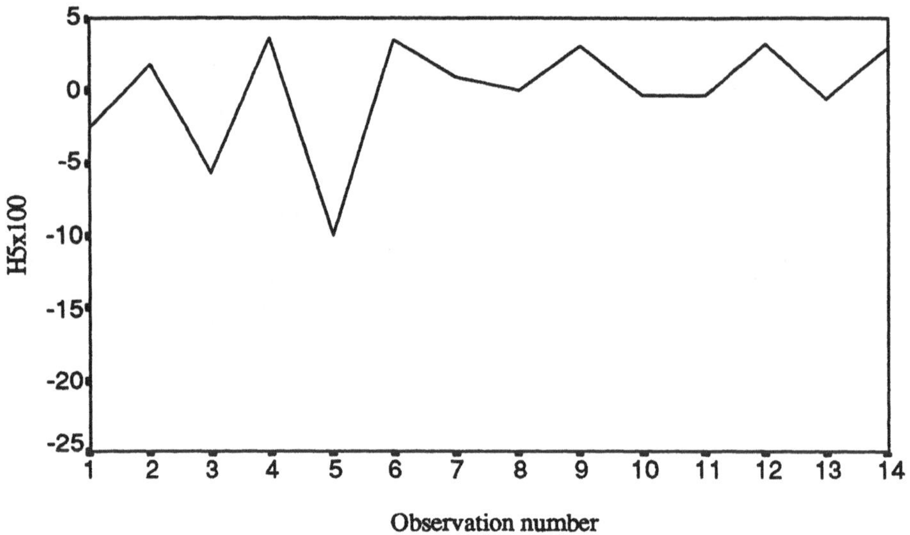

Fig. 1. Index plot of H_{5i}^E for Cement Data.

TABLE III
Collinearity diagnostics and leverage values for Cement Data

Obs.	$H_{2i}^E \times 100$	$H_{3i}^E \times 100$	$H_{4i}^E \times 100$	$H_{5i}^E \times 100$	p_{ii}
1	3.13	-7.44	1.64	-2.54	0.44
2	-1.77	1.80	2.28	1.80	0.22
3	2.68	5.07	-24.62	-5.74	0.99
4	3.46	-2.14	2.97	3.69	0.21
5	2.75	2.87	1.62	-9.91	0.36
6	1.90	3.49	2.62	3.48	0.13
7	2.78	-10.72	2.55	1.02	0.38
8	-3.48	3.43	2.31	0.03	0.30
9	1.52	-1.43	3.10	3.07	0.19
10	-7.54	1.46	-6.06	-0.26	0.72
11	-2.21	1.70	1.84	-0.27	0.33
12	0.85	1.51	3.48	3.28	0.19
13	0.81	0.93	3.26	-0.54	0.25
14	-4.87	-0.51	3.03	2.91	0.31

β_3, remain significant by a t-test. However, R^2- and F-values are still very high. In conclusion, observation 3 is collinearity-influential. The Cement Data suffers from beeing short in one direction, but a removal of observation 3 would make the problems much worse. This conclusion is also clear from the LC plot of ω_{i4}^2 versus ω_{i1}^2, shown in Figure 4.

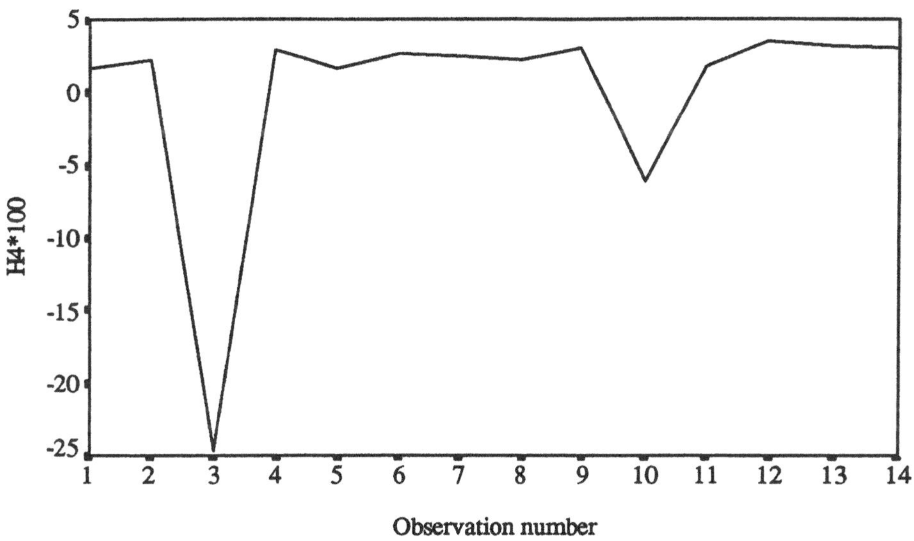

Fig. 2. Index plot of H^E_{4i} for Cement Data.

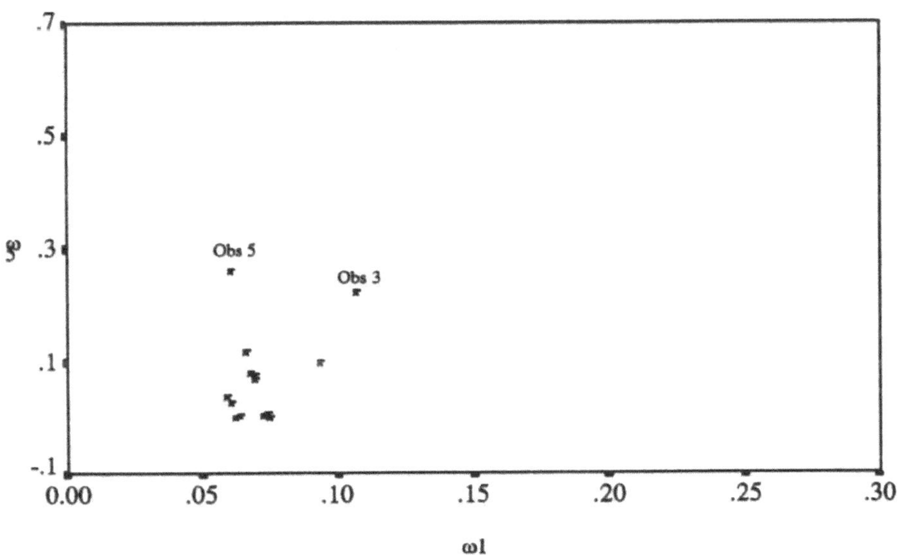

Fig. 3. LC plot of ω^2_{i5} versus ω^2_{i1} for Cement Data.

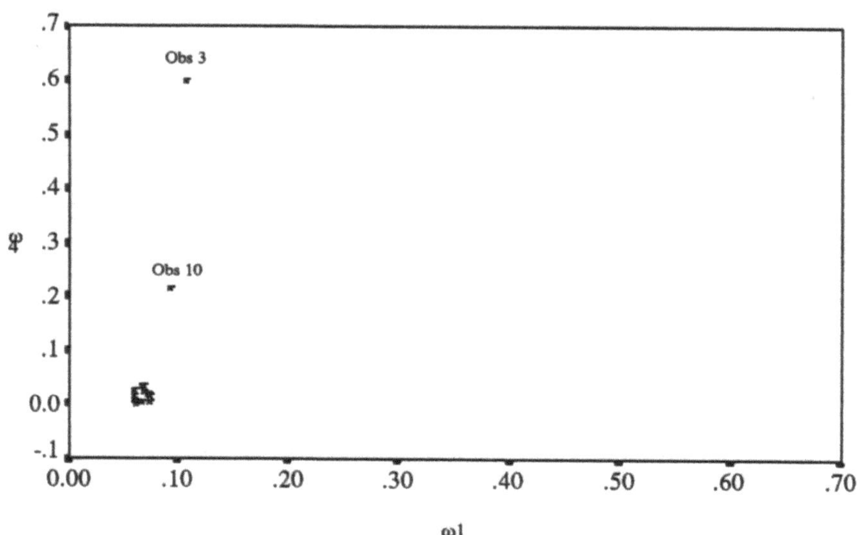

Fig. 4. LC plot of ω_{i4}^2 versus ω_{i1}^2 for Cement Data.

References

Belsley, D.A. (1984). Demeaning conditioning diagnostics through centering (with comments). *The American Statistician* **38**, 73–93.

Belsley, D.A. (1991). *Conditioning Diagnostics, Collinearity and Weak Data in Regression.* Wiley, New York.

Belsley, D.A., Kuh, E., and Welsch, R.E. (1980). *Regression Diagnostics.* Wiley, New York.

Chatterjee, S., and Hadi, A.S. (1988). *Sensitivity Analysis in Linear Regression.* Wiley, New York.

Cook, R.D. and Weisberg, S. (1982). *Residuals and Influence in Regression.* Chapman and Hall, London.

Daniel, C. and Wood, F.S. (1980). *Fitting Equations to Data: Computer Analysis of Multifactor Data*, 2nd ed. Wiley, New York.

Dorsett, D. (1982). Resistant M-estimators in the presence of influential points. Ph.D. dissertation, Southern Methodist University, Dallas.

Hadi, A.S. and Nyquist, H. (1993). Further theoretical results and a comparison between two methods for approximating eigenvalues of perturbed covariance matrices. *Statistics and Computation.* To appear.

Hampel, F.R. (1974). The influence curve and its role in robust estimation. *Journal of the American Statistical Association* **69**, 383–393.

Hampel, F.R., Ronchetti, E.M., Rousseeuw, P.J., and Stahel, W.A. (1986). *Robust Statistics - The Approach Based on Influence Functions.* Wiley, New York.

Hinkley, D.V. (1977). Jackknifing in unbalanced situations. *Technometrics* **19**, 285–292.

Hocking, R.R. and Dunn, M.R. (1982). Collinearity, influential data and ridge regression. Paper presented at Ridge Regression Symposium. University of Delaware.

Mason, R.L., and Gunst, R.F. (1985). Outlier-induced collinearities. *Technometrics*, **27**, 401–407.

Mason, R.L., Gunst, R.F., and Webster, J.T. (1975). Regression analysis and problems of multicollinearity. *Communications in Statistics* **4**, 277–292.

Radhakrishnan, R., and Kshirsagar, A.M. (1981). Influence functions for certain parameters in multivariate analysis. *Communications in Statistics - Theory and Methods A* **10**, 515–529.

Walker, E. (1989). Detection of collinearity-influential observations. *Communications in Statistics*

Theory and Methods **18**, 1675–1690.

Wang, S.G. and Nyquist, H. (1991). Effects on the eigenstructure of a data matrix when deleting an observation. *Computational Statistics & Data Analysis* **11**, 179-188.

Woods, H., Steinour, H.H., and Starke, H.R. (1932). Effect of composition of portland cement on heat evolved during hardening. *Industrial and Engineering Chemistry* **24**, 1207–1214.

USING NONNEGATIVE MINIMUM BIASED QUADRATIC ESTIMATION FOR VARIABLE SELECTION IN THE LINEAR REGRESSION MODEL

STANISLAW GNOT
Agricultural University of Wrocław
Grunwaldzka 53
50-357 Wrocław
Poland

HENNING KNAUTZ
Institute of Statistics and Econometrics
University of Hamburg
Von-Melle-Park 5
D-20146 Hamburg
Germany

and

GÖTZ TRENKLER
Department of Statistics
University of Dortmund
D-44221 Dortmund
Germany

Abstract. The paper is concerned with the problem of variable selection in the linear regression model. We state some properties of the classical procedure known as Mallows' C_p and suggest some new strategies based on nonnegative minimum biased quadratic estimators. Finally we apply these procedures to an example.

Key words: Linear model, Variable selection, Mallows' C_p, Nonnegative minimum biased estimation.

1. Revising Mallows' C_p

Let us consider the conventional regression model

$$y = X\beta + \epsilon, \tag{1.1}$$

where X is the $n \times k$ nonstochastic model matrix with full column rank k, for ϵ, the vector of disturbances, we assume $E(\epsilon) = 0$ and $\text{Cov}(\epsilon) = \sigma^2 I_n$.

Model (1.1) is assumed to be the correct specified regression model including all relevant regressors. Let X be partitioned as $X = (X_1 : X_2)$, where X_1 is of dimension $n \times p$. Correspondingly partition β as $\beta = (\beta_1', \beta_2')'$ with $\beta_1 \in \mathbb{R}^p$. Suppose we leave out the regressors contained in the submatrix X_2 and consider

$$y = X_1\beta_1 + \epsilon. \tag{1.2}$$

T. Caliński and R. Kala (eds.),
Proceedings of the International Conference on Linear Statistical Inference LINSTAT '93, 65–71.

Using the least squares estimator of β_1 based on (1.2), $b_1 = (X_1'X_1)^{-1}X_1'y$, generally induces some bias in the estimation of β_1 and also in the prediction of y, i.e.

$$E(\hat{y}_1 - y) = -(I - P_1)X\beta \neq 0,$$

where $\hat{y}_1 = P_1 y = X_1 b_1$ and $P_1 = X_1(X_1'X_1)^{-1}X_1'$ is the orthogonal projector on the range of $X_1, \mathcal{R}(X_1)$. The problem which arises is to choose the regressors in X_1 such that the (squared) bias of $\hat{y}_1$ under the true model (1.1) is close to zero, i.e. $B^2(\hat{y}_1) = \beta'X'(I - P_1)X\beta \approx 0$. Alternatively, we can also consider the total mean square error of $\hat{y}_1$ given by

$$\begin{aligned} TMSE(\hat{y}_1) &= E(\hat{y}_1 - E(y))'(\hat{y}_1 - E(y)) \\ &= \beta'X'(I - P_1)X\beta + p\sigma^2. \end{aligned} \tag{1.3}$$

A statistic for estimating the relative total mean square error $TMSE(\hat{y}_1)/\sigma^2$, which is used for variable selection, is Mallows' C_p which was introduced in 1964 (see Gorman and Toman, 1966; Daniel and Wood, 1971, p.86; Mallows, 1973; Sen and Srivastava, 1990, p.238). It is based on the residual sum of squares in the model (1.2), given by $RSS_1 = y'(I - P_1)y$, and is defined as

$$C_p = (RSS_1/s^2) - (n - 2p), \tag{1.4}$$

where s^2 is an appropriate estimator for σ^2, usually the unbiased estimator from the full model,

$$s^2 = y'My/(n - k), \tag{1.5}$$

where $M = I - P$ and $P = X(X'X)^{-1}X'$ is the orthogonal projector on the range of $X, \mathcal{R}(X)$. Since $E(RSS_1) = \beta'X'(I - P_1)X\beta + (n - p)\sigma^2$, we have

$$E[RSS_1/\sigma^2 - (n - p)] = B^2(\hat{y}_1)/\sigma^2. \tag{1.6}$$

In view of (1.6), C_p is expected to be close to p if the squared bias of $\hat{y}_1$ is close to zero. Consequently those models are assumed to be appropriate which have C_p-values close to p, the number of regressors which are included in the smaller model.

As it can be seen from the preceding considerations the derivation of this statistic is rather heuristic. In the following we will demonstrate that C_p has some unacceptable properties. We first state the relation between C_p and the unbiased uniformly minimum variance estimator for $TMSE(\hat{y}_1)$ (cf. Gnot, Trenkler and Zmyślony, 1992)

$$\hat{\tau}_1 = b'Hb + s^2[p - \mathrm{tr}H(X'X)^{-1}], \tag{1.7}$$

where $H = X'(I - P_1)X$, b is the least squares estimator in (1.1), i.e. $b = (X'X)^{-1}X'y$, and s^2 is given (1.5).

Theorem 1. *Let C_p and $\hat{\tau}_1$ be defined as in* (1.4) *and* (1.7). *Then it follows that*

$$C_p = \hat{\tau}_1/s^2. \tag{1.8}$$

Proof. Since $\mathcal{R}(X_1) \subset \mathcal{R}(X)$, $(I - P_1)P$ is the projector on $\mathcal{R}^{\perp}(X_1) \cap \mathcal{R}(X)$, where $\perp$ denotes the orthogonal complement, which has dimension $k - p$. Thus we have

$$\begin{aligned} \mathrm{tr} H(X'X)^{-1} &= \mathrm{tr}(I - P_1)P \\ &= \mathrm{rank}(I - P_1)P = k - p, \end{aligned}$$

where tr denotes the trace. Rewriting $\hat{\tau}_1$ we derive

$$\hat{\tau}_1 = y'P(I - P_1)Py + s^2(2p - k). \tag{1.9}$$

Since $\mathcal{R}(X_1) \subset \mathcal{R}(X)$, we also have $PP_1 = P_1 = P_1P$, which implies $P(I - P_1) = P(I - P_1)P$ and $M(I - P_1) = M$. It follows that

$$\begin{aligned} s^2 C_p &= y'(I - P_1)y + (2p - n)s^2 \\ &= \mathrm{tr} P(I - P_1)yy' + \mathrm{tr} M(I - P_1)yy' + (2p - n)s^2 \\ &= y'P(I - P_1)Py + y'My + (2p - n)s^2 \\ &= y'P(I - P_1)Py + (2p - k)s^2, \end{aligned}$$

which in view of (1.9) yields the asserted identity (1.8).□

A direct consequence from the representation of C_p in the above theorem is the following: *despite the fact that $\hat{\tau}_1$ is the unbiased and uniformly minimum variance estimator it can take on negative values for $p < k/2$.* This implies that C_p can be negative whenever we leave out more than half of the regressors of the full model. This is an unacceptable property since the relative total mean square error, which is to be estimated, is always positive. Assuming normal distribution for the errors in (1.1), it can be shown that

$$E(C_p) = \frac{\beta' H \beta}{\sigma^2} + p + \frac{2}{n - k - 2}\left(k - p + \frac{\beta' H \beta}{\sigma^2}\right), \tag{1.10}$$

for $n - k > 2$, and for $n - k > 4$

$$\begin{aligned} \mathrm{Var}(C_p) &= 2\left(\frac{n-k}{n-k-2}\right)^2 \frac{1}{n-k-4}\left[\left(k - p + \frac{\beta' H \beta}{\sigma^2}\right)^2 \right. \\ &\quad \left. + (n - k - 2)\left(k - p + 2\frac{\beta' H \beta}{\sigma^2}\right)\right]. \end{aligned}$$

We conclude from (1.10) that C_p is always positively biased. Moreover, both expectation and bias, heavily depend on the difference $k - p$. Consequently, large numbers of regressors are favoured. It is obvious that at least some of the problems are introduced by estimating the *relative* mean square error of $\hat{y}_1$ and the replacement of the unknown quantity σ^2 by the estimator s^2. Thus we might ask if it were not better to estimate the absolute total mean square error (1.3) by an estimator which is strictly positive. For this we will apply an approach developed by Gnot, Trenkler and Zmyślony (1992), which allows to compute nonnegative minimum biased quadratic estimators for the parametric function (1.3).

2. Nonnegative Minimum Biased Estimation

A parametric function $\beta' H \beta + h\sigma^2$, with an arbitrary n.n.d. matrix H and $h \geq 0$ is nonnegatively estimable in model (1.1) by the uniformly minimum variance unbiased estimator $b'Hb + s^2(h - \mathrm{tr}H(X'X)^{-1})$ if and only if $h \geq \mathrm{tr}H(X'X)^{-1}$. If this condition does not hold we can use an approach of Gnot, Trenkler and Zmyślony (1992) who developed a nonnegative minimum biased estimator. It is given by $b'C_H b + c_H s^2$, where the n.n.d. matrix C_H and the nonnegative scalar c_H are defined as the solution of

$$\min_{C \geq 0, c \geq 0} \mathrm{tr}(H - C)^2 + [\mathrm{tr}C(X'X)^{-1} + c - h]^2.$$

It turns out that $c_H = 0$ and an explicit solution for C_H can be found if H and $X'X$ commute. The general solution for C_H is described by the following two equations:

$$C_H = [H - \lambda(X'X)^{-1}]_+, \qquad (2.1)$$

$$\lambda = \mathrm{tr}C_H(X'X)^{-1} - h, \qquad (2.2)$$

where A_+ is the positive part of a symmetric matrix A which is obtained from the spectral decomposition of A by setting the negative eigenvalues equal to zero. Rewriting (2.1) and (2.2), we obtain

$$\lambda = \mathrm{tr}[H - \lambda(X'X)^{-1}]_+(X'X)^{-1} - h. \qquad (2.3)$$

The unique root of (2.3) can be calculated using conventional derivative free procedures e.g. secant method or regula falsi. However, it is important to mention that the spectral decomposition of a $k \times k$ matrix is necessary in every step of the iteration. The algorithm, which was used for our computations, is described in the Appendix.

Applying this procedure to our problem we have $H = X'(I - P_1)X$ and $h = p$. It can be stated that H and $X'X$ commute if and only if X_1 is orthogonal to X_2, which is not a very realistic situation. Thus we have to apply the numerical procedure as described above. The following strategies for variable selection may be derived from the minimum biased estimation.

Strategy 1. We estimate the total mean square error of $\hat{y}_1$ by the minimum biased estimator $\hat{\tau}_M = b'C_H^M b$ if $p < k/2$. For $p \geq k/2$ we apply the unbiased estimator $\hat{\tau}_1$.

Strategy 2. Estimate $TMSE(\hat{y}_1)$ by $\hat{\tau}_B = b'C_H^B b + ps^2$, where $b'C_H^B b$ is the minimum biased estimator of the squared bias $B^2(\hat{y}_1) = \beta' H \beta$ obtained for $H = X'(I - P_1)X$ and $h = 0$ in the general procedure. This estimator can be motivated by the fact that the total mean square errors (1.3) for different models with the same number of regressors differ only in the squared bias.

To get an impression of the performance of the proposed strategies in comparison with Mallows' C_p we conducted a small computational study. Under normality it is possible to compute expectation and variances of the estimators using Rao and Kleffe

(1988, p.34). We took regressors from a data set which was previously investigated by Gorman and Toman (1966) and Daniel and Wood (1971). It consists of six regressors and a constant term. We used four values for σ^2 : 0.005, 0.01, 0.05, 0.1, while the estimate from the data was $s^2 = 0.0128$. Five values for β were considered: $\beta_{(1)}$ the least squares estimator from the data, $\beta_{(2)}$ was obtained from $\beta_{(1)}$ by setting the two smallest components equal to zero, while $\beta_{(3)}$, $\beta_{(4)}$ and $\beta_{(5)}$ were generated from pseudo-random numbers from the intervals $[0,5], [0,2]$ and $[-1,1]$. To compare the estimators of Strategy 1 and 2 with C_p we computed the relative root mean square error (RRMSE) for each combination of β, σ^2 and each combination of regressors, where the constant term was always included. The RRMSE is defined as

$$RRMSE = \left[E(\hat{\tau} - \tau)^2\right]^{1/2} / \tau$$

for each estimator $\hat{\tau}$ for a positive scalar parameter τ. The results will not be given in detail but can be summarized as follows.

- Mallows' C_p was dominated by Strategy 1 or 2 in nearly every situation (63 combinations of regressors for each combination of β, σ^2).
- Strategy 1 had the best overall performance, dominating C_p in nearly every situation.
- Strategy 2 was quite reliable for the two larger values of σ^2 and β-vectors (1), (2), (4) and (5), while it was worse than C_p in some cases for the two smallest σ^2-values.
- Strategy 2 was superior to Strategy 1 only for $\sigma^2 = 0.05$ and 0.1.

We may conclude from these statements that using the estimates from Strategy 1 for the total mean square error seems to be more precise than using C_p to estimate the relative mean square error. Hence we suggest the latter procedure for variable selection. Computing the estimator in Strategy 2 may give some additional insight in comparing models with the same number of regressors but seems to be inappropriate as a general basis for variable selection. To confirm the above findings similar investigations using different sets of regressors will be necessary. In the following section we apply the proposed procedures to one data set.

3. An Example

Let us consider again the data set of Gorman and Toman (1966) which we used in Section 2. The full model consists of six variables and a constant term with 31 observations. A search of all $2^6 - 1 = 63$ equations (excluding the full model) is possible, including the constant in each model. We computed the estimators given by Strategy 1, 2 and C_p and the five best combinations for each strategy are presented in Table I. *Best* means the smallest values of the estimates for Strategy 1 and 2 while $|C_p - p|$ was considered for Mallows' C_p. It can be seen that all strategies support the choice of the combinations VI, VII and VIII. In addition to C_p, Strategy 1 and 2 indicate that e.g. combinations I and II with small numbers of regressors might deserve further attention.

Even if the new strategies do not share all analytical properties of C_p, as was pointed out by one of the referees, they remove some of its disadvantages and we may finally conclude from our investigations that especially Strategy 1 might serve

TABLE I
Gorman-Toman data with six regressors and a constant. Five best combinations of regressors for each strategy

No.	Regressors	Strat.1	Rank	Strat.2	Rank	C_p	Rank
I	1	0.094	5	0.131	21	20.33	27
II	1 4	0.159	14	0.058	1	20.53	22
III	1 4 5	0.275	23	0.084	5	21.47	21
IV	1 2 6	0.072	4	0.135	23	5.66	4
V	1 2 5 6	0.095	6	0.123	18	7.44	5
VI	1 2 4 6	0.055	1	0.067	2	4.26	3
VII	1 2 4 5 6	0.071	3	0.082	3	5.51	1
VIII	1 2 3 4 6	0.068	2	0.084	4	5.30	2

as an additional useful tool for finding combinations of regressors that have been *overlooked* by C_p.

Appendix

In this section we will briefly describe the algorithm which was used for solving the equation (2.3). We will apply our algorithm to the equation $f(\lambda) = 0$, where

$$f(\lambda) = \text{tr}[H - \lambda(X'X)^{-1}]_+(X'X)^{-1} - h - \lambda.$$

The algorithm consists of two parts: Part 1 is a modified regula falsi to approach the root and the secant method in Part 2 provides a precise solution.

PART 1. Modified Regula Falsi (cf. Conte and de Boor, 1972, p.31)
Starting values are given by $a_0 = 0$ and $b_0 = \text{tr}H(X'X)^{-1} - h$ and it holds that $f(a_0) > 0, f(b_0) < 0$. Now set $F = f(a_0), G = f(b_0)$ and proceed as follows.

Step 1. Compute $w_{n+1} = b_n - G(b_n - a_n)/(G - F)$.

Step 2. If $f(a_n)f(w_{n+1}) \leq 0$ set $a_{n+1} = a_n, b_{n+1} = w_{n+1}$ and $G = f(w_{n+1})$. If also $f(w_n)f(w_{n+1}) > 0$ set $F = F/2$. Otherwise set $a_{n+1} = w_{n+1}, b_{n+1} = b_n, F = f(w_{n+1})$. If also $f(w_n)f(w_{n+1}) > 0$ set $G = G/2$. Then go to Step 1.

Stopping criterion. Stop the iteration if $|a_n - b_n|/\max\{|a_n|, |b_n|\} < \epsilon_1$. For our application $\epsilon_1 = 0.1$ was chosen. The solution for λ is known to be in the interval $[a_n, b_n]$.

PART 2. Secant Method
Starting values are $\delta_0 = a_n, \delta_1 = b_n$ from Part 1, n being the number of the last iteration. Compute $\delta_{n+1} = \delta_n - f(\delta_n)(\delta_n - \delta_{n-1})/[f(\delta_n) - f(\delta_{n-1})]$ until the following stopping criterion is met.

Stopping criterion. Stop the iteration if $\alpha|f(\delta_n)| < \epsilon_2$, where $\alpha = \text{tr}(X'X)^{-2}/[1 + \text{tr}(X'X)^{-2}]$ and ϵ_2 was chosen as 10^{-6}. This criterion is motivated as follows. Equation (2.3) can be equivalently represented as (cf. Gnot, Trenkler and Zmyślony, 1992)

$$\lambda = \frac{1}{d}\left(\mathrm{tr}H(X'X)^{-1} - h + \mathrm{tr}\{[H - \lambda(X'X)^{-1}]_{-}(X'X)^{-1}\}\right) = \tilde{f}(\lambda),$$

where $d = 1 + \mathrm{tr}(X'X)^{-2}$ and $A_{-} = A_{+} - A$ denotes the negative part of A. A reasonable criterion is to compare the functions $f(\lambda) + \lambda$ and $\tilde{f}(\lambda)$, i.e. we stop the iteration if

$$|f(\delta_n) + \delta_n - \tilde{f}(\delta_n)| < \epsilon_2,$$

which can be rewritten in the previous form. As it was already mentioned the spectral decomposition of a $k \times k$ matrix is necessary for computing values $f(\lambda)$, which is no serious problem since this procedure is available in many software packages.

Acknowledgements

This research was supported by grants from Komitet Badań Naukowych, PB 678/2/91, and Deutsche Forschungsgemeinschaft, Grant No. TR 253/1-2. We are grateful to the referees for their helpful comments.

References

Conte, S.D. and de Boor, C. (1972). *Elementary Numerical Analysis*. 2nd ed. Wiley, New York.

Daniel, C. and Wood, F.S. (1971). *Fitting Equations to Data*. Wiley, New York.

Gnot, S., Trenkler, G. and Zmyślony, R. (1992). Nonnegative minimum biased quadratic estimation in the linear regression model. Working Paper.

Gorman, J.W. and Toman, R.J. (1966). Selection of variables for fitting equations to data. *Technometrics* **8**, 27–51.

Mallows, C.L. (1973). Some Comments on C_p. *Technometrics* **15**, 661–676.

Rao, C.R. and Kleffe, J. (1988). *Estimation of Variance Components and Applications*. North-Holland, Amsterdam.

Sen, A. and Srivastava, M.S. (1990). *Regression Analysis: Theory, Methods and Applications*. Springer-Verlag, New York.

PARTIAL LEAST SQUARES AND A LINEAR MODEL

DIETRICH VON ROSEN
Uppsala University
Department of Mathematics
Box 480
S-751 06 Uppsala
Sweden

Abstract. Partial least squares (PLS) is considered from the perspective of a linear model. It is shown that the PLS-predictor is identical to the best linear predictor in a linear model with random coefficients.

Key words: Partial least squares, Gauss-Markov model, Invariant subspaces.

1. Introduction

Partial least squares regression (PLS) has nowadays become a widely used method in some areas of applied statistics, e.g. in chemometrics. One may view PLS as a statistical method and it is often used in situations where one has little knowledge about the structure of data. When applying PLS there are many aspects which can be and also have been questioned, but at the same time there exist many scientists in various fields who are satisfied with PLS. Therefore, we find it important to discuss PLS from a theoretical perspective.

The starting point of this work was Helland's (1988, 1990) papers where various versions of PLS algorithms were studied in detail. Other theoretical aspects and properties have been given by Frank (1987), Höskuldsson (1988), Stone and Brooks (1990), Helland (1991), Frank and Friedman (1993) and von Rosen (1993). For further references to PLS and related subjects we suggest the above cited papers.

In this note we show that an expression obtained from the PLS algorithm is identical to an expression which is based on a weakly singular (see Nordström, 1985) Gauss-Markov model with random coefficients.

2. PLS

Prediction is a major goal in many problems and PLS is one out of several methods which can be applied (see e.g. Frank and Friedman, 1993). In particular, in the presence of collinearity one has suggested PLS as a proper choice. The version of the PLS algorithm used in this paper is based on Helland's (1988, 1990) illuminating works. We will mainly deal with the population version, where all included parameters are supposed to be known. However, one should note that there is a big difference between the sample version and the population version since in the former some stopping rule is implemented which does not automatically follow from

T. Caliński and R. Kala (eds.),
Proceedings of the International Conference on Linear Statistical Inference LINSTAT '93, 73–78.

the algorithm. For details and further references we refer to Helland's papers.

The population version of the PLS algorithm. Let X be a p-dimensional random vector and Y a random variable, where $E[X] = \mu$, $D[X] = \Sigma$ and the covariance $C[X, Y] = w_1$ are known.
(i) Define a starting value $e_0 = X - \mu$. Do the next steps for $a = 1, 2, \ldots$
(ii) Define $w_a = C[e_{a-1}, Y]$, $t_a = e'_{a-1} w_a$.
(iii) Determine $p_a = C[e_{a-1}, t_a]/D[t_a]$.
(iv) Define new residuals $e_a = e_{a-1} - p_a t_a$. Stop if $e_a = 0$.

For our purpose the w_i's are the interesting quantities and it turns out that they generate a certain space which then is used for predicting Y (see formula (8) below). The algorithm implies that

$$\begin{aligned} w_{a+1} &= w_a - D[e_{a-1}]w_a(w'_a D[e_{a-1}]w_a)^- w'_a w_a, \qquad (1)\\ D[e_a] &= D[e_{a-1}] - D[e_{a-1}]w_a(w'_a D[e_{a-1}]w_a)^- w'_a D[e_{a-1}], \end{aligned}$$

where $^-$ denotes a g-inverse in the sense of $G = GG^-G$. Let G_a be any matrix spanning the column space $\mathcal{C}(w_1 : w_2 : \ldots : w_a)$, where : denotes a partitioning. Then

$$D[e_a] = \Sigma - \Sigma G_a(G'_a \Sigma G_a)^- G'_a \Sigma, \qquad (2)$$

$$w_{a+1} = (I - \Sigma G_a(G'_a \Sigma G_a)^- G'_a) w_1. \qquad (3)$$

Since $\mathcal{C}(w_1) \subseteq \mathcal{C}(\Sigma)$, it follows from (2) and (3), that $\mathcal{C}(w_{a+1}) \subseteq \mathcal{C}(D[e_a])$, which implies that the g-inverse in (1) may be replaced by an inverse. The relations (2) and (3) may be proved using an induction argument and straightforward calculations, where it is utilized that $\mathcal{C}(w_1) \subseteq \mathcal{C}(\Sigma)$. An expression for w_{a+1} like the one in (3) has previously been considered by Helland (1988) and others. From (1), (3) and results for projection operators (Shinozaki and Sibuya, 1974), or direct calculations, we obtain

$$\mathcal{C}(w_{a+1}) = \mathcal{C}(w_a)^\perp \cap (\mathcal{C}(D[e_{a-1}]w_a) + \mathcal{C}(w_a)) = \mathcal{C}(G_a)^\perp \cap (\mathcal{C}(\Sigma G_a) + \mathcal{C}(w_1)), \quad (4)$$

and, as a consequence,

$$\mathcal{C}(G_a) = \sum_{i=1}^{a} \mathcal{C}(w_i) = \sum_{i=0}^{a-1} \mathcal{C}(\Sigma^i w_1), \qquad (5)$$

where $\Sigma^i = \underbrace{\Sigma \times \cdots \times \Sigma}_{i\ times}$, $\Sigma^0 = I$.

A subspace $\mathcal{C}(T)$ is Σ–invariant if $\mathcal{C}(\Sigma T) \subseteq \mathcal{C}(T)$. If the PLS algorithm is applied to a p-dimensional space, it has to stop after at most p steps, i.e.

$$\mathcal{C}(w_1 : \Sigma w_1 : \Sigma w_2 : \ldots) = \mathcal{C}(w_1 : \Sigma w_1 : \ldots : \Sigma^{p-1} w_1).$$

Under certain structures in Σ and w_1 the PLS algoritm stops earlier, say after a steps, i.e. $w_{a+1} = 0$; in particular, if Σ is not positive definite (p.d.). Using (4) it is easy to show that the algorithm stops after a steps if and only if $\mathcal{C}(\Sigma G_a) \subseteq \mathcal{C}(G_a)$, i.e. $\mathcal{C}(G_a)$ is Σ–invariant. However, it may be shown (Helland, 1990 and von Rosen, 1993) that the last $p-a$ components in $(w_1 : \Sigma w_1 : \ldots : \Sigma^{p-1}w_1)$ are irrelevant for prediction if and only if $\mathcal{C}(w_1 : \Sigma w_1 : \ldots : \Sigma^{a-1}w_1)$ is Σ–invariant.

Finally note that for each step a of the algorithm

$$X = \mu + p_1 t_1 + \cdots + p_a t_a + e_a, \tag{6}$$

which is a well known relation. Moreover, e_a in (6) is uncorrelated with $t_1, \ldots, t_a$ (Helland, 1990, p. 99). The main object of this note is to reconsider (6) and explain the structure of X.

3. The Linear Model

The general Gauss-Markov model with known dispersion matrix can be written as

$$X = A\beta + \epsilon, \tag{7}$$

where X is a $p \times 1$ random vector, A is a $p \times k$ known matrix, β is a $k \times 1$ vector of unknown parameters and ϵ is a random vector with $E[\epsilon] = 0$ and $D[\epsilon] = \Sigma$, Σ being positive semidefinite (p.s.d.) and known. Furthermore, we will suppose that $\mathcal{C}(A) \subseteq \mathcal{C}(\Sigma)$, which means that (7) is a weakly singular Gauss-Markov model (Nordström, 1985). For the weakly singular Gauss-Markov model there exist many ways to represent the best linear unbiased estimator $\hat{\beta}$. Utilizing the work by Khatri (1968) we obtain

$$A\hat{\beta} = A(A'\Sigma^- A)^- A'\Sigma^- X,$$

which holds for any choice of g-inverse Σ^-.

In the next section we utilize (7) when β is a random vector such that $D[\beta] = (G_a'\Sigma G_a)^-$, $D[X] = \Sigma$ and $A = \Sigma G_a$, where G_a is defined as in the previous section. We do not assume anything about $E[\beta]$ and in particular we may suppose that $E[\beta] = 0$ which means that $E[X] = 0$. Nor do we assume independence between β and ϵ. From a viewpoint of linear models theory, this model is rather peculiar. For example, A is exotic and the interpretation of ϵ in (7) is difficult. Even if ϵ and β are supposed to be independent they are strongly related through the variance structures since $D[X] = \Sigma$ implies that $D[\epsilon] = \Sigma - \Sigma G_a(G_a'\Sigma G_a)^- G_a'\Sigma$. Hence, the interpretation of β as describing a latent structure and ϵ as describing measurements errors is impossible. Furthermore, $D[\epsilon]$ is singular. In the next section it is shown that this model is very strongly connected to the PLS-algorithm.

4. Main Theorem

We still assume that Σ and w_1 are known, i.e. the population version of the PLS algorithm. Without loss of generality we suppose that $E[Y] = 0$ and $E[X] = 0$. According to Helland (1988) the PLS predictor equals

$$\hat{Y}_{a,PLS} = w_1' G_a (G_a'\Sigma G_a)^- G_a' X. \tag{8}$$

It is worth observing that (8), under a normality assumption, just stands for the conditional mean $E[Y \mid G_a'X]$.

Following von Rosen (1993) an alternative way to look at the prediction problem is to predict Y via the relation $\hat{Y} = w_1'\Sigma^-\hat{X}$, where $\hat{X}$ is a predictor of X which follows the linear model given by (7) with the above mentioned random β.

From Rao (1973 p. 234, case 1) it follows that the best linear predictor for $w_1'\Sigma^- X$ is identical to $\hat{Y}_{a,PLS}$ given by (8). Hence we have two approaches which give the same predictor. We show now that the approaches in fact are identical. Note, however, that since we are dealing with least squares theory we are just interested in the first two moments.

Theorem 1. *Let X be a $p \times 1$ random vector following* (7), *where β is a random vector such that $D[\beta] = (G_a'\Sigma G_a)^-$, $D[X] = \Sigma$ and $A = \Sigma G_a$. If $\mu = 0$, then X has the same first two moments as X given by* (6).

Proof. The predictor $\hat{Y}_{a,PLS}$ is invariant under any choice of g-inverse. For simplicity we assume that $G_a = (w_1, w_2, \ldots, w_a)$ and that $(G_a'\Sigma G_a)^-$ is p.s.d. The latter assumption implies that there exist matrices U_a such that

$$(G_a'\Sigma G_a)^- = U_a U_a'.$$

Moreover, note that (7) with the random β is equivalent to

$$X = \Sigma G_a U_a \beta_o + \epsilon, \tag{9}$$

where for the new variables β_o, $D[\beta_o] = I$. Furthermore, since $\mu = 0$, (6) equals

$$X = p_1 D[t_1]^{\frac{1}{2}} r_1 + p_2 D[t_2]^{\frac{1}{2}} r_2 + \cdots + p_a D[t_a]^{\frac{1}{2}} r_a + e_a, \tag{10}$$

where $r_i = D[t_i]^{-\frac{1}{2}} t_i$ and hence $D[r_i] = 1$. Our aim is to show that (9) for one particular choice of U_a is identical to (10). Since the t's are uncorrelated (see Helland, 1990, p. 99) the r's are also uncorrelated. Hence, there exist a β_0 such that the first two moments for $(r_1 \ldots r_a)$ and β_o are identical. We are going to use an induction argument to show that

$$(p_1 D[t_1]^{\frac{1}{2}} : p_2 D[t_2]^{\frac{1}{2}} : \cdots : p_a D[t_a]^{\frac{1}{2}}) = \Sigma G_a U_a \tag{11}$$

for some speecific U_a. From the definitions of p_i and t_i follows that

$$p_i D[t_i]^{\frac{1}{2}} = D[e_{i-1}] w_i (w_i' D[e_{i-1}] w_i)^{-\frac{1}{2}}. \tag{12}$$

For $a = 1$ (11) follows immediately. Now suppose that (11) is true for $a - 1$. Hence the theorem is verified if we are able to show that for some U_a

$$(\Sigma G_{a-1} U_{a-1} : D[e_{a-1}] w_a (w_a' D[e_{a-1}] w_a)^{-\frac{1}{2}}) = \Sigma G_a U_a.$$

Let

$$T_a = \begin{pmatrix} U_{a-1} & -U_{a-1} U_{a-1}' G_{a-1}' \Sigma w_a C^{-\frac{1}{2}} \\ 0 & C^{-\frac{1}{2}} \end{pmatrix},$$

where $C = w'_a D[e_{a-1}]w_a = w'_a(\Sigma - \Sigma G_{a-1}(G'_{a-1}\Sigma G_{a-1})^- G'_{a-1}\Sigma)w_a$. Since

$$U_a U'_a = T_a T'_a = (G'_a \Sigma G_a)^-, \tag{13}$$

we may choose $U_a = T_a$. The last equality in (13) follows from a standard result for g-inverses (see e.g. Srivastava and Khatri p. 34);

$$\begin{aligned}(G'_a\Sigma G_a)^- \\ = \begin{pmatrix} (G'_{a-1}\Sigma G_{a-1})^- & 0 \\ 0 & 0 \end{pmatrix} + \begin{pmatrix} -(G'_{a-1}\Sigma G_{a-1})^- G'_{a-1}\Sigma w_a \\ I \end{pmatrix} \times \\ (w'_a(\Sigma - \Sigma G_{a-1}(G'_{a-1}\Sigma G_{a-1})^- G'_{a-1}\Sigma)w_a)^{-1})(-w'_a\Sigma G_{a-1}(G'_{a-1}\Sigma G_{a-1})^- : I)\end{aligned}$$

and since, by assumption, $U_{a-1}U'_{a-1} = (G'_{a-1}\Sigma G_{a-1})^-$. Some straightforward calculations then show that

$$\Sigma G_a U_a = (\Sigma G_{a-1}U_{a-1} : \Sigma w_a C^{-\frac{1}{2}} - \Sigma G_{a-1}U_{a-1}U'_{a-1}G'_{a-1}\Sigma w_a C^{-\frac{1}{2}})$$

and using (2) and (12) we see that this expression is identical to (11). Finally, we note that if ϵ is independent of β_0 in (9) then e_a in (10) must be identical to ϵ in (9) because $D[X] = \Sigma$. Thus, if $U_a = T_a$, (10) can be written of the form given by (9) and the theorem is proved.□

5. Discussion

Using the least squares theory, we have found that the PLS solution is obtained when X follows a very artificial linear model. It also follows that our approach does not just give a mathematical coincidence between PLS (population version) and linear models theory because if we apply the PLS algorithm X equals (10) which is shown to be identical to the model (9).

Why should one consider PLS at all? It would be very easy to reject the PLS approach on the basis of linear models theory. However, and this is important, in applications PLS has made great success and simulations indicate that PLS performs better than other regression methods, for example principal component regression.

As previously noted, the interpretation of ϵ is not straightforward, in particular if ϵ and β are independent. One may view ϵ as a measure of the performances of PLS, i.e. if the variation in ϵ, namely $\Sigma - \Sigma G_a(G'_a\Sigma G_a)^- G'_a\Sigma$, is large when PLS stops, then it has been advantageous to apply PLS. Another conclusion, also somewhat negative, is that if we have a material with measurement errors which can not be neglected, PLS will not be an appropriate method.

Finally we note that the linear models approach provides us with possibilities to evaluate model assumptions and discuss alternative models. From linear models theory we can adopt approaches for studying influential observations. Thus, we are now, in principle, in a position were we can evaluate the performance of the PLS algorithm in a straightforward manner. However, it should be emphasized that in this note we are just dealing with the population version of the algorithm. In practice, we have to estimate w_1 and Σ as well as find suitable stopping criterias and then everything is much more difficult.

Acknowledgements

The research was supported by the Swedish Natural Research Council. Comments made by two anonymous referees were important and improved the material. The author is also grateful to prof. Allan Gut for helpful comments.

References

Frank, I.E. (1987). Intermediate least squares regression method. *Chemometrics and Intelligent Laboratory Systems* **1**, 233–242.

Frank, I.E. and Friedman (1993). A statistical view of some chemometrics regression tools (with discussion). *Technometrics* **35**, 109–148.

Helland, I.S. (1988). On the structure of partial least squares regression. *Communications in Statistics - Simulation and Computation* **17**, 581–607.

Helland, I.S. (1990). Partial least squares regression and statistical models. *Scandinavian Journal of Statistics* **17**, 97–114.

Helland, I.S. (1991). Maximum likelihood regression on relevant components. *Journal of the Royal Statistical Society B* **54**, 637–647.

Höskuldsson, A. (1988). PLS regression methods. *Journal of Chemometrics* **2**, 211–228.

Khatri, C.G. (1968). Some results for the singular normal multivariate regression models. *Sankhyā A* **30**, 267–280.

Nordström, K. (1985). On a decomposition of the singular Gauss-Markov model. In: T. Caliński and W. Klonecki, Eds., *Linear Statistical Inference. Lecture Notes in Statistics* **35**. Springer-Verlag, Berlin, 231–245.

Rao, C.R. (1973). *Linear Statistical Inference and Its Applications.* Wiley, New York.

von Rosen, D. (1993). PLS, linear models and invariant spaces. *Scandinavian Journal of Statistics.* To appear.

Shinozaki, N. and Sibuya, M. (1974). Product of projectors. Scientific center report, IBM Japan, Tokyo.

Srivastava, M.S. and Khatri, C.G. (1979). *An Introduction to Multivariate Statistics.* North-Holland, New York.

Stone, M. and Brooks, R.J. (1990). Continuum regression: Cross-validated sequentially constructed prediction embracing ordinary least squares, partial least squares, and principal components regression (with discussion). *Journal of the Royal Statistical Society B* **52**, 237–269.

ONE-WAY ANALYSIS OF VARIANCE UNDER TUKEY CONTAMINATION: A SMALL SAMPLE CASE SIMULATION STUDY

RYSZARD ZIELIŃSKI
Institute of Mathematics of the Polish Academy of Sciences
Śniadeckich 8, P.O.Box 137
00-950 Warsaw
Poland

Abstract. Under some violations of assumptions of normality the size and the power function of the one-way analysis of variance test change remarkably.

Key words: One-way analysis of variance, Tukey model, Stability, Robustness.

1. Introduction

As long ago as in 1931, E.S.Pearson (1931) considered the problem of stability of statistical inference in the one-way analysis of variance under violations of the assumption of normality and had shown that "the sampling distribution of η^2 is remarkably insensitive to change in population form" (p. 118); here η^2 was the classical statistic of the F-test.

Since then the problem has been considered by many statisticians. In a very important paper by Atiqullah (1962) the concept of quadratically balanced experimental designs was introduced. Similarly as in Pearson (1931), symmetric (throughout our note we confine ourselves to the symmetrical violating distributions) nonnormality was measured by the value of γ_2 denoting the kurtosis (cf. Scheffé, 1959, p. 331). Concerning robustness of the one-way analysis of variance, Atiqullah's conclusions were: 1) F-test is robust "for a wide range of problems concerning balanced setups" (p. 88) and 2) "in other cases we may judge the influence of γ_2 by numerical calculation... The effect of γ_2 on Z is negligible, except in very extreme cases" (p. 89); here $Z = \log \eta^2$.

Common opinion today is that one can adopt the classical F-test even when random variables under consideration are not normally distributed. The opinion is supported by what one can find in first-class textbooks. Here are three examples.

Scheffé (1959) wrote: "The conclusion is that the effect of violation of the normality asumption is slight on inferences about means..." (p. 337), "Nonnormality has little effect on inferences about means" (p. 345), and "For the one-way layout with more than two equal groups sampling experiments have again indicated that nonnormality of the errors has small effect on the F-test of the hypothesis that the means are equal" (p. 347).

Seber (1977) states that "quadratically balanced F-tests are robust with regard to departure from the assumption of normality. ... one-way classification with equal

T. Caliński and R. Kala (eds.),
Proceedings of the International Conference on Linear Statistical Inference LINSTAT '93, 79–86.

numbers of observations per mean enjoys this robustness" (p. 249). In the context of multiple confidence intervals he states that "F-statistic is no longer robust with respect to nonnormality (because F is no longer quadratically balanced)" (p. 251).

Lehmann (1986) opens his Section 7.8 with a rather categorical statement "The F-test for the equality of a set of means was shown to be robust against nonnormal errors in Section 3", though in Section 3 one can find asymptotic arguments only.

In the present note we consider the linear model

$$Y_{ij} = \mu_i + \eta_{ij}, \quad i = 1, \ldots, k, \ \ j = 1, \ldots, n_i,$$

where η_{ij} are independent identically distributed random errors $N(0, \sigma^2)$, σ^2 being unknown. The problem is to test the hypothesis

$$H : \mu_1 = \mu_2 = \ldots = \mu_k.$$

To this end the classical F-test with the F statistic

$$F = \frac{\sum_{i=1}^{k} n_i(Y_{i\cdot} - \bar{Y})^2/(k-1)}{\sum_{i=1}^{k} \sum_{j}^{n_i}(Y_{ij} - Y_{i\cdot})^2/(N-k)},$$

where $N = n_1 + n_2 + \ldots + n_k$, is used.

We are interested in how much the size and the power of the test are changing when errors η_{ij} are distributed according to

$$\mathbf{T}(\epsilon, \alpha) = (1-\epsilon)N(0, \sigma^2) + \epsilon N(0, (\alpha\sigma)^2)$$

rather than according to $N(0, \sigma^2)$. For obvious reason we assume that $0 \leq \epsilon \leq 0.5$ and $\alpha \geq 1$. These contaminated normal distributions were suggested by Tukey (1960) as a model for observations, which usually follow a normal law but which, where occasionally something goes wrong with the experiment or its recording, might be perturbed by gross errors ('outliers'). This model is referred to as the 'Tukey contamination model'.

In consecutive sections results of a simulation study are presented. We decided to publish the results because they seem to throw some new light on certain well known problems. The Tukey contamination model is really very interesting because in this model we are naturally oriented towards measuring the degree of contamination in terms of ϵ and α while in the traditional approach the degree of nonnormality is typically measured by the value of γ_2, i.e. of kurtosis. Observe however that, given α, the degree of contamination is measured by ϵ, but kurtosis is not changing monotonically with ϵ (see Fig.1) which leads to some unexpected conclusions.

In what follows we report in detail our simulation study for small-sample cases: Case 1 with $k = 5$ and $n_1 = n_2 = n_3 = n_4 = n_5 = 6$ (quadratically balanced set-up, in the sense of Atiqullah, 1962) and Case 2 with $k = 5$ and $n_1 = n_2 = n_3 = n_4 = 2$, $n_5 = 22$. Some partial results for two 'not-very-small-sample-sizes' cases are also presented: for Case 3 with $k = 5$ and $n_1 = n_2 = n_3 = n_4 = n_5 = 10$ and for Case 4 with $k = 5$ and $n_1 = n_2 = n_3 = n_4 = 5$, $n_5 = 30$.

Some of the conclusions are as follows:

1. Even in quadratically balanced set-ups the F-test is not robust. In Case 1 the size of the test is diminishing when, under fixed ϵ, α is growing. One can argue

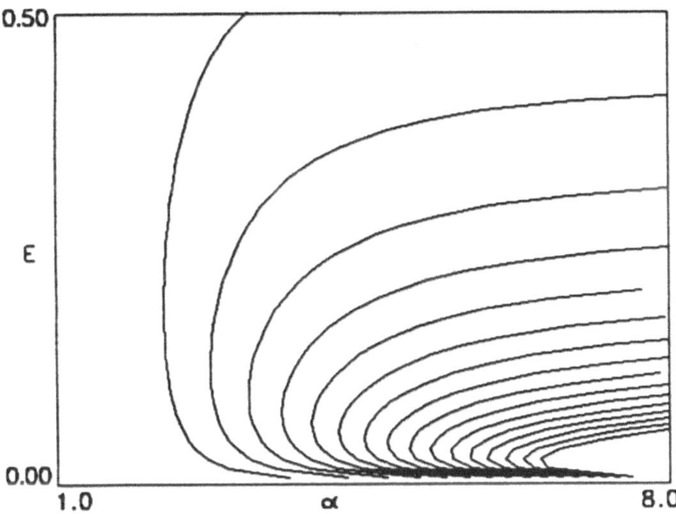

Fig. 1. Level lines of kurtosis on the (α, ϵ)-plane

that this does not make any problem for a statistician because the probability of the Type I error is diminishing. The problem is that the power of the F-test is diminishing also. In Figure 4 (described in Section 4) the power of the F-test under the alternatives

$$K : \mu_2 = \mu_1 + \theta, \mu_3 = \mu_1 + 2\theta, \ \dots \ \mu_k = \mu_1 + (k-1)\theta, \quad \mu_1 \in \mathbb{R}^1, \quad \theta > 0$$

is presented.

2. Given α, the size of the F-test is not changing monotonically with ϵ. See Figure 3 in Section 3. A consequence: the violation of the size of the test might be greater under smaller probability of occuring outliers (under a fixed distribution of the outlier, i.e. under a fixed contaminating distribution).

3. Kurtosis seems to be the main factor responsible for non-robustness of the F-test but apparently it is not the unique factor. To this effect see Figure 5 in Section 5 as well as Figure 3 in Section 3.

Finally we present robustness functions for the problem under consideration and report some simulation results which enable us to asses how robust are two competitors for the classical F-test (Figure 6 in Section 6).

2. Simulation

All simulations have been performed for $N = 10,000$ runs. The aim of the simulation was to estimate some probabilities by appropriate frequencies. To assess the error of that estimation we may assume that the frequencies, for the N chosen, are (almost) normally distributed with mean equal to the estimated value and with standard deviation not greater than $1/(2\sqrt{N}) = 0.005$. When estimating the size of a test with significance level 0.1 the standard deviation of the estimator is approximately equal

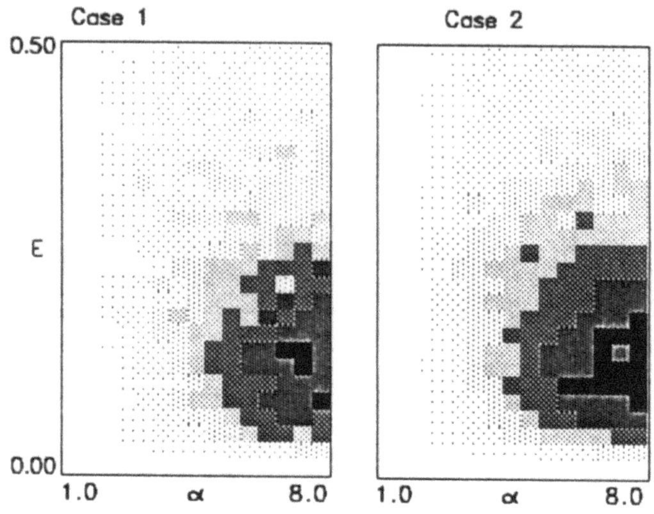

Fig. 2. Simulated size of the F-test

to $\sqrt{0.1 \cdot 0.9/N} = 0.003$. In the report we present simulation results for the F-test with significance level 0.1; simulation results for significance levels 0.05, 0.025, 0.01, and 0.005 are quite similar.

To simulate random numbers uniformly distributed on the unit interval (0, 1) the ULTRA package by A.Zaman and G.Marsaglia (1992) with a random seed generated by Turbo-Pascal 6.0 RANDOM procedure was used.

The following notation is used throughout the paper: size(ϵ, α) denotes the size of the F-test under the Tukey contamination $\mathbf{T}(\epsilon, \alpha)$; similarly, power $(\epsilon, \alpha, \theta)$ denotes the power of that test for a given value of θ.

3. Size of the F-test

In Figure 2 the simulated values of size(ϵ, α) are graphically presented for $\epsilon = 0(0.05)0.5$ and $\alpha = 1(0.5)8$. The observed range of simulated values of size(ϵ, α) was divided into 8 equal subintervals. Consequently, each value of size(ϵ, α) is presented according to which of those intervals it belongs. The smaller values of size(ϵ, α) in Case 1 and the larger values of size(ϵ, α) in Case 2 are presented with more dark rectangles. Generally: more dark areas present higher changes in the size of the test. It is interesting to observe that in the quadratically balanced set-up the size of the test is diminishing under contamination and in the quadratically unbalanced set-up the result is opposite. The smallest observed value was the size $(0.08, 8) = 0.065$ in Case 1 and the largest value was the size $(0.14, 8) = 0.189$ in Case 2 (for the test of the significance level equal to 0.1). For Case 3 the appropriate value is the size $(0.05, 8) = 0.073$ and that in Case 4 is the size $(0.05, 8) = 0.131$. Case 3 and Case 4 represent 'not-very-small-sample' designs so that the observed sensitivity of the F-test to the Tukey contamination is remarkable.

Consider the Tukey contamination with a fixed α. It is interesting to observe

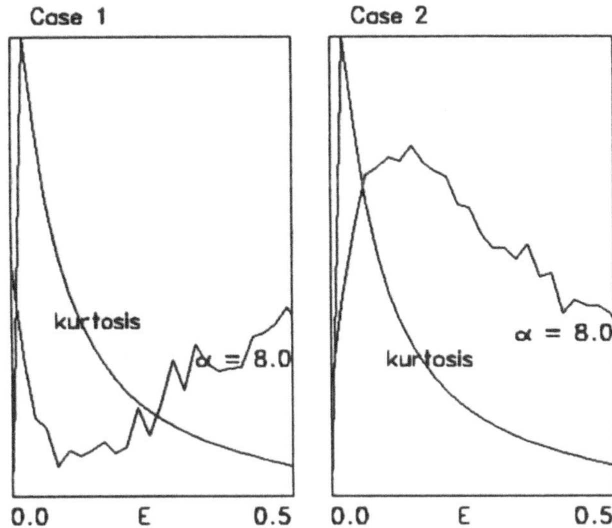

Fig. 3. Simulated size of the F-test and kurtosis

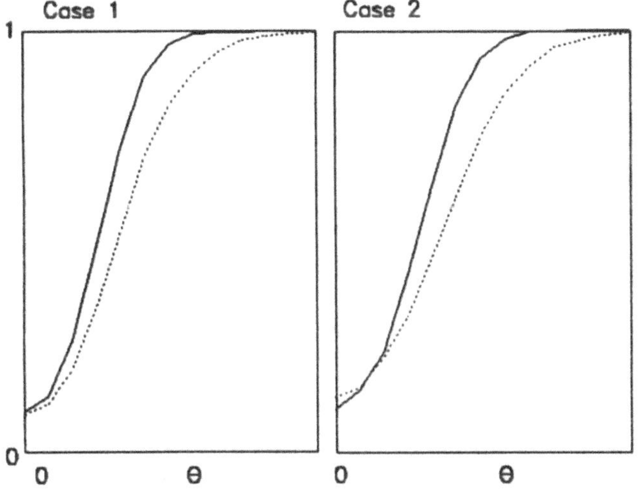

Fig. 4. Simulated power of the F-test

that the size of F-test does not change monotonically with $\epsilon \in (0, 0.5)$ (for $\epsilon \in (0, 1)$ this is obvious). Some results of simulation are presented in Figure 3. One can easly guess that this is a consequence of the fact that kurtosis (which seems to be the main factor responsible for sensitivity of the F-test to nonnormality) does not change monotonically also (see Figure 1). Nevertheles it is rather surprising that the violation of the size of the test may be smaller under a greater probability of the outlier, the distribution of the latter being fixed.

4. Power of the F-test

Statisticians may argue that a loss in the size of a test is not a problem one should worry about because the probability of Type I error is then smaller than assumed. The reasoning is correct unless we look into the power of the test, which typically in such situations is diminishing too. Figure 4 illustrates how much the power is changing: the solid lines give us the power of the F-test under normal errors and the dotted lines give us those for the Tukey contamination with $\epsilon = 0.1$ and $\alpha = 3$.

Even for not very small sample size designs (Case 3 and Case 4) the losses in power seem to be essential. Some numerical values are as follows:

$$\min_{\theta>0} \frac{\text{power}(0.1,3,\theta)}{\text{power}(0,1,\theta)} \leq 70.2 \text{ percent} \quad \text{(Case 1)},$$

$$\min_{\theta>0} \frac{\text{power}(0.1,3,\theta)}{\text{power}(0,1,\theta)} \leq 70.8 \text{ percent} \quad \text{(Case 2)},$$

$$\min_{\theta>0} \frac{\text{power}(0.1,3,\theta)}{\text{power}(0,1,\theta)} \leq 70.4 \text{ percent} \quad \text{(Case 3)},$$

$$\min_{\theta>0} \frac{\text{power}(0.1,3,\theta)}{\text{power}(0,1,\theta)} \leq 70.8 \text{ percent} \quad \text{(Case 4)}.$$

5. Robustnees and kurtosis

We have seen that the kurtosis seems to be the main factor responsible for sensitivity of the F-test to nonnormality. Figure 5 gives us an appropriate comment to that statement. Observe however that curves of kurtosis (represented by solid lines in Figure 5) do not coincide with the size of the test. Figure 3 confirms that remark.

There are two other well-known factors which play here a role. One is the quadratic balancing of the experimental design; some theoretical results concerning that factor can be found in Atiqullah (1962). Another one is the pair $(k-1, N-k)$ of degrees of freedom of the test; theorethical results (see Scheffé, 1959; Lehmann, 1986) are that the test is asymptotically robust. Unfortunately, we are not able as yet to catch quantitatively the influence of those (and perhaps some other) factors. Still we are convinced that it would be possible to give a formula (at least an empirical one) expressing the size of the test as a function of these factors. We will continue this discussion in a separate paper.

6. Robustness function and two competitors for the F-test

Consider the following 'neighbourhood' of the normal $N(0,\sigma^2)$ distribution,

$$\mathbf{T}(\mathcal{E},\mathcal{A}) = \{(1-\epsilon)N(0,\sigma^2) + \epsilon N(0,(\alpha\sigma)^2) : 0 \leq \epsilon \leq \mathcal{E}, 0 \leq \alpha \leq \mathcal{A}\},$$

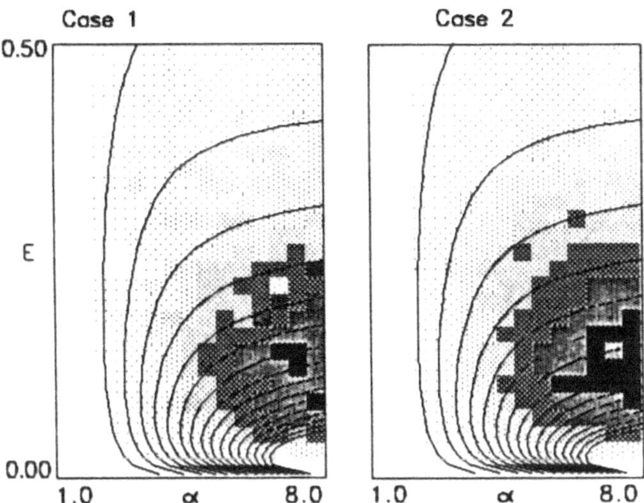

Fig. 5. Simulated size of the F-test and level lines of kurtosis

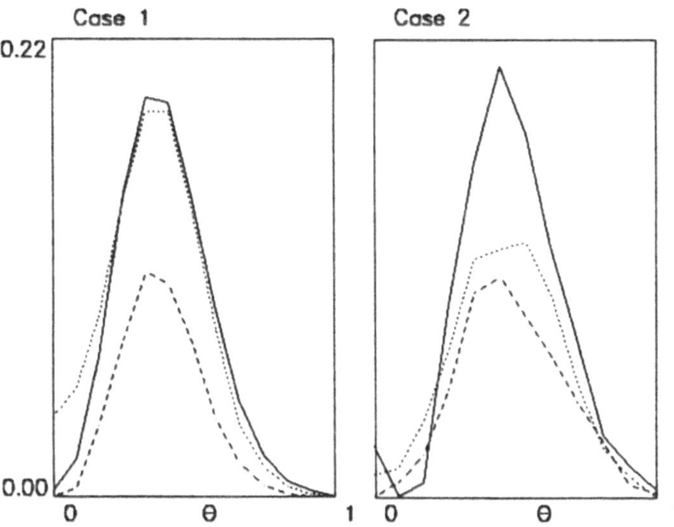

Fig. 6. Robustness functions of F-, L-, and K-test

and define the robustness function

$$r_F(\theta) = \sup\ \text{power}(\epsilon, \alpha, \theta) - \inf\ \text{power}(\epsilon, \alpha, \theta),$$

where sup and inf are taken over $\mathbf{T}(\mathcal{E}, \mathcal{A})$. The robustness function is a function on the parameter space of a given statistical problem and enables us to compare different statistical procedures and to conclude which one (if any) is uniformly more robust than other (for a more complete definition and examples see, e.g., Zieliński, 1983).

The robustness functions $r_F(\theta)$ in Case 1 and Case 2 for $\mathbf{T}(0.1, 3)$ are presented in Figure 6 (solid lines). Parallely with the robustness function $r_F(\theta)$ the robustness functions $r_L(\theta)$ for L-test and $r_K(\theta)$ for K-test were simulated. Here L-test (dotted line) is a test with the statistic which differs from that of F-test by replacing all means by appropriate medians and all squares by absolute values. The K-test (dashed line) is the well know Kruskal-Wallis test.

Acknowledgement

This paper was supported by KBN Grant No. 2P 301 010 04.

References

Atiqullah, M. (1962). The estimation of residual variance in quadratically balanced least–squares problems and the robustness of the F-test. *Biometrika* **49**, 83–91.

Lehmann, E.L. (1986). *Testing Statistical Hypotheses.* Wiley, New York.

Pearson, E.S. (1931). The analysis of variance in case of non-normal variation. *Biometrika* **23**, 114-133.

Scheffé, H. (1959). *The Analysis of Variance.* Wiley, New York.

Seber, G.A.F. (1977). *Linear Regression Analysis.* Wiley, New York.

Tukey, J.W. (1960). A survey of sampling from contaminated distributions. In: I. Olkin, Ed., *Contributions to Probability and Statistics.* Stanford University Press, Palo Alto, CA, 448–485.

Zaman, A. and Marsaglia, G. (1992). ULTRA package, Version 1.01, 18 March 1992.

Zieliński, R. (1983). Robust statistical procudures: a general approach. *Stability Problems for Stochastic Models. Lecture Notes in Mathematics* **982**. Springer-Verlag, New York, 283–295.

A NOTE ON ROBUST ESTIMATION OF PARAMETERS IN MIXED UNBALANCED MODELS

TADEUSZ BEDNARSKI and STEFAN ZONTEK
Institute of Mathematics of the Polish Academy of Sciences
Kopernika 18
51-617 Wrocław
Poland

Abstract. The paper describes a method of robust estimation of shift and scale parameters in a mixed unbalanced interlaboratory model. The emphasis is placed on robust estimation of components of variation. Estimators obtained here result from 'easily computable' Fréchet differentiable functionals.

Key words: Mixed unbalanced model, Robust estimation, Fréchet differentiability.

1. Introduction

A method of robust analysis of random effects in mixed models was introduced in the paper by Rocke (1991) (see also Iglewicz, 1983) and a general tratment of robust estimators of variance components is presented by Fellner (1986). Since these methods rely on estimation of random effects they naturally lead to high dimensional nonlinear equations. In Bednarski, Zmyślony and Zontek (1992) it is shown that Fréchet differentiability of statistical functionals leads quite directly to robust estimators of variance components and treatment fixed effects in a simple interlaboratory model. The basic idea was to construct a functional T that would be Fisher consistent for the parameter $\theta = (\mu_1, \ldots, \mu_a, \sigma_\lambda, \sigma_e)'$ under the model

$$Y_i = \mu_i + \lambda + e_i, \quad \text{for} \quad i = 1, \ldots, a,$$

where the laboratory effect λ and errors e_i are independent normal variables with distributions $\mathcal{N}(0, \sigma_\lambda^2)$ and $\mathcal{N}(0, \sigma_e^2)$, respectively. Considering then the random vector $\mathcal{Y} = (Y_1, \ldots, Y_a)'$ as a single observation, we obtain an estimator $T(\mathcal{F}_n)$ for $\mathcal{F}_n$, the empirical distribution function based on the sample of random vectors $\mathcal{Y}_1, \ldots, \mathcal{Y}_n$. From Bednarski (1992) it follows that important robustness properties of $T(\mathcal{F}_n)$ are a consequence of 'proper' smoothness of T; more precisely, of Fréchet differentiability of T for the supremum norm in the space of distribution functions.

When we let the above model to be unbalanced in the sense that the number of treatment replications may depend on both laboratory and treatment, then a similar point of view is possible. However we need to deal with a number of subpopulations $F^1(\cdot|\theta), \ldots, F^p(\cdot|\theta)$ defined for different designs of treatment replications at various laboratories. The application of von Mises (1947) methodology requires then modifications which partly are subject of this paper.

In the following section a general information about Fréchet differentiability is presented along with modifications specific to its application to the mixed unbal-

T. Caliński and R. Kala (eds.),
Proceedings of the International Conference on Linear Statistical Inference LINSTAT '93, 87–95.

anced model. Section 3 gives the construction of the Fréchet differentiable functional and the asymptotic properties of the resulting estimator. The asymptotic distributions and approximaton to the limiting covariance matrix are given under both, the model assumptions and small departures from the model. The last section provides results of a simulation experiment where the behavior of the maximum likelihood and the robust estimator are compared.

2. Fréchet Differentiability: General Information

Denote by $\mathcal{G}$ the set of distribution functions defined on $\mathbb{R}^r$ and let $\mathcal{D}$ be the convex cone spanned by the differences $F - G$ from $\mathcal{G}$. Here the distance between the distributions F and G from $\mathcal{G}$ will always be defined by the supremum norm

$$||F - G|| = \sup_x |F(x) - G(x)|.$$

A statistical functional T defined from $\mathcal{G}$ to $\mathbb{R}^k$ is said to be Fréchet differentiable at $F \in \mathcal{G}$ for the supremum norm, when there exists a linear functional T' on $\mathcal{D}$ such that

$$|T(G) - T(F) - T'(G - F)| = o(||F - G||),$$

where $| \cdot |$ stands for the Euclidean distance. The use of the supremum norm in the context of robust statistics is shown in Bickel (1981). For description of other concepts of differentiability we refer in particular to Reeds (1976), Fernholtz (1983) and Gill (1989).

Clarke (1983, 1986) and Bednarski, Clarke and Kołkiewicz (1991) give sets of conditions implying Fréchet differentiability for M-functionals. In the model studied later in this paper we consider differentiability of an M-functional, given by some function Ψ, at a product distribution $\mathcal{F}_\theta = F^1(\cdot|\theta) \times ... \times F^p(\cdot|\theta)$. If Ψ satisfies Clarke's conditions and $\int \Psi(\cdot|\theta) d\mathcal{F}_\theta(\cdot) = 0$ (this can always be assumed without loss of generality), then the derivative is

$$T'(G - \mathcal{F}_\theta) = -M^{-1}(\theta) \int \Psi(x|\theta) dG(x),$$

where

$$M(\theta) = \int \left[\frac{\partial \Psi(x|\theta)}{\partial \theta} \right] d\mathcal{F}_\theta(x).$$

To obtain the asymptotic distribution of the estimator resulting from the functional T, we need to approximate $\mathcal{F}_\theta$ on the basis of a random sample. This can be done via $\mathcal{F}_n = F^1_{n_1} \times ... \times F^p_{n_p}$, where $F^i_{n_i}$ is the empirical distribution function resulting from the subpopulation given by $F^i(\cdot|\theta)$, and $n = \sum n_i$. Another feature to be used later is a specific structure of Ψ, namely the representation

$$\Psi(Y_1, ..., Y_p|\theta) = \Psi_1(Y_1|\theta) + ... + \Psi_p(Y_p|\theta),$$

where Y_i is observed in the population distributed according to $F^i(\cdot|\theta)$. Then, under the assumption that n increases to infinity while min $\{n_1/n, ..., n_p/n\}$ tends to a positive constant, we obtain

$$\sqrt{n}[T(\mathcal{F}_n) - \theta] = -M^{-1}(\theta) \sum_{i=1}^{p} \sqrt{n} \int \Psi_i(y_i|\theta) dF^i_{n_i}(y_i) + o_{G^{\otimes n}_n}(1),$$

under the product of G_{n_i}, where $\sqrt{n}||G_{n_i} - F^i(.|\theta)||$ stays bounded. The asymptotic distribution is then normal with expectation zero (under the model) and the covariance matrix

$$M^{-1}(\theta) \left[\sum_{i=1}^{p} \frac{1}{q_i} \int \Psi_i(y_i|\theta) \Psi_i(y_i|\theta)' dF^i(y_i|\theta) \right] M^{-1}(\theta)',$$

where for $i = 1, ..., p$ we have $\lim_{n\to\infty} n_i/n = q_i$, under the whole infinitesimal model given by the supremum norm. Further we will sometimes call the parameter corresponding to the center of the infinitesimal neighbourhood, the 'true' parameter.

One of the essential implications of the Fréchet differentiability given here is that the above covariance matrix can be approximated consistently over the infinitesimal neighbourhoods by

$$\widehat{M}^{-1}(\mathcal{F}_n) \left[\sum_{i=1}^{p} \frac{1}{q_i} \int \Psi_i(y_i|T(\mathcal{F}_n)) \Psi_i(y_i|T(\mathcal{F}_n))' dF^i_{n_i}(y_i) \right] \widehat{M}^{-1}(\mathcal{F}_n)', \tag{1}$$

where

$$\widehat{M}(\mathcal{F}_n) = \sum_{i=1}^{p} \int \left[\frac{\partial}{\partial\theta} \Psi_i(y_i|\theta)|_{\theta=T(\mathcal{F}_n)} \right] dF^i_{b_i}(y_i).$$

In the following we shall give a proposition for Ψ designed for the mixed unbalanced model.

3. The Mixed Unbalanced Model

A single observation y_{ijk} for $i = 1, ..., a, j = 1, ..., b, k = 1, ..., d_{ij}$ is expressible in the model by

$$y_{ijk} = \mu_i + \lambda_j + e_{ijk}, \tag{2}$$

where μ_i are treatment fixed effects, λ_j are independent laboratory random effects and e_{ijk} are independent (also of λ_j) random errors with $\mathcal{N}(0, \sigma^2_\lambda)$ and $\mathcal{N}(0, \sigma^2_e)$ distributions, respectively.

To adopt the already described differentiability approach to this model, we need to single out model distributions for subpopulations for which i.i.d. observations are available. If $\{d_{ij}\}$ denotes the incidence matrix for the mixed model with i and j corresponding to the treatments and laboratory effects, respectively, then the number p of different subpopulations will correspond to the number of distinct columns of the incidence matrix. Denote these columns further by $N_1, ..., N_p$ and let N be the matrix formed by the columns. If b_s is the number of repetitions of the experiment corresponding to $N_s \in \mathbb{R}^a$, where $N_s = (n_{1s}, ..., n_{as})'$, then we have

$$\{d_{ij}\} = (N_1 1'_{b_1} : ... : N_p 1'_{b_p}) \quad \text{and} \quad b = \sum_{s=1}^{p} b_s.$$

To ensure the identifiability of the parameters in the model we assume further that

(I1) every row in N has a nonzero element,

(I2) there is a column in N for which the sum of its elements exceeds 1.

Referring to our earlier notation we can now write

$$F^s(.|\theta) \sim \mathcal{N}(X_s\mu,\ \sigma_\lambda^2 1_{n_{.s}} 1'_{n_{.s}} + \sigma_e^2 I_{n_{.s}}),$$

where $n_{.s} = N_s' 1_a$ for $s = 1, ..., p$, $\theta' = (\mu', \sigma_\lambda, \sigma_e)$, while $\mu = (\mu_1, ..., \mu_a)'$. The matrix $X_s = \text{diag}(1_{n_{1s}}, ..., 1_{n_{as}})$ is a partitioned matrix with the column vectors of ones, $1_{n_{1s}}, ..., 1_{n_{as}}$, on the main diagonal and with zeros elsewhere. To validate the asymptotic argumentation we will need the following assumption,

$$\lim_{b\to\infty} (b_s/b) = q_s > 0, \tag{3}$$

for $s = 1, ..., p$.

For fixed s let P_s be an $n_{.s} \times n_{.s}$ matrix with columns $P_{1s}, ..., P_{n_{.s}s} \in \mathbb{R}^{n_{.s}}$, normed and orthogonal, and such that $P_{1s} = (\sqrt{n_{.s}})^{-1} 1_{n_{.s}}$. Then the random vector Y_s corresponding to $F^s(.|\theta)$ is transformed into

$$P_s'Y_s \sim \mathcal{N}\left(P_s'X_s\mu,\ \text{diag}(n_{.s}\sigma_\lambda^2 + \sigma_e^2, \sigma_e^2, ..., \sigma_e^2)\right).$$

The objective function for the subpopulation s, taken as a simple modification of the loglikelihood function, can be written

$$\Phi_s(y_s|\theta) = \sum_{i=1}^{n_{.s}} \left[\ln(\delta_{is}) + \phi\left(\frac{1}{c\delta_{is}} P_{is}'(y_s - X_s\mu)\right)\right], \tag{4}$$

with a function ϕ properly chosen and δ_{is} as square roots of consecutive elements on the diagonal of the covariance matrix of $P_s'Y_s$. The function Φ_s becomes proportional to the loglikelihood function when ϕ is taken as a quadratic function and $c = 1$. For a given ϕ we can frequently choose c such to make the functional Fisher consistent.

Definition 3.1 (The functionals). Define the functional $T^*(G)$ to be the parameter θ for which

$$\int \Phi(y_1, ..., y_p|\theta) dG(y_1, ..., y_p) \tag{5}$$

attains the minimum value, where $\Phi(y_1, ..., y_p|\theta) = \sum_{s=1}^p \Phi_s(y_s|\theta)$.

Define $T(G)$ to be the solution of the equation

$$\int \Psi(y_1, ..., y_p|\theta) dG(y_1, ..., y_p) = 0, \tag{6}$$

where Ψ is the vector of partial derivatives of Φ with respect to θ.

We state now assumptions concerning ϕ which imply the Fisher consistency of the functional T^* and Fréchet differentiability of T.

(A1) The function ϕ defined on the real line is symmetric about 0, strictly increasing for positive arguments and it is twice continously differentiable.

(A2) The function $x\phi'(x)$ has a nonnegative derivative for $x \geq 0$. Moreover, $x\phi'(x)$ and $x^2\phi''(x)$ are bounded functions.

Theorem 3.1 (Fisher consistency). *Let $F^i(.|\theta_0)$ be the distribution of Y_i for $i = 1, ..., p$, where $\theta_0' = (\mu_1^0, ...\mu_a^0, \sigma_\lambda^0, \sigma_e^0)$ and assume that the model is identifiable* (I1, I2).

(i) *If* (A2) *is satisfied, then there is a unique $c > 0$ defining Φ which satisfies*

$$\int \{\frac{x}{c}\phi'(\frac{x}{c}) - 1\} dF(x) = 0, \tag{7}$$

where F is the distribution function for the standard normal distribution.

(ii) *If ϕ satisfies* (A1) *and* (A2), *then*

$$\int \Phi(y_1, ..., y_p|\theta) d(F^1(y_1|\theta_0) \times ... \times F^p(y_p|\theta_0)) \tag{8}$$

attains the global minimum if and only if $\theta = \theta_0$.

A proof of the theorem is in Bednarski and Zontek (1993) and it will be published elswhere.

It results now from Clarke(1983) that under Assumptions (A1) and (A2) the M-functional T is Fréchet differentiable. We phrase this in the following.

Theorem 3.2 (Fréchet differentiability). *Under* (A1) *and* (A2) *we have*

$$[T(G) - \theta] = \int IF(y_1, ..., y_p|\theta) dG(y_1, ..., y_p) + o(||G - \mathcal{F}_\theta||),$$

where

$$IF(y_1, ..., y_p|\theta) = \left[\sum_{s=1}^{p} \int \Delta_s(y_s|\theta) dF^s(y_s|\theta)\right]^{-1} \Psi(y_1, ..., y_p|\theta),$$

and $T(G)$ is the functional given in Definition 3.1, *while Δ_s is the matrix of second partial derivatives of Φ_s with respect to θ.*

A function ϕ defined by its following derivative,

$$\phi'(x) = \begin{cases} x, & |x| \leq t, \\ -x - 4t - 2t^2/x, & -2t < x < -t, \\ -x + 4t - 2t^2/x, & t < x < 2t, \\ 2t^2/x, & |x| \geq 2t, \end{cases} \tag{9}$$

satisfies the assumptions of Theorem 3.2. It is in fact a smoothed version of Huber's proposition (Huber, 1981, p.137, formula (4.14)).

Bednarski and Zontek (1993) have proved that the covariance matrix for the Fréchet differentiable estimator generated by (6) is of the form

$$\begin{bmatrix} w_1 V_1(\theta) & 0 \\ 0 & w_2 V_2(\theta) \end{bmatrix},$$

where $V_1(\theta)$ and $V_2(\theta)$ are respective asymptotic covariance matrices for the shift and scale of the maximum likelihood estimator. The positive constants w_1 and w_2 are given by the formulas

$$w_1 = \frac{c^2 E\left[\phi'(X/c)\right]^2}{E^2\phi''(X/c)},$$

$$w_2 = \frac{2E[(X/c)\phi'(X/c) - 1]^2}{[E(X/c)^2\phi''(X/c) + 1]^2},$$

where X is a standard normal random variable. For each t in (9) we choose c in (7) so to make the functional Fisher consistent.

In the following section simulation results for a functional induced by this function ϕ are presented and compared with the behavior of the maximum likelihood estimator.

4. Simulation Results

The robust estimator applied in the simulation was given by (9), with constant $t = 1.253$ selected in such a way that the loss of efficiency for the shift estimator was 10%. More exactly, it gave then $w_1 = 1.1$ and $w_2 = 1.29$ ($c = 0.873$). We considered the following two unbalanced mixed models in our simulation study:

(M1) The shift parameter is two-dimensional and equal to $(-4, 4)$, scales $\sigma_\lambda = 2$, $\sigma_e = 1$, subpopulation sample sizes are $b_1 = 5$, $b_2 = 5$, while the matrix N is

$$N = \begin{bmatrix} 2 & 1 \\ 1 & 2 \end{bmatrix}.$$

(M2) Except for $b_1 = 15$ and $b_2 = 15$ other quantities defining the model remain the same as in M1.

Models (M1) and (M2) differ in sample sizes only and they were chosen to see how the number of laboratories affects a precision of estimation of the covariance matrix via the influence function both under the pure and contaminated model for the maximum likelihood estimator and robust estimator. Contaminations of the data simulated from the model were taken to imitate: (i) exchange of samples in a laboratory, (ii) wrong scaling of an instrument in a laboratory, (iii) higher occasional variability in laboratory effects, (iv) higher occasional effects in error effects, (v) occasional interaction.

Only the first type of contamination may be identified with gross errors. For the other types of discrepancies the 'erroneous' data were mildly contaminated. Exactly one laboratory was randomly chosen to be 'contaminated' for the model (M1) and exactly two laboratories were chosen in the case of (M2). The following modifications of the original model distribution were applied to the randomly selected laboratories:

(i) the mean was taken $(4, -4)$ instead of $(-4, 4)$,

(ii) the mean was taken $(-3, 5)$,

(iii) σ_λ was taken equal to 4,

(iv) σ_e was taken equal to 2,

(v) an interaction random effect between a laboratory and a sample was taken to be $\mathcal{N}(0, 2.25)$.

TABLE I
Simulation results

	Model (M1) Estimates for parameters				Model (M2) Estimates for parameters			
	μ_1	μ_2	σ_λ	σ_e	μ_1	μ_2	σ_λ	σ_e
m.l.e.	-4.01	4.01	1.83	0.96	-3.99	4.00	1.93	0.99
	(0.68)	(0.69)	(0.50)	(0.16)	(0.38)	(0.40)	(0.27)	(0.09)
	(0.65)	(0.65)	(0.39)	(0.14)	(0.39)	(0.39)	(0.26)	(0.09)
	(0.69)	(0.69)	(0.49)	(0.16)	(0.40)	(0.40)	(0.28)	(0.09)
r.e.	-4.02	4.00	1.85	0.97	-3.99	4.00	1.93	0.99
	(0.70)	(0.72)	(0.57)	(0.17)	(0.40)	(0.42)	(0.31)	(0.11)
	(0.77)	(0.78)	(0.68)	(0.19)	(0.42)	(0.42)	(0.33)	(0.10)
	(0.72)	(0.72)	(0.55)	(0.18)	(0.42)	(0.42)	(0.32)	(0.10)
m.l.e.	-3.19	3.19	1.04	2.90	-3.47	3.48	1.55	2.51
	(0.69)	(0.71)	(0.79)	(0.21)	(0.40)	(0.40)	(0.40)	(0.13)
(i)	(1.03)	(1.02)	(1.28)	(1.04)	(0.55)	(0.55)	(0.53)	(0.68)
	(0.69)	(0.69)	(0.49)	(0.16)	(0.40)	(0.40)	(0.28)	(0.09)
r.e.	-3.97	3.97	1.94	1.06	-3.98	4.01	2.05	1.05
	(0.73)	(0.74)	(0.60)	(0.20)	(0.42)	(0.42)	(0.33)	(0.12)
(i)	(0.81)	(0.82)	(0.74)	(0.23)	(0.44)	(0.44)	(0.35)	(0.12)
	(0.72)	(0.72)	(0.55)	(0.18)	(0.42)	(0.42)	(0.32)	(0.11)
m.l.e.	-3.98	4.02	2.10	0.95	-3.98	4.02	2.13	0.99
	(0.70)	(0.71)	(0.63)	(0.15)	(0.41)	(0.41)	(0.34)	(0.09)
(iii)	(0.74)	(0.74)	(0.50)	(0.14)	(0.42)	(0.42)	(0.33)	(0.09)
	(0.69)	(0.69)	(0.49)	(0.16)	(0.40)	(0.40)	(0.28)	(0.09)
r.e.	-4.00	4.01	2.00	0.96	-3.98	4.02	2.05	0.99
	(0.72)	(0.72)	(0.63)	(0.17)	(0.41)	(0.41)	(0.33)	(0.10)
(iii)	(0.82)	(0.82)	(0.75)	(0.18)	(0.44)	(0.44)	(0.35)	(0.10)
	(0.72)	(0.72)	(0.55)	(0.18)	(0.42)	(0.42)	(0.32)	(0.10)

Note: Each result is the average of 500 estimates. Sample standard deviation for the estimates, the average of the standard deviations for the estimates obtained via the influence functions, and the 'true' asymptotic value for the standard deviation are given, consecutively, in parentheses.

In Table I typical results of the simulation are given. In each case the estimation was repeated 500 times. The initial values for the estimation were randomly chosen (according to the uniform distribution) from the intervals of lenght one and centered at the true parameter value. Table I lets us judge how rapidly the number of 'observations' improves the credibility of the standard error estimator obtained via the influence function [models (M1) and (M2)]. The table is set up in two-row blocks named by the model and its modification. The first row in the block is always for the maximum likelihood estimator while the second one for the robust estimator. Each model row consists of 4 subraws: the first one is for the average of the 500 estimates, the second one gives sample standard deviation for the estimates, the third one shows the average of the standard deviations for the estimates obtained via the influence functions, while the last subrow gives the 'true' asymptotic value for the standard deviation.

Under the model distribution the robust estimator has about 5% higher standard deviation for the shift and about 10% for the scale estimators. The robust estimator shows, however, much higher performance under small contaminations in both, the accuracy of estimation in terms of unbiasedness and in terms of the assessment of estimate's variability via the influence function. In cases (M1)(i) and (M2)(i) these effects are specially visible. Apart from a clear superiority of the robust estimator under model violations let us also notice an amazing coincidence of the sample standard deviations and the standard deviations obtained via the influence function in the case of the robust estimator. This is a consequence of Fréchet differentiability of the functional generating the robust estimator.

Simulation results for models (M1) and (M2) cases (ii), (iv), (v) are similar to those of (iii) and are not presented. They all show, as in the case (iii), a very mild influence of the assumed discrepancy from the model on estimation effects. Mixtures of different contamination patterns do not much change the picture.

Acknowledgement

This paper was supported by Komitet Badań Naukowych, Grant No. 21052 9101.

References

Bednarski, T., Clarke, B.R. and Kołkiewicz, W. (1991). Statistical expansions and locally uniform Fréchet differentiability. *Journal of the Australian Mathematical Society A* **50**, 88–97.

Bednarski, T. (1992). Freécht differentiability of statistical functionals and implications to robust statistics. In: *Proceedings of the Ascona Workshop on Data Analysis and Robustness.* To appear.

Bednarski, T., Zmyślony, R. and Zontek, S. (1992). On robust estimation of variance components via von Mises functionals. Preprint 504, Institute of Mathematics, Polish Academy of Sciences, Warszawa.

Bednarski, T. and Zontek, S. (1993). Robust estimation of parameters in mixed unbalanced model. Unpublished manuscript.

Bickel, P.J. (1981). Quelques aspects de la statistique robuste. *Ecole d'Eté de Probabilités de St. Flour IX. Lecture Notes in Mathematics* **876**. Springer-Verlag, New York, 1–72.

Clarke, B.R. (1983). Uniqueness and Fréchet differentiability of functional solutions to maximum likelihood type equations. *The Annals of Statistics* **11**, 1196–1206.

Clarke, B.R. (1986). Nonsmooth analysis and Fréchet differentiability of M-functionals. *Probability Theory and Related Fields* **73**, 197–209.

Fernholz, L.T. (1983). *Von Mises Calculus for Statistical Functionals. Lecture Notes in Statistics* **19**. Springer-Verlag, New York.

Fellner, W.H. (1986). Robust estimation of variance components. *Technometrics* **28**, 51–60.

Gill, R.D. (1989). Non- and semi-parametric maximum likelihood estimators and the von Mises method. Part 1. *Scandinavian Journal of Statistics* **16**, 97–128.

Huber, P.J. (1981). *Robust Statistics.* Wiley, New York.

Iglewicz, B. (1983). Robust scale estimators and confidence intervals for location. In: D.C. Hoagling, F. Mosteller and J.W. Tukey, Eds., *Understanding Robust and Exploratory Data Analysis.* Wiley, New York, 404–431.

Reeds, J.A. (1976). On the definition of von Mises functions. Ph. D. thesis, Department of Statistics, Harvard University, Cambridge.

Rocke, D.M. (1991). Robustness and balance in the mixed model. *Biometrics* **47**, 303–309.

Von Mises, R. (1947). On the asymptotic distributions of differentiable statistical functionals. *The Annals of Mathematical Statistics* **18**, 309–348.

OPTIMAL BIAS BOUNDS FOR ROBUST ESTIMATION IN LINEAR MODELS

CHRISTINE H. MÜLLER
Freie Universität Berlin
1. Mathematisches Institut
Arnimallee 2-6
D-14195 Berlin
Germany

Abstract. A conditionally contaminated linear model $Y(t) = x(t)'\beta + Z(t)$ is considered where the errors $Z(t)$ may have different contaminated normal distributions for different experimental conditions t. Estimating the unknown parameter β or a linear aspect $\varphi(\beta) = C\beta$ in such a model, an asymptotic bias will appear. Bounding the maximum asymptotic bias by some bias bound b, optimal robust estimators and optimal designs can be derived by minimizing the trace of the asymptotic covariance matrix (see Bickel, 1984; Rieder, 1987; Kurotschka and Müller, 1992; Müller, 1992a). While the optimal designs, which are the classical A-optimal designs, do not depend on the bias bound b, the optimal robust estimators depend strongly on b and the trace of their asymptotic covariance matrix increases if the bias bound decreases. In this paper optimal bias bounds are derived by minimizing the asymptotic mean squared error or its generalization. In particular at A-optimal designs the optimal bias bounds are easy to compute. For two examples the optimal bias bounds are given.

Key words: Conditionally contaminated linear models, Robust estimation, A-optimal designs, Optimal bias bounds, Mean squared error.

1. Introduction

A general linear model

$$Y_{nN} = x(t_{nN})'\beta + Z_{nN}, \qquad n = 1, \ldots, N,$$

is considered, where Y_{nN} are observations, $t_{nN} \in T$ are experimental conditions, $x : T \to \mathbb{R}^p$ is a known 'regression' function, $\beta \in \mathbb{R}^p$ is an unknown parameter vector, and Z_{nN} are error variables. In classical linear models it is assumed that the error variables $Z_{1N}, \ldots, Z_{NN}$ are independent and identically distributed. Usually it is assumed that they are normally distributed with mean 0 and known or unknown variance σ^2, i.e.

$$\frac{1}{\sigma} Z_{nN} \sim P = \mathcal{N}(0, 1).$$

But if some outlying observations (gross errors) may appear, the normal distribution is not correct. Then, even for designed experiments, a conditionally contaminated linear model is adequate (see Bickel, 1984; Rieder, 1987; Kurotschka and Müller, 1992; Müller, 1992b). In such a model it is assumed that the error variables $Z_{1N}, \ldots, Z_{NN}$ are independent and are distributed according to a contaminated normal distribution, where the contamination may be different for different experimental

T. Caliński and R. Kala (eds.),
Proceedings of the International Conference on Linear Statistical Inference LINSTAT '93, 97–102.

conditions, i.e.

$$\frac{1}{\sigma} Z_{nN} \sim Q_{nN}(dz) = [1 - N^{-1/2}\,\epsilon(t_{nN})]\; P(dz) + N^{-1/2}\,\epsilon(t_{nN})\; g(z, t_{nN})\; P(dz),$$

with $\sum_{n=1}^{N} \epsilon(t_{nN}) \leq N$ for almost all $N \in \mathbb{N}$, $\int g(z,t)\, P(dz) = 1$, $g(z,t) \geq 0$ for all $z \in \mathbb{R}$, $t \in T$. Thereby the markov kernel $g(\cdot, t)P$ models the form and $\epsilon(t) \geq 0$ the proportion of contamination. The set $\mathcal{P}$ of all sequences $(Q^N = \bigotimes_{n=1}^{N} Q_{nN})_{N \in \mathbb{N}}$ defines a conditional contamination neighbourhood around the classical model $(P^N)_{N \in \mathbb{N}}$.

To estimate in this model a linear aspect $\varphi(\beta) = C\beta$, $C \in \mathbb{R}^{s \times p}$, one can use a one-step-M-estimator. An estimator $\widehat{\varphi}_N : \mathbb{R}^N \times T^N \to \mathbb{R}^s$ is called a one-step-M-estimator for $\varphi(\beta) = C\beta$ with a score function $\psi : \mathbb{R} \times T \to \mathbb{R}^s$, an initial estimator $\widehat{\beta}_N^0 : \mathbb{R}^N \times T^N \to \mathbb{R}^p$ *for* β and a variance estimator $\widehat{\sigma}_N^2 : \mathbb{R}^N \times T^N \to \mathbb{R}^+$ for σ^2, if

$$\widehat{\varphi}_N(y_N, d_N) = C\,\widehat{\beta}_N^0(y_N, d_N) + \frac{1}{N} \sum_{n=1}^{N} \psi\left(\frac{y_{nN} - x(t_{nN})'\widehat{\beta}_N^0(y_N, d_N)}{\widehat{\sigma}_N}, t_{nN} \right) \widehat{\sigma}_N$$

(see Bickel, 1975; Rieder, 1985; and Müller, 1992b). Thereby the initial and the variance estimators can be robust or non-robust estimators. For example the initial estimator can be the least squares estimator or some M-estimator and the variance estimator can be the mean squared residuals or Huber's Proposal 2 (Huber, 1964). Because the asymptotic behaviour of the one-step-M-estimators does neither depend on the initial estimator nor on the variance estimator and here the robustness property is derived from the asymptotic distribution we also can use non-robust initial and variance estimators. But if the estimator should also satisfy a finite sample robustness property then the initial and variance estimator should also satisfy the finite sample robustness property.

If we assume that the design $d_N = (t_{1N}, ..., t_{NN})$ converges to an asymptotic design measure δ in the following sense

$$\lim_{N \to \infty} \frac{1}{N} \sum_{n=1}^{N} e_{t_{nN}}(t) = \delta(t) \text{ for all } t \in \operatorname{supp}(\delta), \tag{1}$$

then the one-step-M-estimator with the score function $\psi_\infty(z,t) = C\,I(\delta)^- x(t) z$ behaves asymptotically like the Gauss-Markov estimator for $\varphi(\beta) = C\beta$. Here e_t denotes the Dirac measure on t and $I(\delta) = \int x(t)\, x(t)'\, \delta(dt)$ is the information matrix of the design δ while $I(\delta)^-$ is a g-inverse of $I(\delta)$. In the conditionally contaminated linear model the one-step-M-estimator with the score function ψ_∞, and therefore also the Gauss-Markov estimator, has an unbounded asymptotic bias. But robust estimators should have a bounded asymptotic bias and this is the case for all one-step-M-estimators with bounded score function ψ. This was shown by Bickel (1984) and Rieder (1985, 1987) for estimating the whole parameter vector β and by Kurotschka and Müller (1992) for estimating a linear aspect $\varphi(\beta) = C\beta$. Moreover, they derived optimal robust estimators by minimizing the trace of the asymptotic covariance matrix under the side condition that the asymptotic bias is bounded by some bias bound b. Hence the optimal robust estimators depend on the bias bound b.

Similar optimal robust estimators can be also obtained by deriving the influence functions of the estimators (see Hampel et al., 1986). Basing on this approach, Samarov (1985) proposed optimal bounds b for the robust estimators by minimizing an approximated mean squared error at finite samples. But Samarov calculated the optimal bounds only for few estimators which, in particular for designed experiments, are not the optimal estimators.

In Section 3 of this paper we propose optimal bounds for planned experiments and show that for experiments at optimal designs the bounds are very easy to calculate. Because this approach is based on results of Kurotschka and Müller (1992) and Müller (1992a) concerning optimal robust estimators and optimal designs for robust estimation, at first these results will be briefly repeated in Section 2. In Section 4 we will give the optimal bias bounds for the linear regression model and a one-way lay-out with 3 treatments and a control treatment.

2. Optimal Robust Estimators with Given Bias Bound

Under some regularity conditions (see Müller, 1992b), in particular under condition (1) and the condition $\psi \in \Psi(\delta, C)$, where

$$\Psi(\delta, C) = \{\psi\colon \mathbb{R}\times T \to \mathbb{R}^s;\ \int \psi(z,\cdot)\, P(dz) = 0,\ \int \psi(z,t)\, x(t)'z\, P(dz)\delta(dt) = C\},$$

a one-step-M-estimator with a score function ψ is asymptotically normally distributed for all contaminated error distributions, i.e.

$$\mathcal{L}(\sqrt{N}(\widehat{\varphi}_N - \varphi(\beta))|Q_\beta^N) \overset{N\to\infty}{\longrightarrow} \mathcal{N}(b(\psi, (Q^N)_{N\in\mathbb{N}}), \sigma^2\, V(\psi, \delta)) \tag{2}$$
$$\text{for all } (Q^N)_{N\in\mathbb{N}} \in \mathcal{P}.$$

Thereby the asymptotic covariance matrix is given by

$$V(\psi, \delta) = \int \psi\, \psi'\, d(P\otimes\delta)$$

and the maximum asymptotic bias satisfies

$$\sup\{|b(\psi, (Q^N)_{N\in\mathbb{N}})|;\ (Q^N)_{N\in\mathbb{N}} \in \mathcal{P}\} = \sigma\, ||\psi||_\infty \tag{3}$$

(see Bickel, 1984; Rieder, 1985, 1987; Kurotschka and Müller, 1992; and Müller, 1992b). This shows that the asymptotic behaviour of a one-step-M-estimator is completely determined by its score function.

Optimal robust estimators for a given bias bound b for the maximum asymptotic bias at a given design δ are those which have a score function $\psi_{b,\delta}$ solving

$$\min\{\operatorname{tr} V(\psi, \delta);\ \psi \in \Psi(\delta, C),\ ||\psi||_\infty \le b\}.$$

The solution $\psi_{b,\delta}$ is $P\otimes\delta-$unique and exists if and only if b is greater than or equal to $b_0(\delta) = \min\{||\psi||_\infty;\ \psi \in \Psi(\delta, C)\}$. A general, but very implicit characterization of these optimal score functions $\psi_{b,\delta}$ was given by Hampel (1978) and Krasker (1980). More explicit characterizations for special designs were given by Kurotschka and Müller (1992) and Müller (1992a). In Müller (1992a) it was also shown that the classical A-optimal designs are optimal for these optimal robust estimators. It means that a design δ^* solving

$$\min\{\operatorname{tr} C\, I(\delta)^- C';\ \delta \in \Delta\}$$

is also a solution of

$$\min\{\operatorname{tr} V(\psi_{b,\delta}, \delta);\ \delta \in \Delta \text{ with } b_0(\delta) \leq b\},$$

and this holds for all bias bounds $b \geq b_0(\delta^*)$. Thereby we have $b_0(\delta^*) \leq b_0(\delta)$ for all $\delta \in \Delta$.

3. Optimal Bias Bounds

According to (2), the asymptotic mean squared error of a one-step-M-estimator with a score function ψ at a design δ is equal to

$$\sigma^2 \left(b(\psi, (Q^N)_{N \in \mathbb{N}})^2 + \operatorname{tr} V(\psi, \delta)\right),$$

and, according to (3), the maximum asymptotic mean squared error is equal to

$$MSE(\psi, \delta) \;=\; \sigma^2\, [\|\psi\|_\infty^2 + \operatorname{tr} V(\psi, \delta)].$$

To give more or less weight on the maximum asymptotic bias, the mean squared error criterion can be generalized in the following sense

$$GMSE(\psi, \delta) \;=\; \alpha\, w(\|\psi\|_\infty) + \gamma \operatorname{tr} V(\psi, \delta),$$

where $\alpha,\ \gamma \geq 0$ and $w : \mathbb{R}^+ \to \mathbb{R}^+$ is a convex function.

Now the optimal bias bound can be defined as the solution b^* of

$$\min\{GMSE(\psi_{b,\delta^*}, \delta^*);\ b \geq b_0(\delta^*)\}. \tag{4}$$

In particular, for $\alpha = \gamma$ and $w(b) = b^2$, the solution of (4) will be denoted by b_2. It is the optimal bias bound for the mean squared error criterion, i.e. is a solution of

$$\min\{MSE(\psi_{b,\delta^*}, \delta^*);\ b \geq b_0(\delta^*)\}.$$

The solution of (4) for $\alpha = \gamma$ and $w(b) = b$ will be denoted by b_1.

In Müller (1992a) it was shown that the score functions ψ_{b,δ^*} of the optimal robust estimators with bias bound b at the A-optimal design δ^* have a very simple form and that the trace of the corresponding asymptotic covariance matrix, $\operatorname{tr} V(\psi_{b,\delta^*}, \delta^*)$, is a decreasing, convex function of b with a known first and second derivative. Hence $GMSE(\psi_{b,\delta^*}, \delta^*)$ is also a convex function of b and the minimum point can be calculated by Newton's method.

If we want to minimize $GMSE(\psi_{b,\delta}, \delta)$ with respect to b for an arbitrary, non-A-optimal design, then the problem becomes very complicated, since no explicit formulas for the first and the second derivative are known. The advantage of the above definition of an optimal bias bound follows also from the fact that it provides the estimator and the design which minimizes the generalized mean squared error within all estimators and designs, i.e. $(\psi_{b^*,\delta^*},\ \delta^*)$ is the solution of

$$\min\{GMSE(\psi,\delta);\ \psi \in \Psi(\delta, C) \text{ and } \delta \in \Delta\}.$$

4. Examples

4.1. LINEAR REGRESSION

In a linear regression model

$$Y(t) = \beta_0 + \beta_1 t + Z(t) \text{ with } t \in [-1, 1],$$

the A-optimal design for estimating the whole parameter vector β is $\delta^* = \frac{1}{2}(e_{-1}+e_1)$. According to Müller (1992a), the score function of the optimal robust estimator for β with the bias bound $b > b_0(\delta^*) = \sqrt{\pi}$ at δ^* has the form

$$\psi_{b,\delta^*}(z,t) = \frac{\operatorname{sgn}(z)}{y_b}\ \frac{\min\{|z|, b\, y_b\}}{\sqrt{2}}\ (1,t)',$$

where y_b satisfies $y_b = [2\Phi(b\, y_b) - 1]/\sqrt{2} > 0$, while Φ denotes the standard normal distribution function.

As the optimal bias bound with respect to the mean squared error criterion we get $b_2 = 1.8289$. For this bias bound the fixed point y_b is equal to $y_{b_2} = 0.2386$. The optimal bias bound for the generalized mean squared error criterion with $\alpha = \gamma$ and $w(b) = b$ is $b_1 = 1.9934$, where $y_{b_1} = 0.4317$.

4.2. ONE-WAY LAY-OUT

In a one-way lay-out model

$$Y(i) = \beta_i + Z(i), \text{ for } i = 1, \ldots, 4,$$

with a control treatment, say 1, and three other treatments, say 2, 3, 4, let consider a linear aspect of the form $\varphi(\beta) = (\beta_1, \beta_2 - \beta_1, \beta_3 - \beta_1, \beta_4 - \beta_1)'$. The A-optimal design for estimating this aspect is $\delta^* = \frac{1}{5}(2\, e_1 + e_2 + e_3 + e_4)$. According to Müller (1992a) the score function of the optimal robust estimator for $\varphi(\beta)$ with the bias bound $b > b_0(\delta^*) = 5\sqrt{\pi/2}$ at δ^* has the form

$$\psi_{b,\delta^*}(z,t) = \operatorname{sgn}(z)\, \frac{\min\{|z|, b\, y_b\}}{y_b} \cdot \begin{cases} \frac{1}{2}(1,-1,-1,-1)' & \text{for } t = 1 \\ (e_1(t), \ldots, e_4(t))' & \text{for } t \neq 1 \end{cases},$$

where y_b satisfies $y_b = [2\Phi(b\, y_b) - 1]/5 > 0$.

As the optimal bias bound with respect to the mean squared error criterion we get $b_2 = 6.4660$ with $y_{b_2} = 0.0675$. The optimal bias bound for the generalized mean squared error criterion with $\alpha = \gamma$ and $w(b) = b$ is $b_1 = 8.1950$ with $y_{b_1} = 0.1644$.

Acknowledgements

The author thanks the referees for their comments and suggestions which improved the presentation of the paper.

References

Bickel, P.J. (1975). One-step Huber estimates in the linear model. *Journal of the American Statistical Association* **70**, 428–434.

Bickel, P.J. (1984). Robust regression based on infinitesimal neighbourhoods. *The Annals of Statistics* **12**, 1349–1368.

Hampel, F.R. (1978). Optimally bounding the gross-error-sensitivity and the influence of position in factor space. In: *Proceedings of the ASA Statistical Computing Section*. ASA, Washington, D.C., 59–64.

Hampel, F.R., Ronchetti, E.M., Rousseeuw, P.J. and Stahel, W.A. (1986). *Robust Statistics - The Approach Based on Influence Functions.* Wiley, New York.

Huber, P.J. (1964). Robust estimation of a location parameter. *The Annals of Mathematical Statistics* **35**, 73–101.

Krasker, W.S. (1980). Estimation in linear regression models with disparate data points. *Econometrica* **48**, 1333–1346.

Kurotschka, V. and Müller, Ch.H. (1992). Optimum robust estimation of linear aspects in conditionally contaminated linear models. *The Annals of Statistics* **20**, 331–350.

Müller, Ch.H. (1992a). Optimal designs for robust estimation in conditionally contaminated linear models. *Journal of Statistical Planning and Inference.* To appear.

Müller, Ch.H. (1992b). One-step-M-estimators in conditionally contaminated linear models. Preprint No. A-92-11, Freie Universität Berlin, Fachbereich Mathematik. Submitted for publication.

Samarov, A.M. (1985). Bounded-influence regression via local minimax mean squared error. *Journal of the American Statistical Association* **80**, 1032–1040.

Rieder, H. (1985). Robust estimation of functionals. Technical Report, Universität Bayreuth.

Rieder, H. (1987). Robust regression estimators and their least favorable contamination curves. *Statistics and Decisions* **5**, 307–336.

GEOMETRICAL RELATIONS AMONG VARIANCE COMPONENT ESTIMATORS

LYNN ROY LAMOTTE
Department of Experimental Statistics
Louisiana State University and A and M College
Baton Rouge, LA 70803-5606
USA

Abstract. In linear models with two variance components, mean squared errors (MSEs) of invariant quadratic estimators of linear combinations of the variance components are quadratics in the variance components. However, it is revealing to note that MSEs are linear in convex combinations of the squares and products of the variance components, so that surfaces of MSEs are subsets of planes. Lower bounds on MSEs form a concave surface. MSEs of admissible invariant quadratic estimators are tangent to this surface, while MSEs of inadmissible estimators are not. There are different bounds when attention is restricted to unbiased invariant quadratic estimators or the more general class obtained by dropping the restriction to unbiasedness. The purpose of this paper is to develop these simple relations and to illustrate them with graphs.

Key words: Invariant quadratic estimation, Mixed models with two variance components.

1. Introduction

If Y is normally distributed and follows a mixed linear model with variance components γ (a vector of nonnegative parameters), then mean squared errors (MSEs) of invariant quadratic estimators (IQEs) of these components are quadratic functions of γ. Such MSEs can be expressed as linear functions of $\gamma\gamma'$. Thus MSE functions are subsets of hyperplanes. This fact provides a particularly simple and informative way to view and compare MSEs of invariant quadratic estimators of variance components.

In such a model with two variance components, the dimension of $\gamma\gamma'$ is three, so ($\gamma\gamma'$,MSE) is in four dimensions. However, using an appropriate projection that preserves relations among MSEs, MSEs can be plotted against $\gamma\gamma'$ and viewed in three dimensions. The primary purpose of this paper is to illustrate some characteristics and relations among MSEs with such three-dimensional graphs.

2. The Model, Notation, and Background

Consider a model with fixed effects β and one set of random effects α, plus an error term, of the form

$$Y = X\beta + U\alpha + e,$$

where Y is an n-vector of observations of a response variable, X and U are fixed and known matrices, β is a vector of unknown fixed effects, α and e are uncorrelated

T. Caliński and R. Kala (eds.),
Proceedings of the International Conference on Linear Statistical Inference LINSTAT '93, 103–111.

vector-valued random variables such that Y is normally distributed with $\mathrm{E}(Y) = X\beta$ and

$$\mathrm{Var}(Y) = \sigma_1^2 I + \sigma_2^2 UU'. \tag{1}$$

In the expressions shown here, dimensions will not be denoted unless it is necessary. Assume that they are unrestricted, except that they must make sense for the expressions in which they appear. With $\gamma = (\gamma_1, \gamma_2)' \equiv (\sigma_1^2, \sigma_2^2)'$, consider estimating $p'\gamma = p_1\gamma_1 + p_2\gamma_2$ with invariant quadratics $Y'AY$. Invariance requires that $AX = 0$.

Suppose $Q_i = Y'A_iY$, $i = 1, \ldots, k$, are k invariant quadratics. Let $Q = (Q_1, \ldots, Q_k)'$. Then $\mathrm{E}(Q) \equiv \mu = B\gamma$ where the $k \times 2$ matrix B has as its i-th row $(\mathrm{tr}(A_i), \mathrm{tr}(A_iUU'))$. The variance-covariance matrix of Q has the form

$$\mathrm{Var}(Q) = 2(\mathrm{tr}(A_i \mathrm{Var}(Y) A_j \mathrm{Var}(Y))). \tag{2}$$

With this structure, Q follows a general linear model, as Pukelsheim (1976) and others have noted. LaMotte (1982) established fundamental relations among MSEs of linear estimators in a very general framework that includes the present one as a special case. To correspond with that framework, we are considering estimation of $p'\gamma = c'(B\gamma) = c'\mu$ by estimators $\ell'Q$ that are linear in Q. MSE of such an estimator is

$$\mathrm{MSE}(\ell'Q) \equiv \mathrm{MSE}_\ell(\gamma\gamma') = \mathrm{Var}(\ell'Q) + (\ell - c)'(B\gamma)(B\gamma)'(\ell - c). \tag{3}$$

It is important at this point to recognize that MSE_ℓ is linear in $v \equiv \gamma\gamma'$. Because $\mathrm{MSE}_\ell(av) = a\mathrm{MSE}_\ell(v)$ for any scalar $a \geq 0$, we can characterize MSE_ℓ in terms of its restriction to any appropriate cross-section, such as $\{v = (v_{ij}) \colon v_{11} + v_{22} = 1\}$, by Lemma 3.2 in LaMotte (1982).

2.1. The OSB Partition

Olsen, Seely and Birkes (1976) (OSB) described a minimal sufficient statistic for the family of distributions of a maximal invariant statistic. It is a set of k independent, invariant quadratics $Q_1, \ldots, Q_k$, with $Q_i \sim (\gamma_1 + \lambda_i\gamma_2)\chi^2_{r_i}$. To define r_i and λ_i, let H be a matrix such that its columns form an orthonormal basis for $\{z : X'z = 0\}$. Denote the distinct eigenvalues of $W \equiv H'UU'H$ by $\lambda_i, i = 1, \ldots, k$, with $0 \leq \lambda_1 < \ldots < \lambda_k$, and denote their respective multiplicities by $r_i, i = 1, \ldots, k$. Then $Q_i = Y'HP_iP_i'H'Y$, where the r_i columns of P_i are orthonormal eigenvectors of W corresponding to λ_i.

We will work with the OSB partition $Q = (Q_1, \ldots, Q_k)'$ in the remainder of this paper. Thus we will consider relations among those estimators of $p'\gamma$ which are of the form $\ell'Q = \sum \ell_i Q_i$, where Q_i, $i = 1, \ldots, k$, are as defined above.

Depending on X and UU', the least eigenvalue of W may be zero or not. In common ANOVA models $\lambda_1 = 0$. However, in other two-variance-components models, λ_1 is positive. LaMotte and McWhorter (1978), for example, used the OSB partition to develop a test that $\gamma_2 = 0$ in a model in which λ_1 is typically positive. Whether λ_1 is positive affects, among other things, whether a nonnegative unbiased invariant quadratic estimator of γ_1 exists, and whether powers of tests that $\gamma_2 = 0$ go to 1 as γ_2 goes to ∞.

2.2. The ANOVA Estimator

An analysis of variance (ANOVA) table for Y typically has the following form.

ANOVA Table

Source	df	SS	E(SS)
Fixed (X)			
Residual		SSR	
Random (U)		SSU	$g_1\gamma_1 + g_2\gamma_2$
Error		SSE	$r_1\gamma_1$

SSR, SSU, and SSE are invariant quadratics and g_1, g_2, and r_1 are constants. Residual SS (SSR) is partitioned into SSU and SSE, which are proportional to respective independent central chi-squared random variables. The ANOVA estimator of $p'\gamma$ is ℓ_1SSE+ℓ_2SSU, where ℓ_1 and ℓ_2 are chosen such that the estimator is unbiased.

The invariant quadratics in the ANOVA table may be related to the OSB partition. If the least eigenvalue λ_1 is 0 then SSE is Q_1. If λ_1 is greater than 0 then there does not exist an invariant quadratic with expected value proportional to γ_1. SSR is $Q_1 + \cdots + Q_k$ and, if $\lambda_1 = 0$, SSU is $Q_2 + \cdots + Q_k$.

2.3. Best IQEs and Attainable Lower Bounds on MSE

We shall illustrate relations among linear (in Q) estimators in two classes of estimators. One is the set of all linear estimators, with $\ell \in \mathbb{R}^k$. The other is the set of all linear estimators that are unbiased for $p'\gamma$, with $\ell \in \{\ell : B'\ell = p\}$. Let v be a point in the minimal closed convex cone containing $\{\gamma\gamma' : \gamma \geq 0\}$. In each class of estimators, there exists a vector ℓ_v that minimizes MSE at v among all estimators in the class. We shall say that such an estimator is *best at* v in its class. Different points v may have different best estimators. An attainable lower bound on MSE of estimators in a class is $\text{MSE}_{\ell_v}(v)$. See LaMotte (1982) for details.

3. The Graphs

For the graphs that follow we will use $k = 5$ invariant quadratics, with eigenvalues 0, 1, 2, 3, 4 and multiplicities 5, 3, 3, 1, 3. Having $k > 2$ guarantees that there is no uniformly best unbiased IQE. We will estimate the second variance component γ_2, so $p = (0,1)'$.

Figure 1 shows the MSE of the ANOVA estimator of γ_2 and the lower bound on MSEs (variances) of all unbiased IQEs of γ_2, both plotted against $\gamma_1^2/(\gamma_1^2 + \gamma_2^2) = v_{11}/(v_{11} + v_{22})$. Note the vertical space between the two curves, indicating that the ANOVA estimator is not best at v for any v.

Figure 2 shows $\gamma\gamma'$ in three dimensions. The set $\{(\gamma_1^2, \gamma_2^2, \gamma_1\gamma_2) : \gamma_1 \geq 0, \gamma_2 \geq 0\}$ is the top surface of the cone $\{(v_{11}, v_{22}, v_{12}) : v_{11} \geq 0, v_{22} \geq 0, 0 \leq v_{12} \leq \sqrt{v_{11}v_{22}}\}$. The intersection of this surface and the hyperplane $\{(v_{11}, v_{22}, v_{12}) : v_{11} + v_{22} = 1\}$ is the semicircular arc. The line in the $v_{11} \times v_{22}$ plane across the bottom of the arc is the horizontal axis in Figure 1.

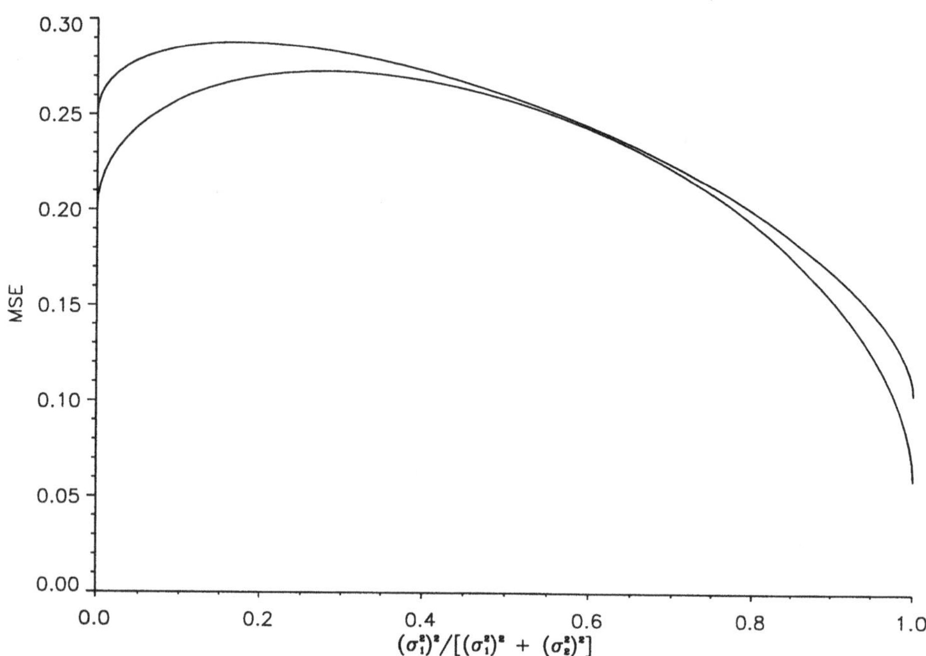

Fig. 1. MSE of ANOVA estimator and bound on MSEs of unbiased IQEs.

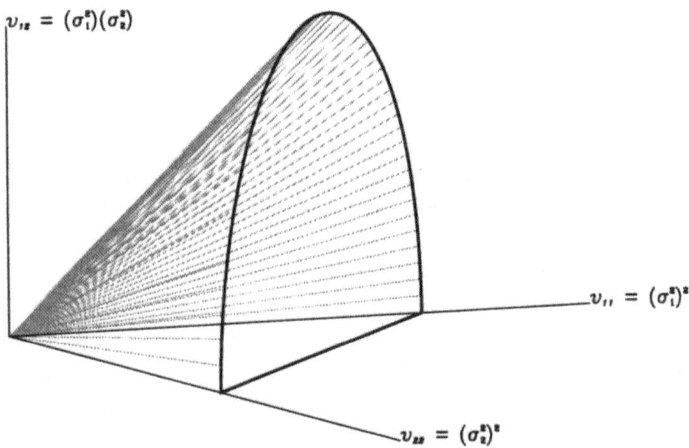

Fig. 2. The parameter set.

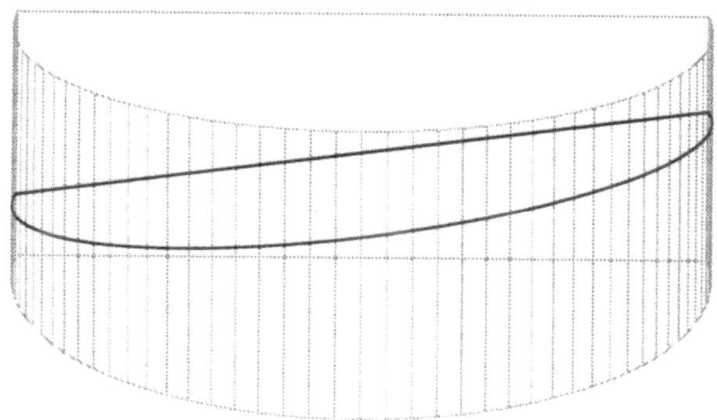

Fig. 3. MSEs of the ANOVA estimator and the IQUE best at $\gamma_1^2 = 0.6, \gamma_2^2 = 0.4$.

In Figure 2, imagine rotating the top of the arc 90° toward the origin so that the arc lies flat on the horizontal plane. Erect a vertical axis for MSE. At this point, MSE is defined only on the outer rim of a half-cylinder rising vertically from the arc on the horizontal plane.

Figure 3 shows MSEs of the ANOVA estimator and a best unbiased IQE (in particular, the one best at $\gamma_1^2/(\gamma_1^2 + \gamma_2^2) = 0.6$). In fact, MSEs of the two estimators are so close that they are practically indistinguishable.

In order to see relations among MSEs of IQEs fully, it is necessary to extend the domain from $\{\gamma\gamma' : \gamma \geq 0\}$ to the minimal closed convex cone containing this set. See Olsen et al. (1976) and LaMotte (1982). In Figure 2 the resulting set is the interior and surface of the cone, i.e. $\{(v_{11}, v_{22}, v_{12}) : v_{11} \geq 0, v_{22} \geq 0, 0 \leq v_{12} \leq \sqrt{v_{11}v_{22}}\}$. Now, instead of looking only on the proximal rim of the half-circle in Figure 3, we will extend the domain of MSE to the entire closed convex half-circle. Keep in mind, though, that, even after this extension, MSE for any particular IQE is the intersection of a hyperplane and the (now solid) half-cylinder.

Figure 4 shows MSEs of three unbiased IQEs of γ_2. Each one is best at γ_0 for some γ_0. That is, each one touches the lower bound on MSEs of unbiased IQEs somewhere on the surface of the cylinder. However, the main purpose of Figure 4 is to illustrate the variety of MSEs possible, even among the very restricted class of unbiased IQEs. With such variety possible, it is natural to ask whether some estimators are dominated by others. That is, given one, can we find another estimator whose MSE fits entirely under the first?

The answer is easier to see by showing the bound on MSEs extended to the entire half-circle. See Figure 5. Olsen et al. (1976) showed for unbiased IQEs that, if the MSE plane does not touch the bound, there exists another unbiased IQE whose MSE plane fits between the bound and the first MSE plane. They also proved that best estimators - those whose MSE planes touch the bound - form a minimal essentially complete class. Thus we may say, only somewhat imprecisely, that for any MSE plane that does not touch the bound, there is one that touches the bound that is everywhere under the first. LaMotte (1977) showed how to identify such a better, best estimator.

Now consider the class of all IQEs, without the restriction to unbiased estima-

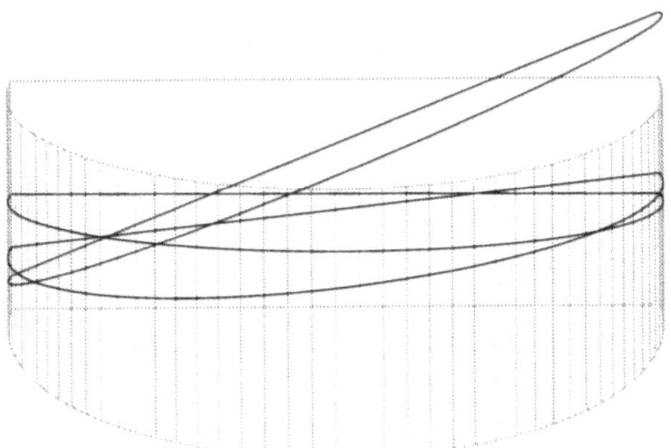

Fig. 4. MSEs of several best unbiased IQEs.

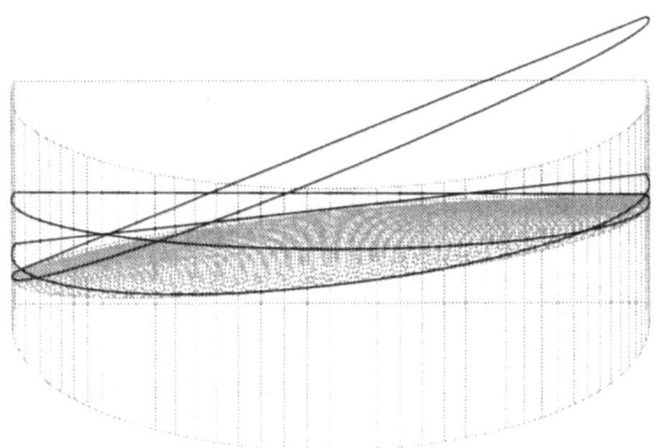

Fig. 5. Bound on MSEs of unbiased IQEs and MSEs of several unbiased IQEs.

tors. Figure 6 shows the lower bound on MSEs of IQEs. The MSE of the ANOVA estimator is included for reference. Given the discussion in the previous paragraph we see, from observing that the ANOVA MSE plane lies well above the surface of the bound, that the ANOVA estimator is not admissible among IQEs. In fact, no unbiased IQE is admissible among IQEs of its expected value (LaMotte, 1980).

Because of the great curvature of the MSE bound, which occurs in the interior of the domain, there is tremendous variety in the shapes of MSEs of best IQEs. Figure 7 and Figure 8 illustrate some of that variety. In those figures the MSE of the ANOVA estimator is sketched lightly for reference. In Figure 7, the three estimators are best at three points along the line joining $(v_{11} = 0, v_{22} = 1, v_{12} = 0)$ and $(v_{11} = .5, v_{22} = .5, v_{12} = .5)$, i.e., the back-right to the front-center. In Figure 8 the points at which the MSE planes are tangent to the bound surface are shown. See how specific is the IQE best at the front-center point. Although it has minimum MSE at that point, its MSE rises extremely rapidly at parameter values different

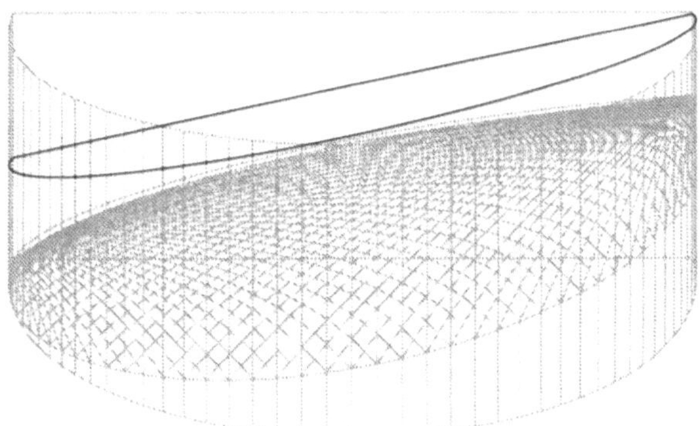

Fig. 6. Bound on MSE of IQEs and MSE of the ANOVA estimator.

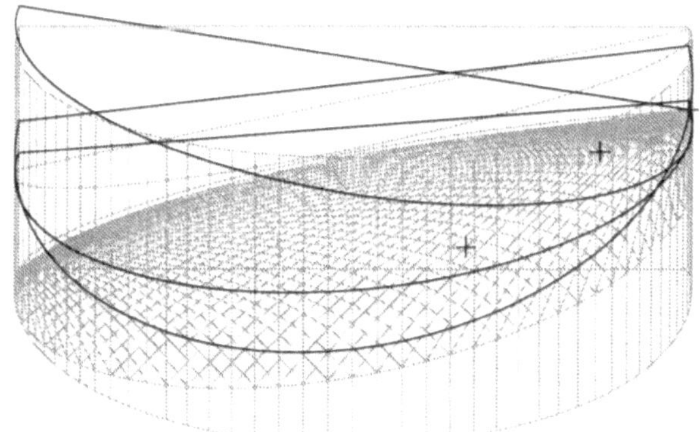

Fig. 7. MSEs of four best IQEs.

from that point. Note also that MSEs of the other three estimators are uniformly below MSE of the ANOVA estimator.

Figure 9 illustrates an interesting special case. At ($v_{11} = 0, v_{22} = 1, v_{12} = 0$), many IQEs are best. In the two-variance-components model, this is the only point that has multiple best estimators, and then only when $\lambda_1 = 0$. There are many MSE planes that touch the bound at that point. Figure 9 shows MSEs of five such estimators.

The fact that there are many IQEs that are best at $v_{11} = 0$ corresponds to the fact that there is a discontinuity in the bound shown in Figure 6 at $v_{11} = 0$. LaMotte (1973, p. 326) showed that for any γ at which $\mathrm{Var}(Y)$ is nonsingular, an attainable lower bound on MSEs of IQEs at γ is $2(p'\gamma)^2/(\nu + 2)$, where $\nu = \sum_{i=1}^{k} r_i$. Points γ correspond to the proximal rim of the half-circle in the horizontal plane of Figure 6. Thus the points that describe the bound along the front surface of the cylinder are coplanar. In the example, $\nu = 15$ and we are estimating γ_2, so the bound is $2\gamma_2^2/17$, and this holds for all $\gamma_2 > 0$. However, at $v = (0, 1, 0)$, the bound on MSE is $2\gamma_2^2/(\sum_{i=2}^{k} r_i + 2) = 2/12$.

The IQEs best at (0,1,0) are of the form $\ell'Q$, with $\ell \in \mathcal{L}_0 \equiv \{\ell_0 + Nz : z \in \mathbb{R}\}$,

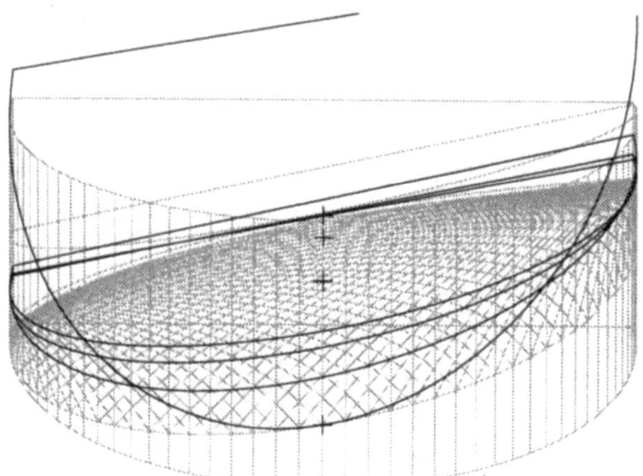

Fig. 8. MSEs of four more best IQEs.

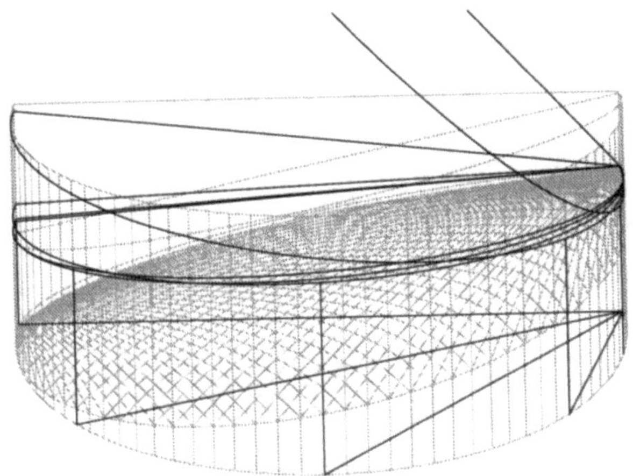

Fig. 9. MSEs of five IQEs best at $v_{11} = 0$.

where $N = (1, 0, \ldots, 0)'$ and

$$\ell_0 = (0, 1/\lambda_2, \ldots, 1/\lambda_k)'/(\sum_{i=2}^{k} r_i + 2). \tag{4}$$

All such estimators have MSE $= 2/(\sum_{i=2}^{k} r_i + 2)$ at (0,1,0). Following LaMotte (1982), it may be shown that IQEs in $\mathcal{L}_0$ that are admissible must be best among $\mathcal{L}_0$ at a nonzero point (v_{11}, v_{22}, v_{12}) with $v_{11} \geq 0$ and $v_{12} \geq 0$. This requires that

$$z = [2v_{12}/v_{11} - \sum_{i=2}^{k} r_i/\lambda_i]/[(r_1 + 2)(\sum_{i=2}^{k} r_i + 2)]. \tag{5}$$

(No point v with $v_{12} > 0$ and $v_{11} = 0$ permits a best estimator among $\mathcal{L}_0$.)

Let ℓ_* denote an estimator that is best among $\mathcal{L}_0$ at v_* with $v^*_{12}/v^*_{11} \geq 0$. Let ℓ_v denote the estimator best among all IQEs at v. Let $v_0 = (0,1,0)$. Then it may be shown that

$$\ell_* = \lim_{\delta \to 0+} \ell_{[(1-\delta)v_0 + \delta v_*]}. \tag{6}$$

Acknowledgements

I am grateful to the organizers and sponsors of LINSTAT'93 for the opportunity to participate in such a well-organized and interesting conference. Research for this paper was supported in part by Grant DMS-9104811 from the National Science Foundation.

References

LaMotte, L. (1973). Quadratic estimation of variance components. *Biometrics* **29**, 311–330.

LaMotte, L. (1977). On admissibility and completeness of linear unbiased estimators in a general linear model. *Journal of the American Statistical Association* **72**, 438–441.

LaMotte, L. (1980). Some results on biased estimation applied to variance component estimation. In: W. Klonecki, A. Kozek and J. Rosiński, Eds., *Mathematical Statistics and Probability Theory. Lecture Notes in Statistics* **2**. Springer-Verlag, New York, 266–274.

LaMotte, L. (1982). Admissibility in linear estimation. *Annals of Statistics* **10**, 245–255.

LaMotte, L. and McWhorter, A. (1978). An exact test for the presence of random walk coefficients in a linear regression model. *Journal of the American Statistical Association* **73**, 816–820.

Olsen, A., Seely, J. and Birkes, D. (1976). Invariant quadratic estimation for two variance components. *The Annals of Statistics* **4**, 878–890.

Pukelsheim, F. (1976). Estimating variance components in linear models. *Journal of Multivariate Analysis* **6**, 626–629.

ASYMPTOTIC EFFICIENCIES OF MINQUE AND ANOVA VARIANCE COMPONENT ESTIMATES IN THE NONNORMAL RANDOM MODEL

PETER H. WESTFALL
Texas Tech University
Lubbock, TX 79409-2101
U.S.A.

Abstract. The following question is addressed: For which quadratic unbiased estimates of variance components, and under what asymptotic assumptions, are the estimates as efficient as estimates based on *the random effects themselves*, with or without the normality assumption? Westfall and Bremer (1993) have identified sufficient asymptotic conditions under which such an efficiency property holds for the 'cell means estimates' in the general k-way classification model. In this paper, the asymptotic behavior of MINQUE, ANOVA, and cell means estimates is considered in the one-way random model.

Key words: Cell means estimates, Kurtosis, Quadratic estimates, Unbalanced data.

1. Introduction

The general mixed ANOVA model has the form

$$\begin{aligned} Y &= \textstyle\sum_{r\in F} U_r\theta_r + \sum_{s\in R} U_s\xi_s + \epsilon \\ &= X\theta + U\xi + \epsilon, \end{aligned} \tag{1}$$

where Y is the column vector of observations $y(i_1, ..., i_k, i_{k+1})$, arranged lexicographically with right-most indices varying fastest. The factors θ_r and ξ_s are fixed and random effects, respectively, associated with the factor combinations r and s. We assume that the ξ_s and ϵ are jointly independent mean zero random vectors, and that the elements within vectors are independent, with common variances $\phi_s \geq 0$ and $\phi_e > 0$, respectively. This model is discussed by Seifert (1979), Khuri (1990), Hocking (1985), and Westfall and Bremer (1993). The terms r and s are subsets of the set $S = \{1, \ldots, k\}$, and are used to refer to different interaction terms in the model.

In his pioneering work on unification of variance component estimation, Rao (1971) considered estimates that would be used if the *random effects themselves* were known. Assuming finite fourth moments, and assuming the effects are observable, the following have minimum variance (Westfall, 1987a) within the class of quadratic unbiased estimates:

$$\hat{\phi}_{so} = \begin{cases} \xi_s'\xi_s/a_s, & \text{for } s \in R \\ \epsilon'\epsilon/n, & \text{for } s = e. \end{cases}$$

Here the 'o' subscript denotes 'optimal', and $a_s = \dim\xi_s$. We call these estimates 'optimal nonparametric estimates', or simply 'optimal estimates'. It is important to reiterate that such estimates are unobservable under the model (1).

T. Caliński and R. Kala (eds.),
Proceedings of the International Conference on Linear Statistical Inference LINSTAT '93, 113–119.

Like the optimal nonparametric estimates, any invariant quadratic estimates based on the observable data are also quadratic functions of the underlying random effects. In Rao's minimum norm invariant quadratic unbiased estimation (MINQUE) procedure, one minimizes the difference between an invariant quadratic unbiased estimator (IQUE) and the optimal estimate by minimizing the (possibly weighted) norm of the difference between the defining matrices.

The MINQUE procedure is 'nonparametric' in that no distributional form is required for the underlying effects. Nevertheless, when the effect distributions are normal, the resulting estimates become locally minimum variance quadratic unbiased estimates for suitably chosen weights (e.g., Rao and Kleffe, 1988, Chapter 11).

We investigate conditions under which commonly-used estimates are asymptotically as efficient as the optimal estimates. Exploiting the simple structure of the 'cell means' estimates, Westfall and Bremer (1993) demonstrate that such estimates are as efficient as the optimal estimates when both $\min\{a_r\} \rightarrow \infty$ and $\min\{n(i_1, ..., i_k)\} \rightarrow \infty$ in model (1), where $n(i_1, ..., i_k)$ is the sample size in a k-fold cell. Because the MINQUE estimates are constructed to be close (in the minimum norm sense) to the optimal estimates, one would expect a similar efficiency result to hold for the MINQUE estimates. The form of the estimates is generally quite complicated, however. Therefore, this paper considers only the one-way random model

$$y_{ij} = \xi_i + \epsilon_{ij}, \tag{2}$$

or in matrix form, $Y = U\xi + \epsilon$. The MINQUE estimates depend on a weight $\gamma \in [0, \infty)$, which is the analyst's fixed *a priori* guess of the variance ratio ϕ_1/ϕ_e. An explicit representation of the resulting MINQUE estimates is given below in the proof of Result 1.

2. Results

2.1. Notation and Definitions

The indices in model (2) are $i = 1, \ldots, a$ and $j = 1, \ldots, n_i$. The random variables are defined as follows: the ξ_i are i.i.d. with mean zero, variance ϕ_1, and finite fourth moment μ_1 with $\mu_1 - \phi_1^2 > 0$; the ϵ_{ij} are i.i.d. with mean zero, variance ϕ_e, and finite fourth moment μ_e with $\mu_e - \phi_e^2 > 0$; and the ξ_i are independent of the ϵ_{ij}. Define $n = \sum n_i$, and let $n_0 = \min\{n_i\}$.

In this model, the optimal estimates have variances $\mathrm{Var}(\hat{\phi}_{1o}) = (\mu_1 - \phi_1^2)/a$, and $\mathrm{Var}(\hat{\phi}_{eo}) = (\mu_e - \phi_e^2)/n$.

2.2. Asymptotic Sequence

The model (2) is indexed by a parameter τ, $\tau = 1, 2 \ldots$. (The random variables are defined on a 'triangular array', e.g., Chung, 1974, p. 196.) All terms in (2), including the random variables, a, and the n_i, are functions of τ. The dependence of these terms on τ will usually be implicit, rather than explicit, in the notation. Specifically, we assume that as $\tau \rightarrow \infty$, $n_0 \rightarrow \infty$, and $a \rightarrow \infty$.

Comments are in order concerning the choice of an asymptotic sequence.

Comment 1. To apply this asymptotic formulation to a given finite-sample design, one must imagine only that the design is part of a sequence of possibly *hypothetical* designs (indexed by τ) for which the given conditions hold. In particular, it is not necessary to assume that the minimum within-cell sample size will increase when more data is collected.

Comment 2. There are many choices of asymptotic sequences available. Westfall (1987b, 1988a, 1988b) considered the case where the n_i are uniformly bounded, but with $a \to \infty$. This type of asymptotic sequence is better for realism, but worse for simplification.

Seely (1979) and Westfall (1987a) considered the case $cn^\alpha < n_i < Cn^\alpha$, for $0 < c < C < \infty$ and $\alpha \in (0,1)$, which represents a special case of the present asymptotic scheme.

Comment 3. The present asymptotics are sufficient, but not necessary, for the results to hold. Nevertheless, they are weaker than Seely's (1979) and Westfall's (1987a) asymptotics. Note also that n_0 and a are allowed to diverge at possibly different rates.

2.3. Efficiency of MINQUE when $\gamma \in (0,\infty)$

Now let us state

Result 1. *Consider the MINQUE estimates $\hat{\phi}_{1\gamma}$ and $\hat{\phi}_{e\gamma}$ which use a priori variance ratio equal to $\gamma \in (0,\infty)$. Under the model (2), with the asymptotic scheme given above,*

$$\mathrm{Var}(\hat{\phi}_{1\gamma})/\mathrm{Var}(\hat{\phi}_{1o}) \to 1 \text{ and } \mathrm{Var}(\hat{\phi}_{e\gamma})/\mathrm{Var}(\hat{\phi}_{eo}) \to 1.$$

Proof. For particular choice γ of *a priori* variance ratio, let $V_\gamma = \gamma UU' + I$. The quadratic forms used to define the MINQUE estimates are $q_{1\gamma} = Y'V_\gamma^{-1}UU'V_\gamma^{-1}Y$, and $q_{e\gamma} = Y'V_\gamma^{-1}V_\gamma^{-1}Y$. Letting $q_\gamma = (q_{1\gamma}, q_{e\gamma})'$, we have $E(q_\gamma) = T_\gamma\phi$, so the MINQUE estimates are $\hat{\phi}_\gamma = T_\gamma^{-1}q_\gamma$.

Consider

$$T_\gamma = \begin{pmatrix} \sum n_i^2/(1+\gamma n_i)^2 & \sum n_i/(1+\gamma n_i)^2 \\ \sum n_i/(1+\gamma n_i)^2 & n - 2\gamma\sum n_i/(1+\gamma n_i) + \gamma^2\sum n_i^2/(1+\gamma n_i)^2 \end{pmatrix}$$

$$= \begin{pmatrix} (a/\gamma^2)\{1+o(1)\} & (a/\gamma^2)o(1) \\ (a/\gamma^2)o(1) & n\{1+o(1)\} \end{pmatrix}.$$

Thus,

$$\hat{\phi}_\gamma = \begin{pmatrix} (\gamma^2/a)\{1+o(1)\} & (1/n)o(1) \\ (1/n)o(1) & (1/n)\{1+o(1)\} \end{pmatrix} q_\gamma, \tag{3}$$

implying $\hat{\phi}_{1\gamma} = (\gamma^2/a)\{1+o(1)\}q_{1\gamma} + (1/n)o(1)q_{e\gamma}$ and $\hat{\phi}_{e\gamma} = (1/n)o(1)q_{1\gamma} + (1/n)\{1+o(1)\}q_{e\gamma}$.

Using Lemma 1 of Westfall (1987b), we have

$$\mathrm{Var}(q_{1\gamma}) = (\mu_1 - \phi_1^2)\sum n_i^4/(1+\gamma n_i)^4 + 4\phi_1\phi_e\sum n_i^3/(1+\gamma n_i)^4$$

$$+2\phi_e^2 \sum n_i^2/(1+\gamma n_i)^4 + (\mu_e - 3\phi_e^2)\sum n_i/(1+\gamma n_i)^4$$
$$= (\mu_1 - \phi_1^2)(a/\gamma^4)\{1+o(1)\} + o(a), \tag{4}$$

and similarly

$$\mathrm{Var}(q_{e\gamma}) = (\mu_e - \phi_e^2)n + o(n). \tag{5}$$

Combining (3), (4), (5), and using $|\mathrm{Cov}(X,Y)| \leq \mathrm{Var}^{1/2}(X)\mathrm{Var}^{1/2}(Y)$, the desired efficiency results follow for $\hat{\phi}_{1\gamma}$ and $\hat{\phi}_{e\gamma}$. □

2.4. Efficiency of MINQUE when $\gamma = 0$: $\hat{\phi}_{10}$

In the following sections, please distinguish the '0' subscript (referring to MINQUE estimates with $\gamma = 0$) from the 'o' subscript (referring to optimal nonparametric estimates). Also note that when 'o' and 'O' are not subscripts, they refer to asymptotic order.

A key element in deriving the $o(1)$ terms in (3), (4), and (5) is the inequality $1/(1+\gamma) \leq n_i/(1+\gamma n_i) < 1/\gamma$; when $\gamma = 0$, the upper bound fails. The MINQUE quadratic forms with $\gamma = 0$ are $q_{10} = Y'UU'Y$ and $q_{e0} = Y'Y$. Thus $E(q_0) = T_0\phi$, where

$$T_0 = \begin{pmatrix} \sum n_i^2 & n \\ n & n \end{pmatrix},$$

and

$$\hat{\phi}_0 = \frac{1}{\sum n_i^2 - n}\begin{pmatrix} 1 & -1 \\ -1 & \sum n_i^2/n \end{pmatrix} q_0.$$

Noting that $n/\sum n_i^2 \leq 1/n_0 = o(1)$, we have $\hat{\phi}_{10} = (q_{10} - q_{e0})\{1+o(1)\}/\sum n_i^2$ and $\hat{\phi}_{e0} = -\{1+o(1)\}q_{10}/\sum n_i^2 + \{1+o(1)\}q_{e0}/n$.

Now,

$$\mathrm{Var}(q_{10} - q_{e0}) = (\mu_1 - \phi_1^2)\sum(n_i^2 - n_i)^2 + 4\phi_1\phi_e \sum n_i(n_i-1)^2 + 2\phi_e^2 \sum n_i(n_i-1)$$
$$= (\mu_1 - \phi_1^2)\{1+o(1)\}\sum n_i^4.$$

(The variance of $q_{10} - q_{e0}$ does not depend on the fourth moment μ_e; Westfall, 1987b, 1988a, used this fact to demonstrate certain robustness properties of related statistics.)

Thus,

$$\frac{\mathrm{Var}(\hat{\phi}_{10})}{\mathrm{Var}(\hat{\phi}_{1o})} - 1 = \{1+o(1)\}\left(a\sum n_i^4 - \left(\sum n_i^2\right)^2\right)/\left(\sum n_i^2\right)^2$$
$$= \{1+o(1)\}\mathrm{CV}_2(\tau),$$

where $\mathrm{CV}_2(\tau)$ is the coefficient of variance (variance divided by the squared mean) of the n_i^2 for model τ in the asymptotic sequence. This proves

Result 2. *If $CV_2(\tau)$ converges to CV_2 as $\tau \to \infty$, then $Var(\hat{\phi}_{10})/Var(\hat{\phi}_{1o}) \to 1 + CV_2$.*

If $\mathrm{CV}_2 = 0$, then we say that the design is *asymptotically balanced.* Of course, this will occur when all n_i are equal, as well as in situations where there is a decreasingly small fraction of uncommon cell sizes.

Example 1. Suppose $n_i \equiv k_1\tau$ for $1 \le i \le a/2$, and suppose $n_i \equiv k_2\tau$ for $a/2+1 \le i \le a$, for any positive integers k_1 and k_2. [Assume wlog that $a = a(\tau)$ is always even.] Then $\mathrm{CV}_2 = \{(k_1^2 - k_2^2)/(k_1^2 + k_2^2)\}^2$.

2.5. Efficiency of MINQUE when $\gamma = 0$: $\hat{\phi}_{e0}$

Consider the representation $\hat{\phi}_{e0} = -\{1 + o(1)\}q_{10}/\sum n_i^2 + \{1 + o(1)\}q_{e0}/n$. Upon careful consideration it is seen that the $o(1)$ terms will not affect the asymptotic order of the variance, hence it will suffice to consider $\mathrm{Var}(-q_{10}/\sum n_i^2 + q_{e0}/n)$ for asymptotic comparisons. The variance is bounded below by the term involving $\mu_1 - \phi_1^2$ as follows:

$$\begin{aligned}
\mathrm{Var}(-q_{10}/\textstyle\sum n_i^2 + q_{e0}/n) &\ge (\mu_1 - \phi_1^2)\left(\textstyle\sum_i n_i^2(\sum_j n_j^2 - n_i n)^2\right)/\left(n^2(\textstyle\sum n_i^2)^2\right) \\
&\ge (\mu_1 - \phi_1^2)n_0\left(\textstyle\sum_i n_i(\sum_j n_j^2 - n_i n)^2\right)/\left(n^2(\textstyle\sum n_i^2)^2\right) \\
&= (\mu_1 - \phi_1^2)n_0\left(n\textstyle\sum n_i^3 - (\sum n_i^2)^2\right)/\left(n(\textstyle\sum n_i^2)^2\right).
\end{aligned}$$

Thus,

$$\mathrm{Var}(\hat{\phi}_{e0})/\mathrm{Var}(\hat{\phi}_{eo}) \ge (\mathrm{Constant})n_0\frac{n\sum n_i^3 - (\sum n_i^2)^2}{(\sum n_i^2)^2}\{1 + o(1)\}. \qquad (6)$$

The fraction in (6) is always positive, and represents a term similar to the coefficient of variance found in Result 2. Thus we may state

Result 3. *If*

$$\frac{n\sum n_i^3 - (\sum n_i^2)^2}{(\sum n_i^2)^2}$$

is bounded away from 0, then $\mathit{Var}(\hat{\phi}_{e0})/\mathit{Var}(\hat{\phi}_{eo})$ *diverges.*

Example 2. Consider the n_i-configuration of Example 1. Then

$$\frac{n\sum n_i^3 - (\sum n_i^2)^2}{(\sum n_i^2)^2} = \frac{k_1k_2(k_1 - k_2)^2}{(k_1^2 + k_2^2)^2},$$

which is clearly greater than zero when $k_1 \ne k_2$.

2.6. Efficiency of ANOVA Estimates

The usual ANOVA quadratic forms are given by $q_{1A} = Y'U(U'U)^{-1}U'Y$ and $q_{eA} = Y'(I - U(U'U)^{-1}U')Y$; equating to expectations and solving yields $\hat{\phi}_{1A} = q_{1A}/n - aq_{eA}/\{n(n-a)\}$ and $\hat{\phi}_{eA} = q_{eA}/(n-a)$.

Since $\hat{\phi}_{eA}$ is the cell means estimate defined by Westfall and Bremer (1993), their results imply that $\mathrm{Var}(\hat{\phi}_{eA})/\mathrm{Var}(\hat{\phi}_{eo})$ converges to unity under the present asymptotic scheme.

Consider the result for $\hat{\phi}_{1A}$: Note that $\text{Var}(q_{1A}) = (\mu_1 - \phi_1^2)\{1 + o(1)\} \sum n_i^2$ and that $\text{Var}(q_{eA}) = O(n)$. These results, together with the covariance inequality, imply the equality $\text{Var}(\hat{\phi}_{1A}) = \{1 + o(1)\}(\mu_1 - \phi_1^2) \sum n_i^2/n^2$, so

$$\frac{\text{Var}(\hat{\phi}_{1A})}{\text{Var}(\hat{\phi}_{1o})} - 1 = \{1 + o(1)\} \frac{a \sum n_i^2 - (\sum n_i)^2}{(\sum n_i)^2},$$

which is $\{1 + o(1)\}$ times the coefficient of variance of the n_i, $\text{CV}_1(\tau)$. Thus the following result may be stated:

Result 4. *If $CV_1(\tau)$ converges to CV_1 as $\tau \to \infty$, then $Var(\hat{\phi}_{1A})/Var(\hat{\phi}_{1o}) \to 1 + CV_1$.*

3. Concluding Remarks

Let us conclude with the following remarks.

Remark 1. What has been defined as 'efficiency' in this paper would ordinarily be called 'inefficiency', since large values are considered bad. The present choice was made to simplify the results.

Remark 2. The ratio of variance of MINQUE estimate to the variance of the optimal nonparametric estimate does not converge uniformly in $\gamma \in [0, \infty)$, the ratio converges to the value $1 + \text{CV}_2$ for the estimate $\hat{\phi}_{10}$.

It may be shown that $\text{CV}_2 \geq \text{CV}_1$; hence, we have the interesting conclusion that the ANOVA estimate is positioned firmly 'between' estimates in the MINQUE class (between $\hat{\phi}_{10}$ and $\hat{\phi}_{1\gamma}$, for $\gamma > 0$).

Remark 3. Excluding the overall mean term in model (2) limits the applicability of the results. Ongoing research following Westfall (1987a, 1987b, and 1988b) considers the case where the mean is in the model.

Remark 4. Other efficiency results of interest follow as corollaries to the results given in this paper. For example, the efficiency of $\hat{\phi}_{10}$ relative to $\hat{\phi}_{1\gamma}$ is also $1 + \text{CV}_2$, for any fixed $\gamma > 0$. Further, the efficiency of $\hat{\phi}_{e0}$ relative to any other estimate of ϕ_ϵ considered in this paper becomes infinite.

Remark 5. Because all MINQUEs have efficiencies approaching unity when considered relative to the optimal nonparametric estimates, and because ML estimates are in some cases asymptotically equivalent to iterated MINQUE, it is reasonable to postulate that maximum likelihood (ML) estimates will have similar efficiency properties.

Remark 6. Seely (1979) has proven related asymptotic results for the case of normally distributed models. The present paper extends Seely's results by (i) allowing a less restrictive asymptotic scheme, and (ii) obtaining stronger efficiency results, ***with or without*** the normality assumption.

References

Chung, K.L. (1974). *A Course in Probability Theory.* Academic Press, New York.

Hocking, R.R. (1985). *The Analysis of Linear Models.* Brooks-Cole, Monterey, CA.

Khuri, A.I. (1990). Exact tests for random models with unequal cell frequencies in the last stage. *Journal of Statistical Planning and Inference* **24**, 177–193.

Rao, C.R. (1971). Estimation of variance and covariance components–MINQUE theory. *Journal of Multivariate Analysis* **1**, 257–275.

Rao, C.R. and Kleffe, J. (1988). *Estimation of Variance Components and Applications.* North-Holland, Amsterdam.

Seely, J. (1979). Large-sample properties of invariant quadratic unbiased estimators in the random one-way model. In: L.D. von Vleck and S.R. Searle, Eds., *Proceedings of a Conference in Honor of C.R. Henderson.* Cornell University, Ithaca, N.Y., 189–201.

Seifert, B. (1979). Optimal testing for fixed effects in general balanced mixed classification models. *Mathematische Operationsforschung und Statistik, Series Statistics* **10**, 237–255.

Westfall, P.H. (1987a). Computable MINQUE-type estimates of variance components. *Journal of the American Statistical Association* **82**, 586–589.

Westfall, P.H. (1987b). A comparison of variance component estimates for arbitrary underlying distributions. *Journal of the American Statistical Association* **82**, 866–874.

Westfall, P.H. (1988a). Robustness and power of tests for a null variance ratio. *Biometrika* **75**, 207–214.

Westfall, P.H. (1988b). Power comparisons for invariant variance ratio tests in mixed ANOVA models. *The Annals of Statistics* **17**, 318–326.

Westfall, P.H. and Bremer, R.H. (1993). Efficiency properties of cell means variance component estimates. *Journal of Statistical Planning and Inference.* To appear.

ON ASYMPTOTIC NORMALITY OF ADMISSIBLE INVARIANT QUADRATIC ESTIMATORS OF VARIANCE COMPONENTS

STEFAN ZONTEK
Institute of Mathematics of the Polish Academy of Sciences
Kopernika 18
51-617 Wrocław
Poland

Abstract. Conditions for a sequence of variance components models are specified so that the appropriate normalized vector of invariant quadratic admissible estimators of variance components, recently proposed by Klonecki and Zontek (1992), has a limiting multivariate normal distribution.

Key words: Variance component model, Admissible invariant quadratic estimator, Asymptotic normality.

1. Introduction and Notation

There is a large literature concerning the problem of estimation of variance components. Using different criteria of estimation (like unbiasedness, admissibility, nonnegativity, robustness and so on) there were proposed several methods of estimation of variance components. An important additional property of estimators (except for those resulting from the above used criteria) is their asymptotic normality. Under some assumptions imposed on a sequence of mixed linear models, asymptotic normality of estimators of variance components was established for the maximum likelihood estimators by Hartley and Rao (1967), Miller (1974), for the MINQUE and I-MINQUE estimators by Brown (1976), for the Henderson's method III estimators by Westfall (1983, 1986), and for robust estimators by Bednarski, Zmyślony and Zontek (1992), Bednarski and Zontek (1994).

Recently, Klonecki and Zontek (1992) elaborated a method of constructing a class of admissible invariant quadratic estimators for the vector of variance components. Here, admissibility is among invariant quadratic estimators under quadratic loss. The method allows construction of admissible estimators within a model with, say, p variance components when an admissible estimator is available in a specified submodel with $p-1$ variance components. Moreover, if an estimator in the submodel possess some additional property like unbiasedness (admissibility among unbiased estimators) or nonnegativity, then the method makes possible construction of an estimator with the same property within the original model. In this paper we show that the method preserves also asymptotic normality. More precisely, we determine conditions under which an appropriate normalized sequence of admissible estimators obtained by using this method has a limiting normal distribution when the corresponding normalized sequence of estimators within submodels is asymptotically normal.

T. Caliński and R. Kala (eds.),
Proceedings of the International Conference on Linear Statistical Inference LINSTAT '93, 121–127.

Some Notation. Let I_n and $\mathbf{0}_{n,m}$ stand for the $n \times n$ unit matrix and the $n \times m$ zero matrix. For any square matrix A, its trace is denoted by tr$A$. For any matrix A, its transpose, the Moore-Penrose generalized inverse, the column space, the null space and the norm are written as A', A^+, $\mathcal{R}(A)$, $\mathcal{N}(A)$ and $||A|| = (\mathrm{tr}A'A)^{1/2}$, respectively. The n-component vector of 1's and 0's are written as $\mathbf{1}_n$ and $\mathbf{0}_n$, respectively. Finally, for any vector $C \in \mathbb{R}^n$ notation $C \geq \mathbf{0}_n$ ($C > \mathbf{0}_n$) means that all coordinates of the vector C are nonnegative (positive). Other notation will be introduced as needed.

2. Preliminaries

Let $X \in \mathbb{R}^n$ be a random vector having the following structure

$$X = D\beta + \sum_{i=1}^{p} J_i \varepsilon_i,$$

where β is the unknown regression parameter, $D, J_1, \ldots, J_p$ are known matrices of orders $n \times r_o, \ldots, n \times r_p$, respectively, while $\varepsilon_1, \ldots, \varepsilon_p$ are independent random vectors normally distributed with zero means and covariance matrices $\sigma_1 I_{r_1}, \ldots, \sigma_p I_{r_p}$ ($\sigma_i \geq 0,\ i = 1, \ldots, p$), respectively. Under this assumption X has n-variate normal distribution $N_n(D\beta,\ \sum_{i=1}^{p} \sigma_i J_i J_i')$. The model is schematically written as

$$\{X,\ N_n(D\beta,\ \sum_{i=1}^{p} \sigma_i J_i J_i')\}. \tag{1}$$

The parameter to be estimated is the vector of variance components $\sigma = (\sigma_1, \ldots, \sigma_p)'$, while the regression vector β is treated as a nuisance parameter. We consider invariant quadratic estimators, i.e., estimators of the following form $\hat{\sigma} = (X'L_1X, \ldots, X'L_pX)'$, where $L_i = ML_iM$ for $i = 1, \ldots, p$, while M is the orthogonal projection matrix onto the intersection of the null space of D' and the space spanned by all columns of $J_1, \ldots, J_p$. The considered estimators are functions of a maximal invariant statistic

$$MX \sim N_n(\mathbf{0}_n,\ \sum_{i=1}^{p} \sigma_i M_i),$$

where $M_i = MJ_iJ_i'M,\ i = 1, \ldots, p$. For convenience we will call the model associated with MX an invariant version of the model (1).

Let S be an $n \times m$ matrix ($m \leq n$) such that $\mathcal{R}(S) = \mathcal{R}(\sum_{i=1}^{p-1} M_i)$. Then the random vector MX can be represented as $MX = X^* + MJ_p\varepsilon_p$, where $X^* = \sum_{i=1}^{p-1} J_i^*\varepsilon_i$, while $J_i^* = S^+MJ_i,\ i = 1, \ldots, p-1$. Following Klonecki and Zontek (1992) we call each model, whose invariant version reduces to

$$\{X^*,\ N_m(\mathbf{0}_m,\ \sum_{i=1}^{p-1} \sigma_i J_i^*(J_i^*)')\}, \tag{2}$$

a submodel of (1) specified by S. In the sequel we assume that $\operatorname{rank}(\sum_{i=1}^{p-1} M_i) < \operatorname{rank}(M)$.

Note that each submodel is also the variance component model but with $p-1$ variance components. Moreover, since the matrix S defining the submodel can be chosen in different ways, we should try to choose it such that the corresponding submodel has a simple structure. Replacing matrix S, leading to the submodel (2), by another $n \times \widetilde{m}$ matrix $\widetilde{S}$ fulfilling condition $\mathcal{R}(\widetilde{S}) = \mathcal{R}(\sum_{i=1}^{p-1} M_i)$, leads to the submodel $\widetilde{X}^* \sim N_{\widetilde{m}}(\mathbf{0}_{\widetilde{m}}, \sum_{i=1}^{p-1} \sigma_i T J_i^* (J_i^*)' T')$, where T is any $\widetilde{m} \times m$ matrix such that $\widetilde{S}^+ = TS^+$.

Let $\hat{\sigma}^* = ((X^*)' A_1^* X^*, \ldots, (X^*)' A_{p-1}^* X^*)'$ be an estimator of the vector $\sigma^* = (\sigma_1, \ldots, \sigma_{p-1})'$ of variance components within a submodel. For the original model define a class of estimators of σ of the following form

$$\hat{\sigma}(C) = (X' A_1 X, \ldots, X' A_p X)' + X' A_p X C, \tag{3}$$

where $A_i = (S^+)' A_i^* S^+$, $i = 1, \ldots, p-1$, $A_p = (1/q)(M - SS^+) M_p^+$, $q = \operatorname{tr}((M - SS^+) M_p^+ M_p)$, while $C = (c_1, \ldots, c_p)' \in \mathbb{R}^p$. Finally, define the $(p-1) \times (p-1)$ matrix K by

$$K = \{\operatorname{tr}(A_i^* J_j^* (J_j^*)')\}$$

and the $(p-1)$ vector W by $W = (\operatorname{tr}(A_1 M_p), \ldots, \operatorname{tr}(A_{p-1} M_p))'$.

Some properties of estimators of the form (3) are formulated in the following theorem (see Klonecki and Zontek, 1992).

Theorem 1. *Assume that M_p and SS^+ commute.*

(i) *If $\hat{\sigma}^*$ is an unbiased estimator of σ^* admissible among unbiased estimators within the submodel* (2), *then $\hat{\sigma}(C)$ with $C = -(W', 0)'$ is an unbiased admissible estimator of σ within the original model.*

(ii) *If $\hat{\sigma}^*$ is an admissible estimator of σ^* within the submodel* (2), *then $\hat{\sigma}(C)$ with*

$$C = [q/(2+q)](\delta'(I_{p-1} - K)' - W', -2/q^2)', \qquad \delta \geq \mathbf{0}_{p-1},$$

form a class of admissible estimators of σ within the original model. Moreover, if in addition $\hat{\sigma}^*$ is nonnegative, then for some $\delta \geq \mathbf{0}_{p-1}$ the resulting estimator of σ is also nonnegative.

The theorem above allows to construct an admissible estimator of σ within the original model when an admissible estimator of σ^* within the submodel is available. In that way the problem of construction of an admissible estimator of the vector of variance components within a model with p variance components is reduced to the problem of construction of an admissible estimator of the vector of variance components within a model with $p-1$ variance components. When we cannot construct an admissible estimator of σ^* within the submodel we can try to find an admissible estimator of variance components within a submodel of the submodel, and so on. After $p-1$ such steps the problem is reduced to a model with 1 variance component. Starting from the admissible estimator within this model and using $p-1$ times Theorem 1, we get an admissible estimator of σ within the original model.

In the next section we will also need the following lemma.

Lemma 1. *Suppose $Y \sim N_k(\mathbf{0}_k, \rho I_k)$ and $Z \sim N_l(\mathbf{0}_l, \tau I_l)$ are independent random vectors. For any non-random symmetric $k \times k$ matrix A, $E(Y'AY) = \rho \mathrm{tr}(A)$ and $\mathrm{Var}(Y'AY) = 2\rho^2||A||^2$, and for any non-random $k \times l$ matrix B, $E(Y'BZ) = 0$ and $\mathrm{Var}(Y'BZ) = \rho\tau||B||^2$.*

3. Main Result

In this section we assume that with the number of observations n tending to infinity, the corresponding number of observations n^*, for a submodel, also tends to infinity.

Let $\Gamma^* = \Gamma^*(n^*) = \mathrm{diag}(\gamma_1, \ldots, \gamma_{p-1})$ be a matrix with normalized factors for an estimator of σ^* within the submodel. Define $\Gamma = \mathrm{diag}(\gamma_1, \ldots, \gamma_p)$, where $\gamma_p = \gamma_p(n) = q^{1/2}$.

Theorem 2. *If*

(i) $\Gamma^*(\hat{\sigma}^* - \sigma^*)$ *has a limiting* $(p-1)$*-variate normal distribution with mean* $\mathbf{0}_{p-1}$ *and covariance matrix* $\Delta = \Delta(\sigma^*)$,

(ii) $\gamma_p \longrightarrow \infty$,

(iii) $\gamma_i \mathrm{tr}(J_p' A_i J_p) \longrightarrow 0, \quad i = 1, \ldots, p-1$,

(iv) $\gamma_i c_i \longrightarrow 0, \quad i = 1, \ldots, p$,

(v) $\gamma_i ||J_j' A_i J_p|| \longrightarrow 0, \quad i = 1, \ldots, p-1, \quad j = 1, \ldots, p$,

then $\Gamma(\hat{\sigma}(C) - \sigma)$ *has a limiting* p*-variate normal distribution with mean* $\mathbf{0}_p$ *and covariance matrix* $\begin{pmatrix} \Delta & \mathbf{0}_{p-1} \\ \mathbf{0}_{p-1}' & 2\sigma_p^2 \end{pmatrix}$.

Proof. For arbitrary p vector $F = (f_1, \ldots, f_p)'$, we have

$$F'\Gamma(\hat{\sigma}(C) - \sigma) = \sum_{i=1}^{p-1} f_i \gamma_i (\hat{\sigma}_i^* - \sigma_i) + f_p \gamma_p (\varepsilon_p' J_p' A_p J_p \varepsilon_p - \sigma_p) + r,$$

where $r = \sum_{i=1}^{p-1} f_i \gamma_i (\varepsilon_p' J_p' A_i J_p \varepsilon_p + c_i \varepsilon_p' J_p' A_p J_p \varepsilon_p) + 2 \sum_{i=1}^{p-1} \sum_{j=1}^{p-1} f_i \gamma_i \varepsilon_j' J_j' A_i J_p \varepsilon_p$. Using Lemma 1, one can easily see that

$$E(r) = \sum_{i=1}^{p-1} f_i \gamma_i (\mathrm{tr}(J_p' A_i J_p) + c_i)$$

and

$$\mathrm{Var}(r) = 2 \sum_{i=1}^{p-1} \sum_{j=1}^{p} f_i^2 \gamma_i^2 \sigma_j \sigma_p ||J_j' A_i J_p||^2 + 2 \sum_{i=1}^{p} f_i^2 \gamma_i^2 c_i^2 \sigma_p^2,$$

which implies that r tends to 0 in probability by assumptions (iii) - (v). Thus the random variable r is asymptotically inessential part of $\hat{\sigma}(C)$.

Since $X' A_p X$ has a limiting $N(0,\ 2\sigma_p^2)$ distribution, the assertion of the theorem follows from the assumption (i) by noting that $\hat{\sigma}_i^*$, $i = 1, \ldots, p-1$, and $\varepsilon_p' A_p \varepsilon_p$ are independent random variables. □

4. Example

The 2-way crossed classification random normal model with interaction and with nonempty cells can be defined in the following way. Let $N = \{n_{ij}\}$ be an $a\times b$ matrix having positive integer elements and let $x_{ijl}, i = 1,\ldots,a,\ j = 1,\ldots,b,\ l = 1,\ldots,n_{ij}$, be random variables having the following structure

$$x_{ijl} = \mu + \varepsilon_{1i} + \varepsilon_{2j} + \varepsilon_{3ij} + \varepsilon_{4ijl},$$

where $\mu \in \mathbf{R}$ is the unknown parameter, while $\varepsilon_{1i}, \varepsilon_{2j}, \varepsilon_{3ij}$ and ε_{4ijl} are unobservable mutually independent random normal variables with means 0 and variances $\sigma_1, \sigma_2, \sigma_3$ and σ_4, respectively. In the notation of (1) this model is specified by

$$X = (x_{111},\ldots,x_{11n_{11}}, x_{121},\ldots,x_{12n_{12}}, \quad \cdots \quad, x_{ab1},\ldots,x_{abn_{ab}})'$$

and by the following matrices:

$$\begin{aligned} D &= \mathbf{1}_n, \quad J_1 = J_3(I_a \otimes \mathbf{1}_b), \quad J_2 = J_3(\mathbf{1}_a \otimes I_b), \\ J_3 &= \operatorname{diag}(\mathbf{1}_{n_{11}}, \mathbf{1}_{n_{12}},\ldots,\mathbf{1}_{n_{ab}}), \quad J_4 = I_n, \end{aligned} \tag{4}$$

where $\otimes$ denotes the Kronecker product, while $n = \sum_{i=1}^{a}\sum_{j=1}^{b} n_{ij}$.

The submodel of the model (4) specified by $S = MJ_3$, where $M = I_n - (1/n)\mathbf{1}_n\mathbf{1}_n'$, is the invariant version of the balanced 2-way crossed classification random normal model without interaction and with one observation per cell. It is given by

$$J_1^* = M^*(I_a \otimes \mathbf{1}_b), \quad J_2^* = M^*(\mathbf{1}_a \otimes I_b) \quad \text{and} \quad J_3^* = M^*,$$

where $M^* = I_{ab} - (1/ab)\mathbf{1}_{ab}\mathbf{1}_{ab}'$. An admissible nonnegative estimator of $\sigma^* = (\sigma_1, \sigma_2, \sigma_3)'$ within the submodel is given by

$$\hat{\sigma}^* = \begin{pmatrix} \frac{1}{a+1}(X^*)'B_1X^* \\ \frac{1}{b+1}(X^*)'B_2X^* \\ \frac{1}{(a-1)(b-1)+2}(X^*)'B_3X^* \end{pmatrix}, \tag{5}$$

where $B_1 = (1/b^2)(J_1^*)'J_1^*$, $B_2 = (1/a^2)(J_2^*)'J_2^*$ and $B_3 = M^* - bB_1 - aB_2$. In fact, this estimator is the limit of a sequence of unique Bayes estimators with respect to a sequence of a prior distributions on $\{\sigma^*(\sigma^*)' : \sigma^* \geq 0_{p-1}\}$ such that

$$E_{\tau_m}[\sigma^*(\sigma^*)'] = \begin{pmatrix} m^2 & 0 & (a-1)/b \\ 0 & m & (b-1)/a \\ (a-1)/b & (b-1)/a & 2 \end{pmatrix}$$

for sufficiently large m. Admissibility of (5) follows now from Theorem 4.2 in Zontek (1989). Admissibility of (5) can be also obtained from Theorem 1.

To derive an admissible estimator of σ within the original model given in Theorem 1 one needs to calculate K and W, which are

$$K = \begin{pmatrix} \frac{a-1}{a+1} & 0 & \frac{a-1}{(a+1)b} \\ 0 & \frac{b-1}{b+1} & \frac{b-1}{a(b+1)} \\ 0 & 0 & \frac{(a-1)(b-1)}{(a-1)(b-1)+2} \end{pmatrix}$$

and

$$W = \frac{1}{ab}\left(\sum_{i-1}^{a}\sum_{j=1}^{b}\frac{1}{n_{ij}}\right)\left(\frac{a-1}{a+1}, \frac{b-1}{b+1}, \frac{(a-1)(b-1)}{(a-1)(b-1)+2}\right)'.$$

For $\delta = (I_3 - K)^{-1}W$ the resulting estimator is nonnegative and is given by

$$\hat{\sigma} = \begin{pmatrix} \frac{1}{a+1}\bar{X}'B_1\bar{X} \\ \frac{1}{b+1}\bar{X}'B_2\bar{X} \\ \frac{1}{(a-1)(b-1)+2}\bar{X}'B_3\bar{X} \\ \frac{1}{n-ab+1}\bar{X}'B_4\bar{X} \end{pmatrix},$$

where $\bar{X} = J_3^+X$, $B_4 = I_n - J_3(J_3^+)'$, while $J_3^+ = \mathrm{diag}^{-1}(n_{11}, n_{12}, \ldots, n_{ab})J_3'$. Note that $\bar{X}$ is cell means statistics.

Since $(X^*)'B_iX^*$, $i = 1, 2, 3$, have independent $(\sigma_1 - \sigma_3/b)\chi^2_{a-1}$, $(\sigma_2 - \sigma_3/a)\chi^2_{b-1}$ and $\sigma_3\chi^2_{(a-1)(b-1)}$ distribution, respectively, one can see that $\Gamma^*(\hat{\sigma}^* - \sigma^*)$, where $\Gamma^* = \mathrm{diag}^{1/2}(a, b, ab)$, has a limiting $N_3(\mathbf{0}_3,\ 2\mathrm{diag}(\sigma_1^2, \sigma_2^2, \sigma_3^2))$ distribution when a and b tend to infinity in such a way that

$$a/b = O(1) \text{ and } b/a = O(1). \tag{6}$$

Now, in addition, assume that

$$\underline{n} = \min\{n_{ij} : i = 1, \ldots, a,\ \ j = 1, \ldots, b\} \longrightarrow \infty \tag{7}$$

and that

$$(ab)^{-1/2}\sum_{i=1}^{a}\sum_{j=1}^{b} n_{ij}^{-1} \longrightarrow 0. \tag{8}$$

Clearly, (7) implies the condition (ii) of Theorem 2. Since

$$\Gamma^*(\mathrm{tr}A_1, \mathrm{tr}A_2, \mathrm{tr}A_3)' = \Gamma^*W \longrightarrow \mathbf{0}_3$$

by (8), the condition (iv) is satisfied. The most restrictive condition among (v) is (for $i = 3, j = 4$)

$$\begin{aligned}(ab)^{1/2}||A_3|| &\leq (ab)^{1/2}\{\underline{n}[(a-1)(b-1)+2]\}^{-1}||B_3|| \\ &= \{ab/[(a-1)(b-1)+2]\}^{1/2}\underline{n}^{-1} \longrightarrow 0\end{aligned}$$

by (7). Thus $\mathrm{diag}^{1/2}(a, b, ab, n-ab)(\hat{\sigma}-\sigma)$ has a limiting $N_4(\mathbf{0}_4, 2\mathrm{diag}(\sigma_1^2, \sigma_2^2, \sigma_3^2, \sigma_4^2))$ distribution when (6)-(8) hold.

References

Bednarski, T., Zmyślony, R. and Zontek, S. (1992). On robust estimation of variance components via von Mises functionals. Preprint 504, Institute of Mathematics, Polish Academy of Sciences, Warszawa.

Bednarski, T. and Zontek, S. (1994). A note on robust estimation of parameters in mixed unbalanced models. In: T. Caliński and R. Kala, Eds., *Proceedings of the International Conference on Linear Statistical Inference LINSTAT'93*. Kluwer Academic Publishers, Dordrecht, 87–95.

Brown, K.G. (1976). Asymptotic behaviour of MINQUE-type estimators of variance components. *The Annals of Statistics* **4**, 746–754.

Hartley, H.O. and Rao, J.N.K. (1967). Maximum likelihood estimation for the mixed analysis of variance model. *Biometrika* **54**, 93–108.

Klonecki, W. and Zontek, S. (1992). Admissible estimators of variance components obtained via submodels. *The Annals of Statistics* **20**, 1454–1467.

Miller, J.J. (1974). Asymptotic properties and computation of maximum-likelihood estimates in mixed model of the analysis of variance. Ph. D. thesis, Stanford University, Stanford, California.

Westfall, P.H. (1983). On the asymptotic normality of the Henderson Method III estimates of components of variance in the mixed linear model. Ph. D. thesis, University of California, Davis.

Westfall, P.H. (1986). Asymptotic normality of the ANOVA estimates of components of variance in the nonnormal, unbalanced hierarchical mixed model. *The Annals of Statistics* **14**, 1572–1582.

Zontek, S. (1989). The minimal complete class for the vector of variance components. *Probability and Mathematical Statistics* **10**, 169–177.

Zontek, S. (1989). Admissible estimation of variance components in crossed classification models with empty cels. Preprint 463, Institute of Mathematics, Polish Academy of Sciences, Warszawa.

ADMISSIBLE NONNEGATIVE INVARIANT QUADRATIC ESTIMATION IN LINEAR MODELS WITH TWO VARIANCE COMPONENTS

STANISŁAW GNOT
Agricultural University of Wrocław
Grunwaldzka 53
50-357 Wrocław
Poland

GÖTZ TRENKLER
Department of Statistics
University of Dortmund
D-44221 Dortmund
Germany

and

DIETMAR STEMANN
Department of Statistics
University of Hagen
Postfach 940 / Roggenkamp 6
D-58009 Hagen
Germany

Abstract. Linear models with two variance components are considered and a new class of nonnegative estimators for singular variance component is presented. It is proved that this class consists of admissible nonnegative invariant quadratic estimators with respect to the mean squared loss function. Numerical results of the mean squared errors of the proposed estimators for one-way ANOVA model are reported.

Key words: Nonnegative invariant quadratic estimator, Variance components, One-way ANOVA model.

1. Introduction

The natural constraint that appears in the theory of estimation of variance components in the general linear model is nonnegativity of estimators for variances, which obviously are nonnegative. Unfortunately, the variance component unbiased estimation standard procedures may produce negative estimators with high probability.

The purpose of this paper is to present a new class of nonnegative estimators for the variance component σ_1^2 in the model $M\{y, X\beta, V(\sigma) = \sigma_1^2 V + \sigma_e^2 I_N\}$. In this model y is an $N \times 1$ normally distributed random vector with

$$E(y) = X\beta, \quad \text{Cov}(y) = V(\sigma),$$

where X is a known $N \times p$ matrix, β is an unknown $p \times 1$ vector of fixed parameters, V is a known nonnegative definite matrix and I_N denotes the $N \times N$ identity matrix. The proposed class is proved to be a class of admissible nonnegative invariant

T. Caliński and R. Kala (eds.),
Proceedings of the International Conference on Linear Statistical Inference LINSTAT '93, 129–137.

quadratic estimators with respect to the mean squared loss function. The results obtained extend that given by Gnot, Kleffe and Zmyślony (1985), where a characterization of the subclass of nonnegative estimators, within all admissible invariant quadratic estimators, was established.

Using the approach of Olsen, Seely and Birkes (1976), we consider estimators that are invariant under the group of translations $g(y) = y + X\alpha$, where α is an arbitrary $p \times 1$ vector. Let $\mathrm{rank}(X) = p - q$, $0 \leq q \leq p$, $r = N - p + q$, and let B be $r \times N$ matrix such that $BB' = I_r$, while $B'B = M = I - XX^+$ is the orthogonal projector on the kernel of X'. Here X^+ denotes the Moore-Penrose inverse of X. Then $t = By$ is a maximal invariant statistic with

$$E(t) = 0, \quad \mathrm{Cov}(t) = W(\sigma) = \sigma_1^2 W + \sigma_e^2 I_r,$$

where $W = BVB'$.

Let $\omega_1 > \omega_2 > ... > \omega_h$ be the distinct positive eigenvalues of W with multiplicities $\nu_1, \nu_2, ..., \nu_h$, respectively, let $W = \sum_{i=1}^h \omega_i E_i$ be the spectral decomposition of W, and, in addition, let $E_{h+1} = I - \sum_{i=1}^h E_i$. We assume that W is singular, i.e., $\nu_{h+1} = \mathrm{tr}(E_{h+1}) > 0$. As it has been proved by Olsen et al. (1976), that the quadratic forms $z_i = t'E_i t/\nu_i$, $i = 1, 2, ..., h+1$, constitute a set of minimal sufficient statistics for the family of distributions of t. Moreover, the z_i's are independent and

$$E(z_i) = \omega_i \sigma_i^2 + \sigma_e^2, \quad \mathrm{Var}(z_i) = 2(\omega_i \sigma_1^2 + \sigma_e^2)^2/\nu_i.$$

The linear space generated by $E_1, ..., E_{h+1}$ will be denoted by $\mathcal{Q}$, i.e. $\mathcal{Q} = \mathrm{sp}\{E_1, ..., E_{h+1}\}$.

2. Bayes Nonnegative Estimation

For a given invariant estimator $\bar{\gamma} = y'Ay$ of $\gamma = f_1\sigma_1^2 + f_2\sigma_e^2$ the mean squared error has the form

$$E_\sigma(\bar{\gamma} - \gamma)^2 = 2\mathrm{tr}[AW(\sigma)AW(\sigma)] + \sigma'(f - \bar{f})(f - \bar{f})'\sigma, \tag{1}$$

where $\sigma = (\sigma_1^2, \sigma_e^2)'$, $f = (f_1, f_2)'$, $\bar{f} = (\bar{f}_1, \bar{f}_2)'$, with $\bar{f}_1 = \mathrm{tr}(AW)$, $\bar{f}_2 = \mathrm{tr}(A)$, and depends only on the unknown parameter vector σ. Let τ be a prior distribution defined on the parameter space

$$\Theta = \{\sigma = (\sigma_1^2, \sigma_e^2)';\ \sigma_1^2 \geq 0,\ \sigma_e^2 > 0\}.$$

Since the Bayes risk $r_\tau(A) = E_\tau E_\sigma(\bar{\gamma} - \gamma)^2$ is linear in $E_\tau\sigma\sigma'$, we can take τ such that

$$E_\tau\sigma\sigma' = T(u,v) = \begin{bmatrix} u^2 + v & u \\ u & 1 \end{bmatrix}, \quad \text{with } u, v \geq 0, \quad \text{or} \quad E_\tau\sigma\sigma' = T_0 = \begin{bmatrix} 1 & 0 \\ 0 & 0 \end{bmatrix},$$

which case is not covered by $T(u,v)$ (cf. Gnot and Kleffe, 1983). Let us denote

$$W(u) = uW + I_r,$$

$$K(u,v) = (u^2+v)W + uI_r = uW(u) + vW,$$
$$\Lambda(u,v) = I_r + 2uW + (u^2+v)W^2 = W^2(u) + vW^2.$$

Following Olsen et al. (1976, Theorem 2.2) (see also Kleffe and Pincus 1974, Theorem 9c) in looking for the Bayes estimators being of the form $t'At$, we can restrict our considerations to $A \in \mathcal{Q}$. Then $t'At = \sum_{i=1}^{h+1} \nu_i \alpha_i z_i$ for some scalars $\alpha_1, ..., \alpha_{h+1}$. Using that fact, it follows from (1) that the Bayes risk $r_\tau(A)$ has the form

$$r_\tau(A) = 2\mathrm{tr}[A\Lambda(u,v)A] + (f-\bar{f})'T(u,v)(f-\bar{f}). \tag{2}$$

Definition 1. A quadratic form $\bar{\gamma}(u,v) = t'A(u,v)t$, $A(u,v) \in \mathcal{Q}$, is called a Bayes nonnegative invariant quadratic (BNIQ) estimator for nonnegative function $\gamma = f_1\sigma_1^2 + f_2\sigma_e^2$ if $A(u,v)$ solves the following problem

$$\min_{A \in \mathcal{Q}^+} r_\tau(A), \tag{3}$$

where $\mathcal{Q}^+$ is the convex cone of all nonnegative definite matrices in $\mathcal{Q}$.

Using an approach similar to Hartung (1981) and Gnot, Trenkler and Zmyślony (1993) we define the Lagrange function

$$L(A,D) = r_\tau(A) - \mathrm{tr}(AD). \tag{4}$$

It can be easily seen that the gradient of $L(A,D)$ with respect to A has the form

$$\frac{\partial L(A,D)}{\partial A} = 8\Lambda(u,v)A - 4\mathrm{diag}[\Lambda(u,v)A] - 2[f_1 - \mathrm{tr}(AW)]\{2K(u,v) - \mathrm{diag}[K(u,v)]\}$$
$$-2[f_2 - \mathrm{tr}(A)]\{2W(u) - \mathrm{diag}[W(u)]\} - 2D + \mathrm{diag}(D),$$

and it vanishes if and only if

$$D = 4\Lambda(u,v)A - 2[f_1 - \mathrm{tr}(AW)]K(u,v) - 2[f_2 - \mathrm{tr}(A)]W(u).$$

The considerations from above together with the arguments used by Hartung (1982, Theorem 3.1 and Lemma 3.1) lead to the following results.

Theorem 1.
(i) *There exists a nonnegative definite Lagrange multiplier $D(u,v)$ such that for a BNIQ estimator $t'A(u,v)t$ we have $\mathrm{tr}[A(u,v)D(u,v)] = 0$ and*

$$r_\tau[A(u,v)] = L[A(u,v), D(u,v)] = \min_{A \in Q^+} L[A, D(u,v)].$$

(ii) *$t'A(u,v)t$ with a nonnegative definite $A(u,v)$ is a BNIQ estimator for nonnegative function $\gamma = f_1\sigma_1^2 + f_2\sigma_e^2$ if and only if*

$$D(u,v) = 4\Lambda(u,v)A(u,v) - 2\{f_1 - \mathrm{tr}[A(u,v)W]\}K(u,v) - 2\{f_2 - \mathrm{tr}[A(u,v)]\}W(u) \tag{5}$$

is nonnegative definite and $\mathrm{tr}[A(u,v)D(u,v)] = 0$.

Nonnegative estimation of the single variance component σ_1^2 appears to have a special meaning. Now using Theorem 1, we characterize a class of admissible nonnegative IQ estimators of σ_1^2. We start with an explicit solution for a BNIQ estimator of the single variance component σ_1^2, i.e. under assumption that $f = (1,0)'$. For this purpose we preceed as follows. For $i = 1, 2, ..., h+1$ denote by

$$w_i(u) = u\omega_i + 1, \quad k_i(u,v) = (u^2+v)\omega_i + u, \quad \lambda_i(u,v) = 1 + 2u\omega_i + (u^2+v)\omega_i^2$$

the eigenvalues of $W(u), K(u,v)$ and $\Lambda(u,v)$, respectively. For a given integer $k \in \{1, 2, ..., h+1\}$ set

$$W_k(u) = \sum_{i\leq k} w_i(u)E_i, \quad K_k(u,v) = \sum_{i\leq k} k_i(u,v)E_i, \quad \Lambda_k(u,v) = \sum_{i\leq k} \lambda_i(u,v)E_i,$$

$$g_{11}(k) = \mathrm{tr}(\Lambda_k^+ K_k W),\ g_{12}(k) = \mathrm{tr}(\Lambda_k^+ W_k W),$$
$$g_{21}(k) = \mathrm{tr}(\Lambda_k^+ K_k), \quad g_{22}(k) = \mathrm{tr}(\Lambda_k^+ W_k),$$

$$G(k) = \{g_{ij}(k)\}, \quad f = (1,0)',$$

$$\phi(k) = [\phi_1(k), \phi_2(k)]' = [2I + G(k)]^{-1}G(k)f, \tag{6}$$

$$\alpha_i(k) = \frac{1}{2}a_i'[f - \phi(k)] = a_i'[2I + G(k)]^{-1}f, \tag{7}$$

where $a_i = a_i(u,v) = [k_i(u,v)/\lambda_i(u,v),\ w_i(u)/\lambda_i(u,v)]'$.

Let k_* be the maximal integer among $1, ..., h+1$, such that $\alpha_{k_*}(k_*) > 0$. A proof of existence and properties of k_* are given in the Appendix.

Theorem 2. *A quadratic form* $\bar{\sigma}_1^2 = t'A(u,v)t$ *is a BNIQ estimator for* σ_1^2 *with respect to the prior distribution* τ, *such that* $E_\tau\sigma\sigma' = T(u,v)$, *if and only if* $A(u,v) = \sum_{i\leq k_*} \alpha_i(k_*)E_i$, *with*

$$\alpha_i(k_*) = a_i'(u,v)[2I + G(k_*)]^{-1}f, \tag{8}$$

and $a_i(u,v) = [k_i(u,v)/\lambda_i(u,v), w_i(u)/\lambda_i(u,v)]'$.

Proof. Let $A(u,v) = \sum_{i=1}^{h+1} \alpha_i E_i$. Then from (5) also D is of the form $D(u,v) = \sum_{i=1}^{h+1} \delta_i E_i$, and from $\mathrm{tr}(AD) = 0$ it follows that $\delta_i = 0$ for i such that α_i is positive. Now from (5) we have

$$4\lambda_i(u,v)\alpha_i - \delta_i = 2[f_1 - \mathrm{tr}(AW)]k_i(u,v) + 2[f_2 - \mathrm{tr}(A)]w_i(u), \tag{9}$$

for $i = 1, ..., h+1$. Putting $\alpha_i = \alpha_i(k_*)$ given by (8) for $i \leq k_*$, $\alpha_i = 0$ for $i > k_*$, $\delta_i = 0$ for $i \leq k_*$ and $\delta_i = -4\lambda_i(u,v)\alpha_i(k_*)$ for $i > k_*$ we easily find that $(\mathrm{tr}AW, \mathrm{tr}A)' = G(k_*)[f - \phi(k_*)] = \phi(k_*)$ and also (9) is valid. From Lemma A.3

in the Appendix it follows that $A(u, v)$ and $D(u, v)$ are nonnegative definite and uniquely given for each pair (u, v). □

Another formula for a BNIQ estimator of σ_1^2 can be obtained if we note that $t'A(u,v)t = l_1 Q_1(k_*) + l_2 Q_2(k_*)$, where

$$Q_1(k_*) = \sum_{i \le k_*} \frac{\nu_i \omega_i z_i}{\lambda_i(u,v)}, \quad Q_2(k_*) = \sum_{i \le k} \frac{\nu_i z_i}{\lambda_i(u,v)}, \quad (l_1, l_2)' = T(u,v)[2I + G(k_*)]^{-1} f.$$

In particular, it can happen that $k_* = h + 1$. Then

$$Q_1(h+1) = y'[M\Lambda(u,v)M]^+ VMy, \quad Q_2(h+1) = y'[M\Lambda(u,v)M]^+ y,$$

and the BNIQ estimator $\bar{\sigma}_1^2$ reduces to the BIQ estimator

$$\hat{\sigma}_1^2 = l_1 y'[M\Lambda(u,v)M]^+ VMy + l_2 y'[M\Lambda(u,v)M]^+ y,$$

with $(l_1, l_2)' = T(u,v)[2I + G(h+1)]^{-1} f$, obtained by Gnot and Kleffe (1983), which in this case is nonnegative. A necessary and sufficient condition for BIQ estimator of $\hat{\sigma}_1^2$ to be nonnegative at (u, v) is

$$2u - vs(u,v) \ge 0, \tag{10}$$

where

$$s(u,v) = \sum_{i=1}^{h} \frac{\omega_i \nu_i}{\lambda_i(u,v)},$$

while $\lambda_i(u,v) = 1 + 2u\omega_i + (u+v)\omega_i^2$ (cf. Gnot at al., 1985, p. 254). For example inequality (10) is fulfiled for each $u \ge 0$ and $v = 0$. However, there is a much richer class of BIQ estimators $\hat{\sigma}_1^2$ which are nonnegative. For a given (u, v) the BNIQ estimator given by Theorem 2 is admissible as a unique Bayes estimator. To obtain further admissible nonnegative estimators separate considerations should be done for prior distributions τ for which $E_\tau \sigma\sigma' = T_0$. In such a case the Baysian risk given by (2) reduces to $r_\tau(A) = 2\text{tr}[AW^2A] + (f_1 - \bar{f}_1)^2$, while

$$\frac{\partial L(A,D)}{\partial A} = 8W^2A - 4\text{diag}(W^2A) - 2[f_1 - \text{tr}(AW)][2W - \text{diag}(W)] - 2D + \text{diag}(D).$$

It vanishes if and only if

$$D_0 = 4W^2A_0 - 2[f_1 - \text{tr}(A_0W)]W. \tag{11}$$

Condition (11) is equivalent to

$$4\omega_i^2\alpha_i - \delta_i = 2[f_1 - \text{tr}(A_0W)]\omega_i. \tag{12}$$

Putting $f_1 = 1$, it is seen that the nonnegative solution of (12) with respect to α_i and δ_i can be presented as

$$\alpha_i = c/\omega_i, \ for \ i = 1, \ldots, h; \quad \alpha_{h+1} \ge 0, \tag{13}$$

where $c = [1 - \mathrm{tr}(A_0 W)]/2$, while α_{h+1} is an arbitrary constant. The values of c can be given explicitly if we note that

$$\sum_{i=1}^{h} \alpha_i \omega_i \nu_i = \frac{1}{2}[1 - \mathrm{tr}(A_0 W)] \sum_{i=1}^{h} \nu_i = \mathrm{tr}(A_0 W).$$

Since $\sum_{i=1}^{h} \nu_i = \mathrm{rank}(W_0)$, it follows that $\mathrm{tr}(A_0 W) = \mathrm{rank}(W)/[2 + \mathrm{rank}(W)]$ and $c^{-1} = 2 + \mathrm{rank}(W)$. Thus we have proved the following theorem.

Theorem 3. *The BNIQ estimator for σ_1^2 with respect to the prior distribution τ, for which $E_\tau \sigma\sigma' = T_0$, has the form $t'A_0 t$, where*

$$A_0 = c \sum_{i=1}^{h} \omega_i^{-1} E_i + \alpha_{h+1} E_{h+1} = cW^+ + \alpha_{h+1} E_{h+1}, \tag{14}$$

$c^{-1} = 2 + \mathrm{rank}(W)$, *while α_{h+1} is an arbitrary nonnegative constant.*

Remark 1. It has been proved by Gnot and Kleffe (1983) that $t'A_0 t$ with A_0 given by (14) and with an arbitrary α_{h+1} forms the entire class of BIQ estimators with respect to τ_0, and among them only that with

$$\alpha_{h+1} \geq -c\mathrm{tr}(W^+)/(2 + \nu_{h+1}) \tag{15}$$

are admissible in the class of all invariant quadratic estimators. These estimators can be obtained in an alternative way as the limiting of BIQ estimators if v tends to infinity. It justifies to use the following notation

$$\hat{\sigma}_1^2(\alpha_{h+1}, \infty) = ct'W^+ t + \alpha_{h+1} t' E_{h+1} t.$$

Since $\mathrm{tr}(W^+)/(2 + \nu_{h+1})$ is positive, it follows that $\hat{\sigma}_1^2(\alpha_{h+1}, \infty)$ with an arbitrary nonnegative α_{h+1} is admissible in the class of nonnegative invariant quadratic estimators.

Remark 2. An interesting property of $\hat{\sigma}_1^2(0, \infty)$ has been given by Mathew, Sinha and Sutradhar (1992) (it is denoted there by $\hat{\sigma}_{11}^2$). They have specified a sufficient condition under which $\hat{\sigma}_1^2(0, \infty)$ has a uniformly smaller mean squared error then every unbiased IQ estimator of σ_1^2.

3. Example

In this section we present the MSE's and squared biases of several estimators of σ_1^2 for the one-way ANOVA model considered by Mathew et al. (1992). In our example, $m = (3, 3, 5, 5, 7, 7)$ denotes the vector of treatment replications,

$$\begin{array}{llllll} \omega_1 = 7, & \nu_1 = 1, & \omega_2 = 5.918, & \nu_2 = 1, & \omega_3 = 5, & \nu_3 = 1, \\ \omega_4 = 3.548, & \nu_4 = 1, & \omega_5 = 3, & \nu_5 = 1, & \omega_6 = 0, & \nu_6 = 24. \end{array}$$

Moreover,

$$\hat{\sigma}_1^2(u,v) = \sum_{i=1}^{h+1} \nu_i \alpha_i(h+1) z_i \quad \text{and} \quad \bar{\sigma}_1^2(u,v) = \sum_{i \le k_*} \nu_i \alpha_i(k_*) z_i,$$

with $\alpha_i(k_*)$ given by (8), denote, the BIQ and BNIQ estimators of σ_1^2, respectively, at a given point (u,v), while

$$\hat{\sigma}_1^2(\alpha_{h+1},\infty) = \frac{t'W^+t}{2+\text{rank}(W)} + \nu_{h+1}\alpha_{h+1}z_{h+1}$$

is a limiting BIQ estimator, which is nonnegative by construction (see Remark 1). In comparing of $\hat{\sigma}_1^2(u,v)$ and $\bar{\sigma}_1^2(u,v)$ the parameters u, v are chosen such that (10) does not hold ($k_* < h+1$). It ensures that $\hat{\sigma}_1^2(u,v)$ and $\bar{\sigma}_1^2(u,v)$ are different.

TABLE I

MSE and the squared bias (within brackets) of $\hat{\sigma}_1^2(u,v)$, $\bar{\sigma}_1^2(u,v)$ and $\hat{\sigma}_1^2(\alpha_{h+1},\infty)$ for $\sigma_e^2 = 1$ and for different values of σ_1^2.

σ_1^2	.10	.30	.50	1.0	4.0	9.0	16.0
$\hat{\sigma}_1^2(0,\infty)$	.0402 (.0175)	.0630 (.0057)	.1088 (.0003)	.3230 (.0155)	4.609 (.9640)	23.18 (5.810)	73.18 (19.45)
$\hat{\sigma}_1^2(0.5,\infty)$	.4435 (.4000)	.3310 (.4092)	.3977 (.2685)	.4692 (.1408)	3.898 (.2322)	21.04 (3.650)	69.04 (15.29)
$\hat{\sigma}_1^2(1,\infty)$	1.338 (1.282)	1.297 (1.156)	1.228 (1.037)	1.157 (.7661)	3.728 (.0003)	19.44 (1.989)	65.44 (11.63)
$\hat{\sigma}_1^2(5,\infty)$	28.45 (26.34)	27.90 (25.76)	27.37 (25.18)	26.16 (23.77)	21.87 (16.15)	26.16 (6.71)	56.16 (.3476)
$\bar{\sigma}_1^2(0.2,1)$	.0303 (.0122)	.0487 (.0020)	.0900 (.0007)	.2970 (.0393)	4.590 (1.506)	23.39 (8.656)	74.17 (28.55)
$\hat{\sigma}_1^2(0.2,1)$	.0215 (.0015)	.0514 (.0007)	.1043 (.0082)	.3375 (.0634)	4.763 (1.491)	23.67 (8.045)	74.34 (25.99)
$\bar{\sigma}_1^2(0.4,5)$	.0369 (.0157)	.0580 (.0042)	.1021 (.0000)	.3125 (.0220)	4.581 (1.128)	23.14 (6.678)	73.18 (22.23)
$\hat{\sigma}_1^2(0.4,5)$	.0301 (.0087)	.0555 (.0011)	.1038 (.0008)	.3247 (.0318)	4.654 (1.174)	23.32 (6.718)	73.48 (22.13)

An inspection of Table I leads to the following conclusions.

(i) In both cases the MSE of the limiting BIQ estimator $\hat{\sigma}_1^2(\alpha_{h+1},\infty)$ very strongly depends on α_{h+1} and it is flatter as a function of σ_1^2 when α_{h+1} is large.

(ii) For small σ_1^2 the BNIQ estimators $\bar{\sigma}_1^2(u,v)$ with v close to zero are strong competitors to $\hat{\sigma}_1^2(\alpha_{h+1},\infty)$ for large α_{h+1} and quite comparable with $\hat{\sigma}_1^2(0,\infty)$.

(iii) For $\alpha_{h+1} = 0$ the estimator $\hat{\sigma}_1^2(0,\infty)$ can be recommended in practice, because of its simple form and small bias.

(iv) Generally the MSE of BNIQ estimators $\bar{\sigma}_1^2(u,v)$ is very close to the MSE of the BIQ estimators $\hat{\sigma}_1^2(u,v)$, which can, however, take on negative values.

Appendix

Three lemmas will be proved below.

Lemma A.1. $a_1(1) > 0$.

Proof. Putting $f = (1,0)'$ and noting that

$$(2I+G)^{-1} = \frac{1}{\det(2I+G)} \begin{bmatrix} 2+g_{22} & -g_{12} \\ -g_{21} & 2+g_{11} \end{bmatrix},$$

we find that an explicite form of $\alpha_i(k)$ given by (7) is $\alpha_i(k) = c_i[2k_i(u,v) + k_i(u,v)g_{22}(k) - w_i(u)g_{21}(k)]$, where $c_i^{-1} = \lambda_i(u,v)\det(2I+G)$ is positive, since $\det(2I+G) > 0$. It follows that $\alpha_i(k) > 0$ if and only if

$$\bar{\alpha}_i(k) = c_i^{-1}\alpha_i(k) = 2k_i(u,v) + k_i(u,v)g_{22}(k) - w_i(u)g_{21}(k) \tag{A.1}$$

is positive. Let

$$g_1(k) = \sum_{i\leq k} \frac{\nu_i\omega_i}{\lambda_i(u,v)}, \quad g_0(k) = \sum_{i\leq k} \frac{\nu_i}{\lambda_i(u,v)}.$$

Using $w_i(u) = 1+u\omega_i$, $k_i(u,v) = uw_i(u)+v\omega_i$, $g_{22}(k) = ug_1(k)+g_0(k)$, $g_{21}(k) = ug_{22}(k) + vg_1(k)$, we find from (A.1) that

$$\bar{\alpha}_i(k) = 2k_i(u,v) + v[\omega_i g_0(k) - g_1(k)], \tag{A.2}$$

while $\bar{\alpha}_1(k) = 2k_1(u,v) > 0$. □

Lemma A.2. *$\bar{\alpha}_{k+1}(k+1) = \bar{\alpha}_{k+1}(k) < \bar{\alpha}_k(k)$ for each $k = 1, ..., h+1$.*

Proof. Proof follows directly from (A.2).

Let k_* be the maximal integer among $1, 2, ..., h+1$, such that $\alpha_{k_*}(k_*) > 0$, or, equivalently, $\bar{\alpha}_{k_*}(k_*) > 0$.

Lemma A.3. *$\alpha_i(k_*) > 0$, for $i \leq k_*$, and $\alpha_i(k_*) \leq 0$, for $i > k_*$.*

Proof. From (A.2) we find that $\alpha_1(k_*) > \alpha_2(k_*) > ... > \alpha_{h+1}(k_*)$. Suppose that $\bar{\alpha}_{k_*+1}(k_*) > 0$. Then from Lemma A.2 it follows that $\bar{\alpha}_{k_*+1}(k_*) = \bar{\alpha}_{k_*+1}(k_*+1) > 0$. But this contradicts the definition of k_*. □

Acknowledgements

The research was supported by Komitet Badań Naukowych, PB 678/2/91 and by Deutsche Forschungsgemeinschaft (DFG), TR 253/2-1. The authors are grateful to the referees for their helpful comments.

References

Gnot, S. and Kleffe, J. (1983). Quadratic estimation in mixed linear models with two variance components. *Journal of Statistical Planning and Inference* **8**, 267–279.

Gnot, S., Kleffe, J. and Zmyślony, R. (1985). Nonnegativity of admissible invariant quadratic estimates in mixed linear models with two variance components. *Journal of Statistical Planning and Inference* **12**, 249–258.

Gnot, S., Trenkler, G. and Zmyślony, R. (1993). Nonnegative estimation of the weighted squared lenght of the parameter vector in the linear regression model. Submitted for publication.

Hartung, J. (1981). Nonnegative minimum biased invariant estimation in variance component models. *The Annals of Statistics* **9**, 278–292.

Kleffe, J. and Pincus. R. (1974). Bayes and best quadratic unbiased estimators for variance components and heteroscedastic variances in linear model. *Mathematische Operationsforschung und Statistik, Series Statistics* **5**, 147–159.

Mathew, T., Sinha, B. K. and Sutradhar, B. C. (1992). Nonnegative estimation of variance components in unbalanced mixed models with two variance components. *Journal of Multivariate Analysis* **42**, 77–101.

Olsen, A., Seely, J. and Birkes, D. (1976). Invariant quadratic unbiased estimation for two variance components. *The Annals of Statistics* **4**, 878–890.

ABOUT THE MULTIMODALITY OF THE LIKELIHOOD FUNCTION WHEN ESTIMATING THE VARIANCE COMPONENTS IN A ONE-WAY CLASSIFICATION BY MEANS OF THE ML OR REML METHOD

V. GUIARD
Research Institute for Biology of Farm Animals
FB Biometrie
18196 Dummerstorf
Wilhelm-Stahl-Allee 2
Germany

Abstract. Maximum Likelihood (ML) and Restricted Maximum Likelihood (REML) estimators of variance components are widely used in animal breeding. Hoeschele (1989) has stated that the likelihood is always unimodal for REML, but this statement is not true in all cases. The surface of the profile likelihood functions for ML and REML are discussed and compared for a general two variance components model and for the one-way ANOVA model. For some cases upper bounds for the probability of multimodality are given.

Key words: Variance components, Maximum likelihood, Restricted maximum likelihood, Multimodality.

1. Introduction

Maximum Likelihood (ML; Hartley and Rao, 1967) and Restricted Maximum Likelihood (REML; Patterson and Thompson, 1971) estimators of variance components are widely used in animal breeding. The iterative maximization of the likelihood function or the restricted likelihood function is difficult if there are multiple maxima. Hoeschele (1989) has stated that the likelihood is always unimodal for REML, but this statement is analytically proved only for a very simple special case. In general, this statement is not true. Dietrich (1991) and Mardia and Watkins (1989) have shown the multimodality for a special class of models. In the following sections the surface of the profile likelihood functions for ML and REML will be discussed and compared for a general two variance components model and for the one-way ANOVA model. For some cases upper bounds for the probability of multimodality will be given.

2. The Model

Let y be an $(N \times 1)$ random vector of the form

$$y = X\beta + \varepsilon, \tag{1}$$

where X is a known $(N \times p)$ matrix of rank p, and β is a $(p \times 1)$ vector of unknown parameters while $\varepsilon \sim \mathbf{N}\ (0, V)$ with $V = \gamma_1 V_1 + \gamma_0 V_0$, γ_1 and γ_0 being unknown

T. Caliński and R. Kala (eds.),
Proceedings of the International Conference on Linear Statistical Inference LINSTAT '93, 139–146.

variance components, V_1 being a positive semidefinite known $(N \times N)$ matrix of rank q and $V_0 = I_N$. Moreover, let r and r_0 be defined as

$$r = \text{rank}(X, V_1) - \text{rank}(X) \quad \text{and} \quad r_0 = N - \text{rank}(X, V_1). \tag{2}$$

It is assumed that $r > 0$ and $r_0 > 0$ and, therefore, $q < N$ (A more general model with $q = N$ has been considered by Dietrich, 1991).

If λ_i, $i = 1, \ldots, s$, are different non-zero (and hence positive) eigenvalues of V_1 with multiplicities q_i, $i = 1, \ldots, s$, and $\lambda_1 < \cdots < \lambda_s$, then the eigenvalues of V have the form $\Delta_i = \gamma_0 + \gamma_1 \lambda_i$, with multiplicities q_i, $i = 0, \ldots, s$, where $\lambda_0 = 0$, $q_0 = N - q$ and $q_1 + q_2 + \cdots + q_s = q$. Since V is positive definite, we have $\Delta_i > 0$ for $i = 0, 1, \ldots, s$. Therefore, the parameter space for (γ_0, γ_1) is given by the inequalities $\gamma_0 > 0$ and $\gamma_1 > -\gamma_0/\lambda_s$, where $\lambda_s = \max(\lambda_i)$. If θ and B are defined as $\theta = \gamma_1/\gamma_0$ and $B = (1/\gamma_0)V = \theta V_1 + I_N$, then we can use the vector (γ_0, θ) instead of (γ_0, γ_1). The parameter space Ω for θ is given by the condition $\theta > -1/\lambda_s = \theta_{min}$. The matrix B has the eigenvalues $1 + \theta\lambda_i$, $i = 0, 1, \ldots, s$, with multiplicities q_i.

The log-likelihood of y differs by only a multiplicative and an additive constant from the function

$$l(\beta, \gamma_0, \theta) = -\ln |B| - N \ln \gamma_0 - Q(\beta, \theta)/\gamma_0,$$

in which

$$Q(\beta, \theta) = (y - X\beta)' B^{-1} (y - X\beta). \tag{3}$$

Maximizing $l(\beta, \gamma_0, \theta)$ with respect to β, yields the unique estimate

$$\hat{\beta} = (X'B^{-1}X)^{-1} X'B^{-1} y. \tag{4}$$

Using (3) and an $N \times (N - p)$ matrix K defined by the equations

$$K'K = I_{N-p}, \quad KK' = I_N - X(X'X)^{-1}X', \quad K'X = 0, \tag{5}$$

we get

$$Q(\theta) = Q(\hat{\beta}, \theta) = y'K(K'BK)^{-1}K'y \tag{6}$$

(see Harville, 1974). Maximizing $l(\hat{\beta}, \gamma_0, \theta)$ with respect to γ_0, we get the unique estimator

$$\hat{\gamma_0} = Q(\theta)/N. \tag{7}$$

Using (6) and (7), the so called profile likelihood function can be expressed as

$$l(\theta) = l(\hat{\beta}, \hat{\gamma_0}, \theta) - \text{const} = -\ln |B| - N \ln Q(\theta) = -\sum_{i=1}^{s} q_i \ln(1 + \theta\lambda_i) - N \ln Q(\theta). \tag{8}$$

In order to describe the representation of $Q(\theta)$ given by Harville and Fenech (1985) assume that Z is an $(N \times q)$ matrix of rank$(Z) = q$, such that $ZZ' = V_1$. Harville and Fenech (1985) defined the $(N \times N)$ matrix $C = Z'(I_N - X(X'X)^{-1}X)Z$. With r, as in (2), we have rank$(C) = r$ (see Marsaglia and Styan, 1974, p. 441).

Now, let η_j, $j = 1, \ldots, t$, are the different positive eigenvalues of C with multiplicities r_j, $r_1 + r_2 + \cdots + r_t = r$, and let $\eta_1 < \cdots < \eta_t$. Then C can be expressed as

$$C = \sum_{j=1}^{t} \eta_j U_j U_j',$$

where the $(N \times r_j)$ matrices U_j fulfil conditions $U_j'U_j = I_{r_j}$ and $U_j'U_i = 0$ if $j \neq i$. Now, we can define the quadratic forms $c_j = y'P_jP_j'y$, $j = 1, \ldots, t$, with $P_j = \eta_j^{-1/2}(I_N - X(X'X)^{-1}X')ZU_j$, $j = 1, \ldots, t$. For $j = 0$ we have the residual sum of squares

$$c_0 = y'\{I_N - (X, Z)[(X, Z)'(X, Z)]^-(X, Z)'\}y.$$

With these definitions we can write (6) in the form (see Harville and Fenech, 1985)

$$Q(\theta) = \sum_{j=0}^{t} \frac{c_j}{1 + \theta\eta_j}, \tag{9}$$

where $\eta_0 = 0$. For $j = 0, 1, \ldots, t$ the random variables $c_j/(\gamma_0(1 + \theta\eta_j))$ are independently distributed as $\chi^2(r_j)$ with r_j degrees of freedom. Moreover, for the maximum eigenvalues λ_s and η_t, the inequality $\lambda_s \geq \eta_t$ holds (see Harville and Fenech, 1985).

3. The ML-Method

The maximization of the profile likelihood function $l(\theta)$ yields the ML-estimator $\hat{\theta}$. Using $\hat{\theta}$ instead of θ, we get the ML-estimators $\hat{\beta}$ and $\hat{\gamma}_0$. Now, we have to study the function $l(\theta)$ given in (8) with $Q(\theta)$ in the form (9). It can be shown that

$$\lim_{\theta \to \infty} l(\theta) = -\infty, \text{ if } c_0 > 0, \qquad (P(c_0 > 0) = 1),$$

$$\lim_{\theta \to \theta_{min}} l(\theta) = \begin{cases} -\infty, \text{ if } \eta_t = \lambda_s \text{ and } c_t > 0, & (P(c_t > 0) = 1), \\ +\infty, \text{ if } \eta_t = \lambda_s,\ c_t = 0 \text{ and } \sum_{j=0}^{t} c_j > 0, & \\ +\infty, \text{ if } \eta_t < \lambda_s \text{ and } \sum_{j=0}^{t} c_j > 0, & (P(\sum_{j=0}^{t} c_j > 0) = 1). \end{cases} \tag{10}$$

Now, let us consider the case $c_j > 0$, $j = 0, \ldots, t$. From (10) it follows that when $\eta_t < \lambda_s$, the ML-estimate $\hat{\theta}$ is given by $\hat{\theta} = \theta_{min} = -1/\lambda_s$ with probability one if the parameter space $\Omega = [\theta_{min}, \infty)$ is admissible. In Section 5.2 it will be shown that this pathological behaviour of the likelihood function occurs in a lot of simple cases. Of course, if only the parameter space $\Omega_+ = [0, \infty)$ is admissible, then this behaviour is of no interest.

With

$$F(\theta) = \sum_{i=1}^{s} \frac{q_i\lambda_i}{1 + \theta\lambda_i} \quad \text{and} \quad Q'(\theta) = \frac{\delta}{\delta\theta} Q(\theta) = -\sum_{j=1}^{t} \frac{c_j\eta_j}{(1 + \theta\eta_j)^2} \tag{11}$$

we get $l'(\theta) = -F(\theta) - NQ'(\theta)/Q(\theta)$.

Now, let us define the sets $L = \{\lambda_1, \ldots, \lambda_s\}$, $E = \{\eta_1, \ldots, \eta_t\}$ and the product $P_A = \prod_{\xi \in A}(1 + \theta\xi)$, where A is a set of eigenvalues ξ. If A is empty then $P_A = 1$. For $\theta > \theta_{min}$, we have $Q(\theta)P_E^2 P_{L-E} > 0$ and therefore $l'(\theta) = 0$ if and only if $H(\theta) = 0$, where $H(\theta)$ is a polynomial in θ of degree $2t - 1 + \text{card}(L - E)$ of the form

$$H(\theta) = l'(\theta)Q(\theta)P_E^2 P_{L-E} = -[F(\theta)P_L][Q(\theta)P_E]P_{E-L} - N[Q'(\theta)P_E^2]P_{L-E}. \quad (12)$$

The highest order coefficient of $H(\theta)$ is negative. Moreover, if $\eta_t = \lambda_s$ then $H(\theta_{min}) > 0$. This means that in the case $\eta_t = \lambda_s$ the polynomial $H(\theta)$ has at least one real zero within Ω.

4. The REML-Method

The restricted likelihood method (REML) is defined as the likelihood method for the data vector $K'y$ instead of y, where K is defined in (5). The appropriate likelihood function is $\tilde{l}(\gamma_0, \theta) = -\ln|K'BK| - (N-p)\ln\gamma_0 - Q(\theta)/\gamma_0$, with $Q(\theta)$ given in (6) or (9). Observe that the function $\tilde{l}(\gamma_0, \theta)$ does not depend on β. With the unique REML-estimate $\hat{\gamma}_0 = Q(\theta)/(N-p)$ we get, apart from an additive constant, the profile likelihood function $\tilde{l}(\theta) = \tilde{l}(\hat{\gamma}_0, \theta) = -\ln|K'BK| - (N-p)\ln Q(\theta)$. Since the matrix $K'BK$ has the same non-zero eigenvalues as the matrix C (see Harville and Callanan, 1988), we get

$$\tilde{l}(\theta) = -\sum_{j=1}^{t} r_j \ln(1 + \theta\eta_j) - (N-p)\ln Q(\theta).$$

This function is defined for $\theta > -1/\eta_t$. Therefore, we can define the extended parameter space $\Omega_e = (\theta_e, \infty)$ with $\theta_e = -1/\eta_t \le \theta_{min} = -1/\lambda_s$.

For $\tilde{l}(\theta)$ we can show that

$$\lim_{\theta\to\infty} \tilde{l}(\theta) = \begin{cases} -\infty, \text{ if } c_0 > 0, & (P(c_0 > 0) = 1), \\ +\infty, \text{ if } c_0 = 0 \text{ and } \sum_{j=1}^{t} c_j > 0, & \end{cases}$$

$$\lim_{\theta\to\theta_e} \tilde{l}(\theta) = \begin{cases} -\infty, \text{ if } c_t > 0, & (P(c_t > 0) = 1), \\ +\infty, \text{ if } c_t = 0 \text{ and } \sum_{j=1}^{t} c_j > 0. & \end{cases}$$

Now, let us consider the case $c_j > 0$, $j = 0, \ldots, t$. With

$$D(\theta) = \sum_{j=1}^{t} \frac{r_j\eta_j}{1 + \theta\eta_j}$$

and $Q'(\theta)$ as in (11), we get $\tilde{l}'(\theta) = -D(\theta) - (N-p)Q'(\theta)/Q(\theta)$. For $\theta > \theta_e$, we have $Q(\theta)P_E^2 > 0$ and, therefore, $\tilde{l}'(\theta) = 0$ if and only if $G(\theta) = 0$, where $G(\theta)$ is a polynomial in θ of degree $2t - 1$ of the form

$$G(\theta) = \tilde{l}'(\theta)Q(\theta)P_E^2 = -[D(\theta)P_E][Q(\theta)P_E] - (N-p)Q'(\theta)P_E^2. \quad (13)$$

The highest order coefficient of $G(\theta)$ is negative. Moreover, we have $G(\theta_e) > 0$. This means that $G(\theta)$ has at least one real zero within Ω_e. If $t = 1$, $G(\theta)$ is linear and $\tilde{l}(\theta)$ is unimodal. But if $t > 1$, $\tilde{l}(\theta)$ can be multimodal.

5. The One-Way Classification

5.1. Model and Eigenvalues

Let us write the model of the one-way analysis of variance with a random factor A in the form

$$y_{ikl} = \mu + a_{ik} + e_{ikl}$$

where i is the index of the set of q_i factor levels having the same sample size n_i, $i = 1, \ldots, s$, k is the index of the levels within the i-th set, $k = 1, \ldots, q_i$, $q = q_1 + q_2 + \cdots + q_s$ is the number of all levels of the factor A, l is the index of the individual experimental units, $l = 1, \ldots, n_i$, and $N = \sum_{i=1}^{s} q_i n_i$ is the total sample size. Moreover, let us assume that $1 \leq n_1 < n_2 < \cdots < n_s$. The random variables a_{ik} and e_{ikl} are distributed independently as $N(0, \sigma_a^2)$ and $N(0, \sigma_e^2)$, respectively. This model can be written in the framework (1), with $\beta = \mu$, $X = \mathbf{1}_N$, $p = 1$,

$$V_1 = ZZ', with Z = \bigoplus_{i=1}^{s} \bigoplus_{k=1}^{q_i} \mathbf{1}_{n_i}, r = q - 1, \gamma_1 = \sigma_a^2 \geq 0, \text{ and } \gamma_0 = \sigma_e^2 > 0.$$

Since $\theta = \sigma_a^2/\sigma_e^2 \geq 0$, the parameter space is now $\Omega_+ = [0, \infty)$. But if we assume that there are non zero covariances $c = \text{cov}(e_{ikl}, e_{ikl'})$, $l \neq l'$, then we have

$$\gamma_1 = \sigma_a^2 + c \text{ and } \gamma_0 = \sigma_e^2 - c \tag{14}$$

(see Smith and Marray, 1984). In this case, the parameter space $\Omega = [\theta_{min}, \infty)$ is more justified.

For the eigenvalues λ_i, $i = 0, 1, \ldots, s$, of V_1 we have $\lambda_0 = 0$ with multiplicity $q_0 = N - q$ and $\lambda_i = n_i$, $i = 1, \ldots, s$, with multiplicities q_i. The set E of positive eigenvalues η_i of the matrix C can be divided in two subsets.

The subset of trivial eigenvalues $\eta_j = \eta_i^t$. For each $i = 1, \ldots, s$ such that $q_i > 1$ there is a trivial eigenvalue $\eta_i^t = n_i$ with multiplicities $r_j = q_i - 1$. The appropriate quadratic form is given by

$$c_i^t = n_i \sum_{k=1}^{q_i} (\overline{y}_{ik\cdot} - \overline{y}_{i\cdot\cdot})^2.$$

The subset of nontrivial eigenvalues $\eta_j = \eta_i^{nt}$. For each $i = 2, \ldots, s$ there is a nontrivial eigenvalue $\eta_j^{nt} \in (n_{i-1}, n_i)$ with multiplicity 1, being a unique solution of

$$N = \sum_{u=1}^{s} \frac{n_u^2 q_u}{n_u - \eta_j^{nt}}.$$

The appropriate quadratic form is given by

$$c_i^{nt} = \eta_i^{nt} \left(\sum_{u=1}^{s} \frac{n_u q_u}{n_u - \eta_i^{nt}} \overline{y}_{u\cdot\cdot} \right)^2 \Bigg/ \left(\sum_{u=1}^{s} \frac{n_u^2 q_u}{(n_u - \eta_i^{nt})^2} \right).$$

In the special case $s = 2$ we have $\eta_2^{nt} = qn_1n_2/N$ with the appropriate quadratic form $c_2^{nt} = n_1n_2q_1q_2(\overline{y}_{1\cdot\cdot} - \overline{y}_{2\cdot\cdot})^2/N$.

Moreover, for $i = 0$ the zero eigenvalue η_0 has the multiplicity $r_0 = N - q$, and the appropriate quadratic form is given by $c_0 = \sum_{ikl}(y_{ikl} - \overline{y}_{ik\cdot})^2$.

5.2. The ML-Method

If the maximum sample size n_s occurs only in one factor level, then the multiplicity of the eigenvalue $\eta_s^t = n_s$ is zero. Therefore, $\eta_s = \max(\eta_i) = \eta_s^{nt} < n_s = \lambda_s$. In this case it was shown in Section 3 that the ML-estimate of θ within $[\theta_{min}, \infty)$ is given by $\hat{\theta} = \theta_{min} = -1/\lambda_s = -1/n_s$. Using (14), we get $\text{var}(\overline{y}_{s\cdot\cdot}) = 0$, which corresponds with the singularity of V in the case $\theta = \theta_{min}$.

Hoeschele (1989) considered the case $s = q = 2$, $q_1 = q_2 = 1$. This is also the very extreme case, mentioned above, with $t = 1$ and $\eta_1 = \eta_2^{nt}$. The degree of the polynomial (12) is then $2 - 1 + \text{card}(L - E) = 3$. Therefore, in view of (10), it can be expected that there are also cases with multimodality in Ω_+. This agrees with the simulation results performed by Hoeschele (1989).

5.3. The REML-Method

For $t = 1$ the polynomial (13) is a linear function having its zero within Ω_e. For a balanced design we have $s = 1$ and, therefore, $t = 1$. It is well known that in this case the REML-method yields the unique ANOVA-estimator. But also in a special unbalanced case we can have $t = 1$. It is so, when $s = q = 2$ and $q_1 = q_2 = 1$. In this case the REML-estimate is identical with the ANOVA-estimate, too. This is again the case studied analytically by Hoeschele (1989). It is clear that there is always unimodality. Hoeschele (1989) has performed some simulation studies for the cases with $s = 2$ and $t = 3$ and did not find multimodality. We will see that multimodality is really very seldom but not impossible. Let us consider the easier case $s = t = 2$, $q_1 = q - 1$ and $q_2 = 1$ or $q_1 = 1$ and $q_2 = q - 1$. In this case the polynomial $G(\theta)$ is of degree 3. There is one real zero $\theta_1 \geq \theta_e$ or three real zeros $\theta_1 \geq \theta_2 \geq \theta_3 \geq \theta_e$. These zeros depend on the functions $h_j = (c_j/r_j) \,/\, (c_0/r_0)$, $j = 1, 2$. The vector (h_1, h_2) has a two-dimensional transformed F-distribution with the density function (note that $r_0 + r_1 + r_2 = N - 1$)

$$f(h_1, h_2) = \frac{\Gamma\left(\frac{N-1}{2}\right) h_1^{\frac{r_1}{2}-1} h_2^{\frac{r_2}{2}-1}}{\Gamma\left(\frac{r_1}{2}\right)\Gamma\left(\frac{r_2}{2}\right)\Gamma\left(\frac{r_0}{2}\right) M}, \tag{15}$$

where

$$M = \left[1 + \frac{h_1}{d_1} + \frac{h_2}{d_2}\right]^{\frac{N-1}{2}} d_1^{\frac{r_1}{2}} d_2^{\frac{r_2}{2}} = M(h_1, h_2, \theta), \text{ while } d_j = (1 + \theta\eta_j) r_0/r_j.$$

Now, for every point (h_1, h_2) in the first quadrant we have to calculate the real zeros of the cubic polynomial $G(\theta)$. If there are three real zeros and if the argument θ_2

of the relative minimum is positive, then we have multimodality in Ω_+. In order to get the probability of multimodality we integrate the density function (15) over the region of points (h_1, h_2) such that $\theta_2 > 0$ (Studying the polynomial $G(\theta)$, it is possible to get some useful bounds of this region). Unfortunately, the density (15) depends on the unknown parameter θ. In order to get an upper bound of this probability in Ω_+ we calculate an upper bound of this density by using the following lower bound for $M(h_1, h_2, \theta)$,

$$M(h_1, 0, \theta^*) \leq M(h_1, 0, \theta) \leq M(h_1, h_2, \theta),$$

where θ^* is defined as $\theta^* = \arg\left(\min_{\theta \in \Omega_+} M(h_1, 0, \theta)\right)$. The essential part of the calculation of θ^* consists in solving a quadratic equation. The results of the integrations are shown in Table I.

TABLE I

Probability of the event $A = \{\theta_2 > 0\}$, which means multimodality, and the event $B = \{[(\tilde{l}(\theta_3) > \tilde{l}(\theta_1)$ *and* $\theta_3 > 0]$ *or* $[\tilde{l}(0) > \tilde{l}(\theta_1)]\}$ for a one-way classification model with q levels, one replicated n times and the rest $q-1$ replicated m times.

	$q = 5$			$q = 20$		
m	n	$P(A)$	$P(B)$	n	$P(A)$	$P(B)$
	2	1.1E−02	2.0E−03	2	6.2E−02	7.0E−03
	6	9.7E−03	2.0E−03	6	7.9E−02	1.1E−02
1	10	8.4E−03	1.8E−03	10	7.5E−02	1.2E−02
	40	6.5E−03	1.5E−03	40	5.9E−02	1.2E−02
	320	5.9E−03	1.5E−03	320	4.9E−02	1.2E−02
	3000	5.7E−03	1.4E−03	3000	4.7E−02	1.1E−02
	1	3.5E−04	7.7E−05	1	3.9E−07	6.5E−08
	$2 \leq n \leq 7$	0.0E+00	0.0E+00	$2 \leq n \leq 9$	0.0E+00	0.0E+00
2	8	2.1E−05	0.0E+00	10	7.0E−05	1.7E−05
	80	5.1E−03	1.2E−03	80	3.9E−02	8.7E−03
	3000	5.8E−03	1.4E−03	3000	4.7E−02	1.1E−02
	1	1.1E−03	4.9E−04	1	2.8E−10	2.8E−10
	4	3.6E−06	9.3E−07	5	1.8E−21	4.5E−22
	$5 \leq n \leq 58$	0.0E+00	0.0E+00	$6 \leq n \leq 65$	0.0E+00	0.0E+00
10	59	1.2E−05	0.0E+00	66	6.0E−07	1.5E−07
	80	3.3E−04	7.5E−05	80	5.7E−04	1.4E−04
	320	3.8E−03	9.1E−04	320	2.6E−02	5.9E−03
	3000	5.6E−03	1.4E−03	3000	4.5E−02	1.1E−02
	1	7.4E−03	4.7E−03	1	2.7E−12	2.7E−12
	10	5.0E−04	2.3E−04	10	2.2E−11	1.2E−11
	40	3.5E−08	1.3E−08	47	2.0E−24	5.0E−25
100	$41 \leq n \leq 625$	0.0E+00	0.0E+00	$48 \leq n \leq 691$	0.0E+00	0.0E+00
	626	3.0E−06	7.5E−07	692	8.8E−11	2.2E−11
	1000	6.4E−04	1.6E−04	1000	2.1E−03	4.9E−04
	3000	3.4E−03	8.4E−04	3000	2.3E−02	5.3E−03

References

Dietrich, C.R. (1991). Modality of the restricted likelihood for spatial Gaussian random fields. *Biometrika* **78**, 833–839.

Harville, D.A. (1974). Bayesian inference for variance components using only error contrasts. *Biometrika* **61**, 383–385.

Harville, D.A. and Callanan, T.P. (1990). Computational aspects of likelihood-based inference for variance components. In: D. Gianola and K. Hammond, Eds., *Advances in Statistical Methods for Genetic Improvement of Livestock*. Springer-Verlag, Berlin, Heidelberg.

Harville, D.A. and Fenech, A.P. (1985). Confidence intervals for a variance ratio, or for heritability, in an unbalanced mixed linear model. *Biometrics* **41**, 137–152.

Hartley, H.O. and Rao, J.N.K. (1967). Maximum likelihood estimation for the mixed analysis of variance model. *Biometrika* **54**, 93–108.

Hoeschele, I. (1989). A note on local maxima in maximum likelihood, restricted maximum likelihood, and Bayesian estimation of variance components. *Journal of Statistical Computation and Simulation* **33**, 149–160.

Mardia, K.V. and Watkins, A.J. (1989). On multimodality of the likelihood in the spatial linear model. *Biometrika* **76**, 289–295.

Marsaglia, G. and Styan, G.P.H. (1974). Rank conditions for generalized inverses of partitioned matrices. *Sankhyā A* **36**, 437–442.

Patterson, H.D. and Thompson, R. (1971). Recovery of inter-block information when block sizes are unequal. *Biometrika* **58**, 545-554.

Smith, D.W. and Murray, L.W. (1964). An alternative to Eisenhart's model II and mixed model in the case of negative variance estimation. *Journal of the American Statistical Association* **79**, 145–151.

PREDICTION DOMAIN IN NONLINEAR MODELS

SYLVIE AUDRAIN
Institut de Recherches Internationales Servier
Division Biométrie
6, Place des Pléiades
92415 Courbevoie Cedex
France

and

RICHARD TOMASSONE
Institut National Agronomique
Département de Mathématique et Informatique
16, rue Claude Bernard
75231 Paris Cedex 05
France

Abstract. Most of the difficulties arising in the interpretation of linear regression are due to collinearity, which is inherent to the structure of the design points (the $\mathcal{X}$ space) in the classical linear model

$$\mathbf{y}_{n\times 1} = \mathbf{X}_{n\times p}\boldsymbol{\Theta}_{p\times 1} + \mathbf{e}_{n\times 1},$$

where the subscripts indicate the dimensions of the vectors and matrices. The structure of the $\mathcal{X}$ space has to be analysed as a warning to limit a correct use of a regression model for prediction purposes; the portion of space where prediction is good was introduced as the 'effective prediction domain' or EPD by Mandel (1985). This notion may be extended when a linear model contains x variables that are nonlinear functions of one or more of the other variables, such as x_j^2 or $x_j x_k$.

In this paper we extend the notion of EPD to nonlinear models, which have the general form

$$\mathbf{y}_{n\times 1} = \eta(\mathbf{X}_{n\times p}, \boldsymbol{\Theta}_{p\times 1}) + \mathbf{e}_{n\times 1}.$$

Key words: Linear and nonlinear regression, Collinearity, Prediction, Effective prediction domain.

1. Introduction and Notation

1.1. The Main Purposes of Regression

One of the most obvious reasons for searching relations between a response Y, considered as a function of $p+1$ factors $X_0, X_1, \cdots, X_p$, and these factors, is to express a quantitative relationship between two sets of variables: the first, Y, is generally a variable of interest, for example a yield, which may be influenced by the values of the second, for example controlled values of temperature, pH, etc. When a *formal model* $\mathcal{M}$ has been chosen, it may contain some *parameters*, such as a and b in the linear relation $Y = aX_0 + bX_1$; then an *initial experiment* $\mathcal{D}$ is made to estimate the values of the parameters of $\mathcal{M}$, such as a and b. As $\mathcal{M}$ is only an approximation to reality, a *stochastic* part is obligatorily included in $\mathcal{M}$, and a *statistical* approach is necessary. The best statistical technique is the 'old regression' analysis, which allows its user to serve two different purposes:

T. Caliński and R. Kala (eds.),
Proceedings of the International Conference on Linear Statistical Inference LINSTAT '93, 147–158.

• to analyse the relation, for selecting the best factors to explain Y;

• to use this relation to estimate Y-values, for new sets of controlled values not included in the initial experiment $\mathcal{D}$.

The literature on the topic of regression is always an important one, but the first of these two purposes seems to be preferred by statisticians. The second is often neglected, even if the aim of *prediction* is of primary interest for a majority of applications, specially in industrial or economical endeavours.

1.2. Notations and Formulas

If we have obtained n Y-values, y_i, in the form of a vector $\mathbf{y}$ associated with a design matrix $\mathbf{X}$, its ith row being of the form

$$\mathbf{x}_i' = [x_{i0}, x_{i1}, \ldots, x_{ip}],$$

we write the linear regression model as

$$\mathcal{M} : y_i = \theta_0 x_{i0} + \theta_1 x_{i1} + \cdots + \theta_p x_{ip} + e_i, \tag{1}$$

or in a more condensed form as

$$\mathcal{M} : \mathbf{y}_{n\times 1} = \mathbf{X}_{n\times(p+1)}\mathbf{\Theta}_{(p+1)\times 1} + \mathbf{e}_{n\times 1}. \tag{2}$$

The components of $\mathbf{e}$ are random variables, with the classical assumptions of independence and Normality, $\mathbf{e} \sim \mathcal{N}(\mathbf{0}, \sigma^2\mathbf{I}_n)$, where σ^2 is often unknown and must be estimated along with the components of the vector $\mathbf{\Theta}$; the Normality assumption is often a bonus, and sometimes may be relaxed. In many cases the first column in $\mathbf{X}$ is constant, so often $x_{i0} = 1$ for all i. Furthermore, it is assumed that $\mathbf{X}$ is of full rank, $p + 1 (< n)$. Then the least squares solution for the vector $\mathbf{\Theta}$ is

$$\hat{\mathbf{\Theta}} = (\mathbf{X}'\mathbf{X})^{-1}\mathbf{X}'\mathbf{y}, \tag{3}$$

and the set of predicted values is

$$\hat{\mathbf{y}} = \mathbf{X}\hat{\mathbf{\Theta}} = \mathbf{X}(\mathbf{X}'\mathbf{X})^{-1}\mathbf{X}'\mathbf{y} = \mathbf{H}\mathbf{y} \tag{4}$$

with

$$\mathrm{Cov}(\hat{\mathbf{y}}) = \sigma^2\mathbf{X}(\mathbf{X}'\mathbf{X})^{-1}\mathbf{X}' = \sigma^2\mathbf{H}. \tag{5}$$

From (5) we see that

$$\mathrm{Var}(\hat{y}_i) = \sigma^2\mathbf{x}_i'(\mathbf{X}'\mathbf{X})^{-1}\mathbf{x}_i = \sigma^2 h_{ii}. \tag{6}$$

The $\mathbf{H} = [h_{ij}]$ matrix is known as the hat-matrix and plays a central role in the influential analysis of individual observations (Belsley, 1990; Tomassone, Audrain, Lesquoy-de Turckheim and Millier, 1992; Tomassone, Dervin and Masson, 1993).

Often it is useful to write a centred version for $\mathcal{M}$; in this case, we have to replace $\mathbf{X}$ by the derived matrix $[\mathbf{1}_n, \mathbf{Z}]$ ($\mathbf{X} \rightarrow [\mathbf{1}_n, \mathbf{Z}]$), where the columns of $\mathbf{Z}$ add to zero. We have to adapt the previous formula; for example,

$$\mathrm{Var}(\hat{y}_i) = \sigma^2[\frac{1}{n} + \mathbf{z}_i'(\mathbf{Z}'\mathbf{Z})^{-1}\mathbf{z}_i]. \tag{7}$$

In (6) and (7), we generally replace σ^2 by its unbiased estimate s^2.

1.3. Collinearity and Model Reformulation

It is well known that the structure of the design points, the $\mathcal{X}$ space, is very important: difficulties in interpretation arise because of the condition known as *collinearity* or more properly *pseudo-collinearity*. In fact, if strict collinearity exists, the computation of $(\mathbf{Z}'\mathbf{Z})^{-1}$ is impossible; but pseudo-collinearity induces difficulties in the inversion and furnishes unstable values.

Pseudo-collinearity is a sort of disease inherent to a bad choice of the design. Sometimes, the experiment $\mathcal{D}$ has been badly designed; sometimes the user has had no other possibilities in collecting his data. A strict analysis of the information provided by the data to the analyst may give a good warning indicating limitations to a correct use of the estimated model. Thus, the examination of pseudo-collinearity will give an indication of the subspaces of $\mathcal{X}$ space where the model may be used with confidence; Mandel (1985) called this subspace the *effective prediction domain* (EPD). In fact, it is useful to use the principal components of $\mathbf{Z}$, and particularly the form of the *singular value decomposition* (SVD) of $\mathbf{Z}$,

$$\mathbf{Z}_{n \times p} = \mathbf{U}_{n \times r}\mathbf{D}_{r \times r}\mathbf{V}'_{r \times p} \tag{8}$$

where, if $\mathbf{Z}$ is of full rank, $r = p$, $\mathbf{D}$ is a diagonal matrix containing the square roots of the nonzero eigenvalues of $\mathbf{Z}'\mathbf{Z}$, being the same as those of the matrix $\mathbf{Z}\mathbf{Z}'$. The columns of $\mathbf{U}$ are the orthonormal eigenvectors of $\mathbf{Z}\mathbf{Z}'$ and the columns of $\mathbf{V}$ are those of $\mathbf{Z}'\mathbf{Z}$; the orthonormality condition implying that $\mathbf{U}'\mathbf{U} = \mathbf{V}'\mathbf{V} = \mathbf{I}_r$. From a computational point of view, excellent algorithms exist for obtaining the SVD of a matrix (Chambers, 1977; Van Huffel and Vandevalle, 1991).

One of the interests of the reformulation of $\mathcal{M}$ in terms of the elements of the SVD of $\mathbf{Z}$ relies in the great simplification of formulae. The two important formulae can be rewritten in the following manner:

$$\begin{aligned} (\mathbf{Z}'\mathbf{Z})^{-1} &= \mathbf{V}\mathbf{D}^{-2}\mathbf{V}', \\ \mathbf{H} &= \mathbf{U}\mathbf{U}'. \end{aligned} \tag{9}$$

Equalities (9) may be written exactly in the same manner for the non-centred version of $\mathcal{M}$, where $\mathbf{Z}$ is replaced by $\mathbf{X}$. In the following text, the set $\{\mathbf{U},\mathbf{D},\mathbf{V}\}$ is calculated for $\mathbf{Z}$, and we shall use $\mathbf{X}$ when we have a non-centred version, for example in the applications to non linear models. The isolation of the constant term of the model leads to the classical formula

$$\mathrm{Var}(\hat{y}_i) = \sigma^2[\frac{1}{n} + \mathbf{u}_i'\mathbf{u}_i] = V_{Fi}\sigma^2. \tag{10}$$

In (10) $\mathbf{u}_i'$ is the ith row of $\mathbf{U}$. The precision of an estimated value is linked with the precision of the model (σ^2 value) and with the position of the observation in the design (V_{Fi}); the latter term is known as the *variance factor*. V_{Fi} is a function of the design $\mathcal{D}$ alone, more precisely of the row number i of $\mathbf{H}$; and we may write it in a condensed form $V_F = f(\mathcal{D})$. From (10), it is easy to show that the average variance of the fitted values $\hat{y}_i$ is

$$\bar{V}(\hat{y}) = \frac{1}{n}\sum_{i=1}^{n}[\frac{1}{n} + \mathbf{u}_i'\mathbf{u}_i]\sigma^2 = \frac{r+1}{n}\sigma^2, \tag{11}$$

with $r = p$ in the full rank case.

1.4. Prediction Domain for a Linear Model

The value of the volume of the parallelotope associated with $\mathbf{Z}$, i.e. with the p columns of this $n \times p$ matrix, is important to have a geometrical interpretation of $\mathbf{Z}$. If we standardize $\mathbf{Z}$ by scaling all its columns to have the same length, say, unity, the associated maximum volume is 1, and it occurs when the columns are mutually orthogonal: we have a rectangular parallelotope. When we make orthogonal transformations on $\mathbf{Z}$, we rotate vectors but their relative lengths and orientations remain unchanged. So the volume associated with $\mathbf{Z}$ is stricly the same as the volume associated with $\mathbf{D}$. But the last one is far easier to calculate: $\mathbf{D}$ being diagonal, its columns are necessarily mutually orthogonal and their length are the singular values of $\mathbf{Z}$, so the volume associated with $\mathbf{D}$ (and with $\mathbf{Z}$) is the product of its singular values.

A practical consequence is that a matrix $\mathbf{Z}$ is ill-conditioned when one side of the associated $\mathbf{D}$ is small relative to the longest side. With a sample, it is possible to visualize this volume by computing for each column of $\mathbf{U}$ the maximal and the minimal values, i.e.

$$u_{M,j} = \max_i u_{i,j} \quad , \quad u_{m,j} = \min_i u_{i,j} \quad , \quad j = 1, \ldots, p. \tag{12}$$

The associated observations define the extreme points of the EPD, where the precision is satisfactory. It is well known (Sen and Srivastava, 1990) that estimates along the first component are quite good and their variances are low. The increase of variance is inversely proportional to the eigenvalues. When one wants to estimate a y value for a vector $\mathbf{x}_0$, all p components of $\mathbf{u}_0$ must be inside the observed parallelotope defined by values in (12), that is to say,

$$u_{m,j} \leq u_{0,j} \leq u_{M,j} \quad , \quad j = 1, \ldots, p. \tag{13}$$

In fact, it is useful to write (13) as inequality constraints in the original $\mathbf{x}_0$ values. For example, for the last component ($j = p$) associated with $\mathbf{v}_p$ (the last column of $\mathbf{v}$) we have

$$d_p u_{m,p} \leq \mathbf{x}_0' \mathbf{v}_p \leq d_p u_{M,p}, \tag{14}$$

and (14) shows clearly that this inequality corresponds to a narrow domain if d_p has a very low value. More technical and applied aspects may be found on this topic in Brook and Arnold (1985), Jackson (1991), Mandel (1985), Tomassone et al. (1992, 1993).

2. Example: Polynomial Regression

It may be interesting to look at the EPD research when we have a simple polynomial regression. As X_1 is non-stochastic, all the former results are applicable; but, doing so, we introduce pseudo-collinearity (sometimes a strong one) in the $\mathcal{X}$ space and the influence has to be analysed.

Using the 8 observations in Table I to fit a parabola, we obtain the fitted model

$$\hat{y} = 17.25 - 4.9x + 0.375x^2.$$

TABLE I
Data to fit a parabola

									Mean	St. dev.
x	2	2	4	4	6	6	8	8	5	2.236
x^2	4	4	16	16	36	36	64	64	30	22.716
y	8	10	4	3	2	1	3	1		
V_F	0.475	0.475	0.275	0.275	0.275	0.275	0.475	0.475		
$s(\hat{y})$	0.696	0.696	0.530	0.530	0.530	0.530	0.696	0.696		

We must notice that the introduction of the x^2 term modifies the classical hyperbolas which limit the confidence region along the straight line when we have a simple linear model. In the latter case, application of (10) gives

$$\begin{aligned}\mathrm{Var}(\hat{y}_i) &= \sigma^2[\tfrac{1}{n} + u_{1i}^2] = V_{Fi}\sigma^2 \\ &= \sigma^2[\tfrac{1}{n} + (x_i - \bar{x})^2 / \textstyle\sum_{i=1}^n (x_i - \bar{x})^2],\end{aligned}$$

and we know that the minimal variance is for the mean point in the $\mathcal{X}$ space, here for the x variable alone. It is also true with parabola, but the minimal variance is then for the mean point in the new $\mathcal{X}$ space, it is to say for the x and for the x^2 variables. This introduces a surprising result: for the mean point of the x variable the variance of the fitted value may be greater than for adjacent points (Figure 1). The result is due to the fact that $s(\hat{y})$ as a function of x (on which the limits of the confidence set are based) is a polynomial of degree 4; and it depends on the design itself, the choice of x_i values.

3. Prediction Domain for a Nonlinear Model

3.1. Extension from Linear to Nonlinear Models

Generally, when we have to fit data to a nonlinear model, we present it in the following manner:

$$\mathcal{M} : y_i = \eta(x_i; \Theta) + \epsilon_i, \tag{15}$$

where $\eta(x; \Theta)$ is a nonlinear function of the p components of the Θ vector. If we have to fit data that start from 0, when X = 0, and attain a maximum value when X is increasing to infinity, one of the simplest function may be the simplified monomolecular function,

$$\eta(x; \Theta) = \theta_1(1 - \theta_2^x), \quad 0 < \theta_1, \;\; 0 < \theta_2 < 1. \tag{16}$$

To estimate Θ, one of the best algorithms is the Marquardt algorithm, used in all classical statistical softwares. In fact, Marquardt algorithm is a compromise between the steepest descent and the Gauss-Newton method: using the respective advantages of both of them, it begins with the first and ends with the second. To

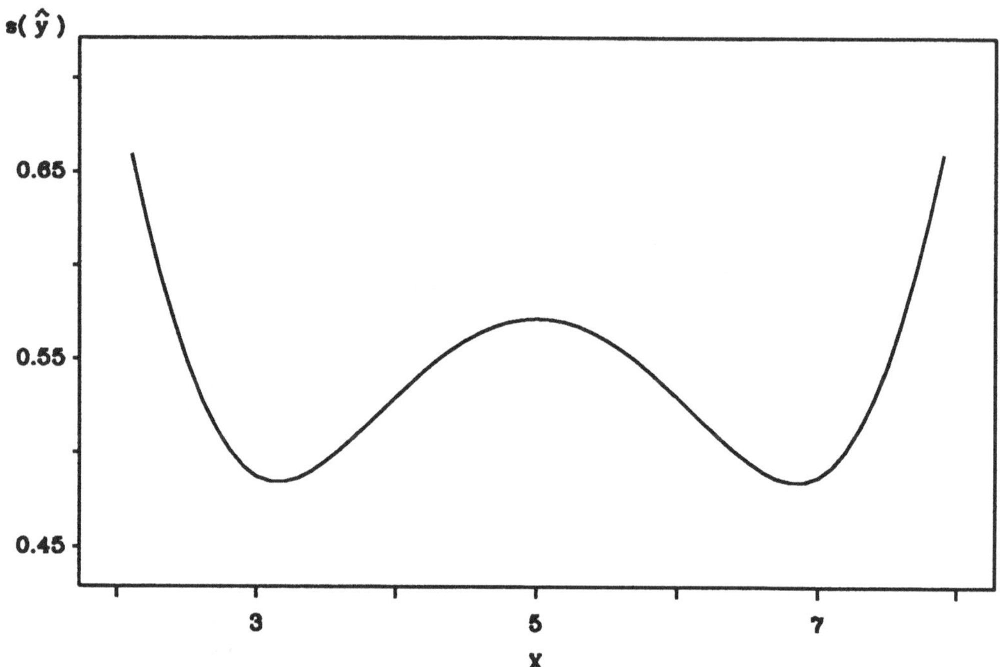

Fig. 1. Polynomial fitting of data presented in table I.

obtain variance estimate of an estimated value, the classical approximation is to use the Gauss-Newton method (i.e. that of Marquardt) at the minimal value of

$$SR(\Theta|x_i; i = 1, \ldots, n) = \sum_{i=1}^{n} (y_i - \eta(x_i; \Theta))^2. \tag{17}$$

This practise is equivalent, when one has found the minimal values of $SR(\Theta|\cdot)$, to performing a formal regression of the form

$$w_i = (y_i - \hat{y}_i) + \sum_{j=1}^{p} \hat{\Theta}_j x_{ij}^* \quad \text{on} \quad x_{ij}^*, \quad \text{with} \quad x_{ij}^* = \frac{\partial \eta(x_i; \Theta)}{\partial \theta_j}|_{\Theta=\hat{\Theta}}. \tag{18}$$

We propose to extend the use of the linear model methodology for locally linearized models. So the SVD decomposition of the $n \times p$ matrix $\mathbf{X}^* = [x_{ij}]$ permits us to obtain the variance factor V_{Fi} for each observation, the estimator of the residual variance σ^2 being chosen as $s^2 = SR(\hat{\Theta}|\cdot)/(n-p)$. But (18), corresponding to the last iteration in the nonlinear model, is formally identical to (1) where y_i is replaced by w_i. The 'pseudo-design matrix' $\mathbf{X}^*$ depends on Θ (as, of course, the matrix $\mathbf{H}^*$ derived from $\mathbf{X}^*$ does), and we may write

$$V_F = f(\mathcal{D}; \Theta). \tag{19}$$

TABLE II
5 observations to fit to the simplified monomolecular model

x	y	$\hat{y}$	$\frac{\partial\eta}{\partial a}$	$\frac{\partial\eta}{\partial b}$	w
1	2	1.8552	0.3434	-5.4026	-1.5474
2	3	3.0734	0.5689	-7.0948	-1.6585
3	4	3.8732	0.7169	-6.9877	-0.5882
4	4	4.3984	0.8141	-6.1176	-0.0169
5	5	4.7432	0.8780	-5.0211	-1.7031

TABLE III
SVD of $[\frac{\partial\eta}{\partial a}, \frac{\partial\eta}{\partial b}]$ matrix

X	=	U	D	V
$\begin{bmatrix} 0.3434 & -5.4026 \\ 0.5689 & -7.0948 \\ 0.7169 & -6.9877 \\ 0.8141 & -6.1176 \\ 0.8780 & -5.0211 \end{bmatrix}$	=	$\begin{bmatrix} 0.3892 & -0.4784 \\ 0.5120 & -0.3843 \\ 0.5054 & -0.0545 \\ 0.4439 & 0.3381 \\ 0.3659 & 0.7115 \end{bmatrix}$	$\begin{bmatrix} 13.8977 & 0.0 \\ 0.0 & 0.4807 \end{bmatrix}$	$\begin{bmatrix} 0.1058 & 0.9944 \\ -0.9944 & 0.1058 \end{bmatrix}$

3.2. Examples

When we try to fit the 5 observations (y, x) appearing in the first two columns of Table II to the the simplified monomolecular function, we obtain the estimated function (where $a = \theta_1$ and $b = \theta_2$)

$$y = 5.403(1 - 0.657^x), \text{ with } s(= \hat{\sigma}) = 0.298.$$

With these values, it is easy to compute

$$x_{1i}^* = \frac{\partial\eta(x_i|\Theta)}{\partial a} = 1 - \hat{b}^x = \frac{\hat{y}}{\hat{a}}$$

and

$$x_{2i}^* = \frac{\partial\eta(x_i|\Theta)}{\partial b} = -\hat{a}x\hat{b}^{x-1} = \frac{x(\hat{y} - \hat{a})}{\hat{b}},$$

to obtain the values of the derivatives (Table II). The SVD decomposition of the 5×2 matrix formed by the $\frac{\partial\eta}{\partial a}$ and $\frac{\partial\eta}{\partial b}$ columns is given in Table III. With these values, we may compute the estimated variances of each estimated fitted value, as indicated in Table IV. For a new observation x_0, we compute $\hat{y}_0 = \hat{a}(1 - \hat{b}^{x_0})$, $\hat{x}_{01} = \frac{\hat{y}_0}{\hat{a}}$, $\hat{x}_{02} = \frac{x_0(\hat{y}_0 - \hat{a})}{\hat{b}}$; further, we use the fact that $\mathbf{X} = \mathbf{UDV}'$, to compute:

TABLE IV
Variance factors and estimated standard errors of fitted values

x	y	u_1	u_2	V_F	$s(\hat{y})1$
1	2	0.3892	-0.4784	0.3803	0.184
2	3	0.5120	-0.3843	0.4098	0.191
3	4	0.5054	-0.0545	0.2584	0.152
4	4	0.4439	-0.3381	0.3114	0.166
5	5	0.3659	-0.7115	0.6401	0.239

TABLE V
United States population from 1790 to 1960 in millions (Snedecor and Cochran, 1971)

Year	Population	Year	Population	Year	Population
1790	3.9	1850	23.2	1910	92.0
1800	5.3	1860	31.4	1920	105.7
1810	7.2	1870	38.6	1930	122.8
1820	9.6	1880	50.2	1940	131.4
1830	12.9	1890	62.9	1950	150.7
1840	17.1	1900	76.0	1960	178.5

$$\mathbf{u}_0' = [\frac{\hat{y}_0}{\hat{a}}, \frac{x_0(\hat{y}_0 - \hat{a})}{\hat{b}}]\mathbf{VD}^{-1}, \quad V_{F_0} = \mathbf{u}_0'\mathbf{u}_0 \quad \text{and} \quad s(\hat{y}_0) = 0.298\sqrt{V_{F_0}}.$$

For illustration, another model may be used to estimate the population of United States in 2100, knowing the 18 values from 1790 to 1960 (Table V). A classical model used to fit these data is the logistic one (Snedecor and Cochran, 1971),

$$\eta(x;\Theta) = \frac{\theta_1}{(1+\theta_2 e^{-\theta_3 x})} = \frac{a}{(1+be^{-cx})}, \tag{20}$$

and the estimated values for parameters are easily obtained (Table VI) as the 3 derivatives:

$$x_{i1}^* = \frac{\hat{y}}{\hat{a}}, \quad x_{i2}^* = \hat{a}e^{-\hat{c}x}(\frac{\hat{y}_i}{\hat{a}})^2, \quad x_{i3}^* = \hat{a}\hat{b}xe^{-\hat{c}x}(\frac{\hat{y}_i}{\hat{a}})^2. \tag{21}$$

Then, it is easy to obtain the SVD of the $\mathbf{X}^*$ matrix whose lines are given by (21). In this case, it could be interesting to compute the estimated value far ahead, say for 2100, and its associated variance. The prediction for $\hat{y}_\infty$ is clearly $\hat{a}$, and $\mathbf{u}_\infty$ is

$$\mathbf{u}_\infty' = [1 \quad 0 \quad 0]'\mathbf{VD}^{-1},$$

TABLE VI
United States population from 1790 to 1960. Parameters estimation for the logistic model

Parameter	Estimate	Asymptotic Standard Error	Asymptotic Confidence Lower	95% Intervall Upper
a	251.030	19.220	210.060	292.000
b	63.180	6.170	50.040	76.320
c	0.273	0.015	0.241	0.305

TABLE VII
United States population from 1790 to 1960. Singular values for D and V, in logistic model

D	V
$\begin{bmatrix} 2471.647 & 0 & 0 \\ 0 & 0.617525 & 0 \\ 0 & 0 & 0.176188 \end{bmatrix}$	$\begin{bmatrix} 0.000558 & 0.001095 & -0.999999 \\ -0.145657 & 0.989335 & 0.001002 \\ 0.989335 & 0.145657 & 0.000712 \end{bmatrix}$

so the variance for $\hat{y}_\infty$ is equal to the estimated variance for $\hat{a}$, the estimated value of the asymptotic variance factor being

$$V_{F_\infty} = \mathbf{u}'_\infty \mathbf{u}_\infty = [\mathbf{V}\mathbf{D}^{-2}\mathbf{V}']_{11} = [(\mathbf{X}'\mathbf{X})^{-1}]_{11} = \frac{\mathrm{Var}(\hat{a})}{s^2}.$$

In fact, it is clear that the interest of this model is only historical, surely not real for practical purposes. It suffices to have a glance at its residuals to verify that their structure it surely not a random one, and we certainly should not use it: the fitted values overestimate the observed ones from 1790 to 1850, in 1870, 1940 and 1950. We must notice that the region defined around the logistic curve (Figure 2) is only a *descriptive* one, it is not a region inside of which we have a certain probability to find the curve.

4. Discussion and Further Developments

In this paper we have proposed a guess; the problem we have presented is only a facet of the prediction problem; three directions seem important to investigate for practical purposes:

• From (10), it is clear that the variance factor V_F is essential to predict the precision for new values. In linear models, we have noticed that this factor is a function of the design $\mathcal{D}$ alone. In non linear ones, we have to prove that the use of linear methodology for locally linearized model is correct. But, the difficulty also

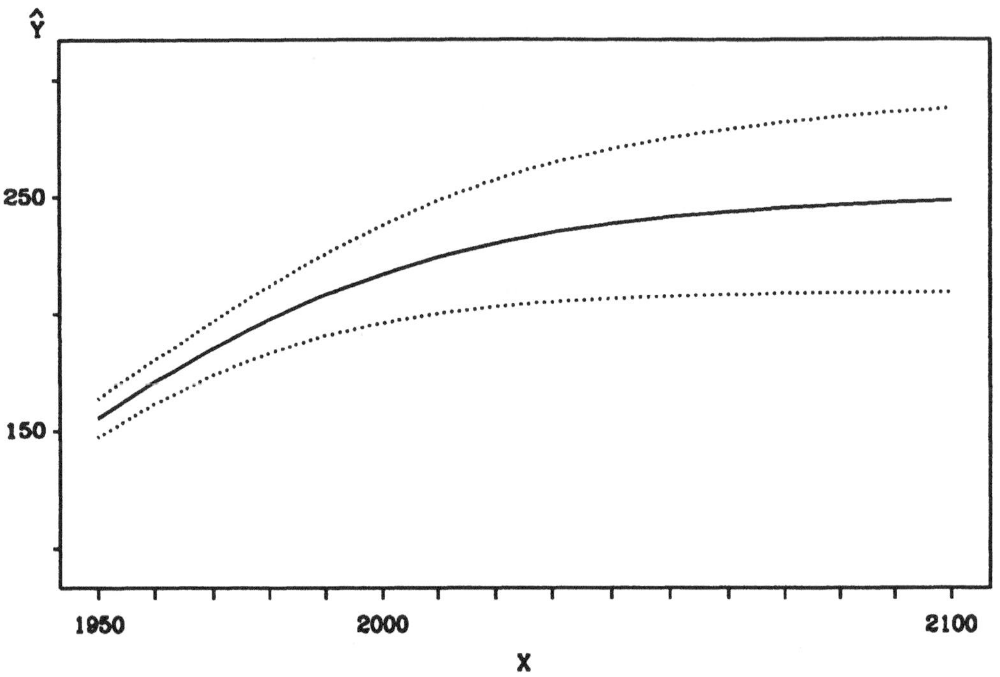

Fig. 2. United states population from 1950 to infinity.

comes from the dependence on parameter values, as we have seen it in (19). So it may be important to analyse the variance factor in terms of the *curvature measures of nonlinearity* (intrinsic curvature and parameter-effects curvature).

• Another problem, also important, is associated with *influential data*: some classical results concerning linear models may be extended to non linear ones; diagnostics as the diagonal terms h_{ii} of the hat matrix $\mathbf{H}$, Dffits$_i$ (the difference between $\hat{y}_i$ and the fitted value the observation i being deleted, $\hat{y}_{(-i)}$, normalized by the standard deviation $s_{(-i)}$), or the Dfbeta$_{ij}$ (the difference between $\hat{\Theta}_j$ and the same coefficient $\hat{\Theta}_{j(-i)}$, also normalized). Of course, a lot of work has to be done to obtain results on *leverage values*; nevertheless, even in a linear model, these values must be considered as approximations. It is interesting to notice that in the logistic model associated with the population of the United States the last observations correspond to an important change in the model. Even if the logistic model has been true, there is surely a rupture in the last two decades(Table VIII), perhaps due to the babyboom after the second world war.

• Of course, the nature of distribution is also important; one of the possibility is to analyse the mean square error prediction (Droge, 1987). Normality assumption is surely a hard one and, even if it is not the most important, it could be interesting to relax it by a computationally intensive approach. This approach is useful for obtaining confidence intervals without distributional assumptions for parameters by reusing the actual sample with jackknife or bootstrap methods: a lot of recent papers

TABLE VIII
United States population from 1790 to 1960. Some diagnostics associated with influential data (only results after 1900 are given)

year	1900	1910	1920	1930	1940	1950	1960
i	12	13	14	15	16	17	18
$y_i - \hat{y}_i$	2.02	3.05	0.55	0.71	-7.76	-5.04	7.21
h_{ii}	0.16	0.19	0.22	0.21	0.20	0.28	0.64
$\hat{y}_i - \hat{y}_{(-i)}$	10.27	0.48	0.09	0.12	-1.60	-1.18	10.68
d^*(i,a)	-0.14	-0.31	-0.06	-0.06	0.27	-0.47	8.05
d^*(i,b)	0.09	0.02	-0.02	-0.04	0.67	0.34	-0.79
d^*(i,c)	-0.08	-0.24	-0.06	-0.07	0.63	-0.10	4.93

[d^*(i,a) in this Table is the same as Dfbeta$_{ij}$ in the text]

are promising (Butler and Rothman, 1980 ; Cooks and Weisberg, 1990 ; Efron 1985, 1987 ; Stine, 1985 ; Thombs and Schucany, 1990 ; Wei, 1992).

Samples of pseudodata can be obtained by sampling with replacement from residuals $\hat{\epsilon}_i = y_i - \eta(x_i; \hat{\Theta})$; we may add sampled residuals to observed y_i and fit again the model. This sampling and fitting is repeated a large number of times, say 1000, to estimate the empirical distribution of $\hat{\Theta}$. As a prediction is a function of estimated parameters, theoretically the transposition of this approach to compute confidence intervals for prediction is possible, surely cumbersome. Even if the problem is similar, it has some strong specific aspects: the statistical behaviour of fitted values does not depend of the parametrization.

Nevertheless, the problem of prediction for future values remains entire. The quality of prediction is a fundamental question linking a real world and a possible set of formal models. Fortunately (or unfortunately?), even if statisticians should take an active part in discussing and exploring epistemological issues, truth is surely not within our grasp!

References

Belsley, D.A. (1990). *Conditionning Diagnostics: Collinearity and Weak Data in Regression.* Wiley, New York.

Brook, R.J. and Arnold, G.C. (1985). *Applied Regression Analysis and Experimental Design.* Dekker, New York.

Butler, R. and Rothman, E.D. (1980). Predictive intervals based on reuse of the sample. *Journal of the American Statistical Association* **75**, 881–889.

Chambers, J.M. (1977). *Computational Methods for Data Analysis.* Wiley, New York.

Cooks, R.D. and Weisberg, S. (1990). Confidence curves in nonlinear regression. *Journal of the American Statistical Association* **85**, 544–551.

Droge, B. (1987). A note on estimating MSEP in nonlinear regression. *Statistics* **18**, 499–520.

Efron, B. (1985). Bootstrap confidence intervals for a class of parametric problems. *Biometrika* **72**, 45–58.

Efron, B. (1987). Better bootstrap confidence interval. *Journal of the American Statistical Association* **82**, 171–185.

Jackson, J.E. (1991). *A User's Guide to Principal Component.* Wiley, New York.

Mandel, J. (1985). The regression analysis of collinear data. *Journal of Research of the National Bureau of Standards* **90**, 465–476.

Sen, A. and Srivastava, M. (1990). *Regression Analysis: Theory, Methods and Applications.* Springer-Verlag, New York.

Snedecor, W.G. and Cochran, G.W. (1971). *Méthodes Statistiques.* ACTA, Paris.

Stine, R.A. (1985). Bootstrap prediction intervals for regression. *Journal of the American Statistical Association* **80**, 1029–1031.

Thombs, L.A. and Schucany, W.R. (1990). Bootstrap prediction intervals for autoregression. *Journal of the American Statistical Association* **85**, 486–492.

Tomassone, R., Audrain, S., Lesquoy-de Turckheim, E. and Millier, C. (1992). *La Régression: Nouveaux Regards sur une Ancienne Méthode Statistique.* Masson, Paris.

Tomassone, R., Dervin, C. and Masson, J-P. (1993). *Biométrie, Modélisation de Phénomènes Biologiques.* Masson, Paris.

Van Huffel, S. and Vandevalle, J. (1991). *The Total Least-Squares Problem: Computational Aspects and Analysis.* SIAM, Philadelphia.

Wei, C.Z. (1992). On predictive least squares principles. *The Annals of Statistics* **20**, 1–42.

THE GEOMETRY OF NONLINEAR INFERENCE: ACCOUNTING OF PRIOR AND BOUNDARIES

ANDREJ PÁZMAN
Department of Probability and Statistics
Faculty of Mathematics and Physics
Comenius University
842 15 Bratislava
Slovakia

Abstract. In nonlinear models with prior, we consider the parameter estimator equal to the modus of the posterior density. The statistical model considered here is the curved exponential family; a particular case is the nonlinear regression. We analyze the importance of 3 concepts: the shift vector, the modified information matrix, and the curvature modified by the prior. A normal approximation of the posterior density is presented. A small-sample approximation of the distribution of the estimator, including the distribution on the boundary of the sample space, is considered. In appendix we consider the nonlinear regression, in examples the linear regression with non-normal prior is discussed.

Key words: Curved exponential families, Geometry in statistics, Posterior distribution, Information matrix, Constraints on parameters.

1. Introduction

Let us suppose that the statistical model is given by a family of densities

$$\{f(y|\vartheta) : \vartheta \in \Theta\} \tag{1}$$

of the sample vector $y \in \mathbb{R}^N$, and that a prior density $\pi(\vartheta)$ on $\Theta \subseteq \mathbb{R}^p$ is given. As is well known, the Bayesian approach to statistical inference requires the consideration of the posterior density

$$\pi(\vartheta|y) = f(y|\vartheta)\pi(\vartheta)/f^*(y) \tag{2}$$

with $f^*(y) = \int_\Theta f(y|\vartheta)\pi(\vartheta)d\vartheta$. A pure Bayesist takes $\pi(\vartheta|y)$ as a final result of the statistical inference. However, a frequentist would like to use an estimator of ϑ, but once knowing prior density he can not neglect it.

A traditional approach of the frequentists to this problem is to consider the estimator which has the minimal mean square error. As is well known, this estimator is equal to the posterior mean (cf. Pilz, 1983, Theorem 5.1),

$$\hat{\vartheta}_M(y) = \int_\Theta \vartheta\, \pi(\vartheta|y)\, d\vartheta.$$

In the particular case of a linear regression model with normal errors,

$$f(y|\vartheta) \approx \exp\{-\frac{1}{2\sigma^2}||y - F\vartheta||^2\}, \tag{3}$$

T. Caliński and R. Kala (eds.),
Proceedings of the International Conference on Linear Statistical Inference LINSTAT '93, 159–170.

and with a normal prior,

$$\pi(\vartheta) \approx \exp\{-\frac{1}{2}\vartheta^T B\vartheta\}, \tag{4}$$

this estimator has the form

$$\hat{\vartheta}_M(y) = \sigma^{-2}(M+B)^{-1}F^T y.$$

Here F and B are given matrices, and $M = \sigma^{-2}F^TF$ is the Fisher information matrix.

In this paper we consider an alternative estimator, namely

$$\hat{\vartheta}(y) = \arg\max_{\vartheta\in\Theta} \pi(\vartheta|y),$$

which is equal to the modus of the posterior density. The main reason why to use $\hat{\vartheta}(y)$ is that it is a 'natural extension' of the maximum likelihood estimator [see below, equation (5)].

In the linear model (3) with the prior (4) we have $\hat{\vartheta}_M(y) = \hat{\vartheta}(y)$. On the other hand, when $\pi(\vartheta)$ is uniform on Θ, the estimator $\hat{\vartheta}(y)$ is equal to the maximum likelihood (ML) estimator $\hat{\vartheta}_{ML}$,

$$\hat{\vartheta}_{ML}(y) = \arg\max_{\vartheta\in\Theta} f(y|\vartheta).$$

In the general case, the estimator $\hat{\vartheta}(y)$ can be considered as 'the ML estimator modified by the prior'. This is evident when we take the logarithm of (2), and write

$$\hat{\vartheta}(y) = \arg\max_{\vartheta\in\Theta} [\ln f(y|\vartheta) - l(\vartheta)]. \tag{5}$$

Here $l(\vartheta) = -\ln\pi(\vartheta)$ is a kind of 'penalty function'. So there is a very close relation between $\hat{\vartheta}(y)$ and the maximum likelihood estimator.

The penalized ML estimator of the form (5) may appear also in some experiments where no prior density is given. Of particular interest are two cases.

(i) A penalty $l(\vartheta)$ is used to approximate the distribution of $\hat{\vartheta}_{ML}$ on the boundary of the parameter space Θ. An elementary example, and a discussion about this are presented in Section 4.

(ii) The penalization of $\hat{\vartheta}_{ML}$ is used when there is a necessity to regularize the model. Then $l(\vartheta)$ is some regularization term. (This is a numerical and not a statistical problem, so we give no details here.)

Another use of the penalty is to decrease the bias of $\hat{\vartheta}_{ML}$ (cf. Firth, 1993). So the penalized ML estimator has many applications, and is worth to be studied.

Numerically, the value of $\hat{\vartheta}(y)$ can be obtained by some iterative algorithm. This problem has been considered in Green (1990). A modification of the Gauss-Newton method has been proposed in Pázman (1992) for the case of a nonlinear regression.

In the present paper we investigate the properties of the estimator (5) exploiting some geometrical concepts. The investigation is restricted to the case where the family (1) is a (regular) curved exponential family (cf. Barndorff-Nielsen, 1979; Efron, 1978), i.e.

$$f(y|\vartheta) = \exp\{-\psi(y) + t^T(y)\gamma(\vartheta) - \kappa[\gamma(\vartheta)]\} \tag{6}$$

with smooth model functions $\gamma(\cdot)$, $\kappa[\cdot]$ and measurable functions $\psi(y)$, $t(y)$. [We note that $\psi(y)$ and $\kappa[\gamma(\vartheta)]$ are scalars, but, as a rule, $t(y)$ and $\gamma(\vartheta)$ are vectors.] A particular but important case of the considered model is the nonlinear regression model with normal errors,

$$f(y|\vartheta) = \frac{1}{(2\pi)^{\frac{N}{2}} det^{\frac{1}{2}}(\Sigma)} \exp\{-\frac{1}{2}[y - \eta(\vartheta)]^T \Sigma^{-1}[y - \eta(\vartheta)]\}. \tag{7}$$

Here Σ is the variance matrix of y, and $\eta(\vartheta)$ is the hypothetical mean of y. Σ is supposed to be known, and regular, the form of the function $\eta(\vartheta)$ is known as well (and sufficiently smooth), but the vector of parameters ϑ is unknown.

By a comparison with (6) we obtain that (7) can be written in the exponential form (6) when

$$\psi(y) = \frac{1}{2} y^T \Sigma^{-1} y, \quad t(y) = y, \quad \gamma(\vartheta) = \Sigma^{-1}\eta(\vartheta),$$

$$\kappa[\gamma(\vartheta)] = \frac{N}{2}\ln(2\pi) + \frac{1}{2}\ln\det(\Sigma) + \frac{1}{2}\gamma^T(\vartheta)\Sigma^{-1}\gamma(\vartheta).$$

Recent results from nonlinear regression (cf. Pázman and Pronzato, 1992; Pázman, 1992) are extended here to the exponential family (6); Proposition 1 is new also for the model (7). On the other hand, some of the presented results may be interesting also for a linear model (3), but with a non-normal prior, or with constraints on parameters. This is presented in examples.

2. The Geometry of the Curved Exponential Family with Prior

As known from the classical theory of exponential families (cf. Lehmann, 1959), the random vector $t(y)$ in (6) is a sufficient statistic, and its mean [denoted by $\eta(\vartheta)$], and variance (denoted by Σ_ϑ) are equal to

$$\eta(\vartheta) = \frac{\partial\kappa(\gamma)}{\partial\gamma}|_{\gamma(\vartheta)} \quad and \quad \Sigma_\vartheta = \frac{\partial^2\kappa(\gamma)}{\partial\gamma\partial\gamma^T}|_{\gamma(\vartheta)}, \tag{8}$$

respectively.

Let us use the notation $\dot{\eta}(\vartheta) = \partial\eta(\vartheta)/\partial\vartheta$, $\ddot{\eta}(\vartheta) = \partial^2\eta(\vartheta)/\partial\vartheta\partial\vartheta^T$. Similar notations will be used also for the derivatives of other functions of the argument $\vartheta : \gamma(\vartheta), l(\vartheta)$, etc. From (8) we have

$$\dot{\eta}(\vartheta) = \Sigma_\vartheta \dot{\gamma}(\vartheta).$$

The Fisher information matrix is equal to

$$M(\vartheta) = -E_\vartheta[\partial^2 \ln f(y|\vartheta)/\partial\vartheta\partial\vartheta^T] = \dot{\gamma}(\vartheta)^T \dot{\eta}(\vartheta).$$

Here the first equality is the definition of the information matrix, the second is obtained by a straightforward derivation of the logarithm of (6), taking into account equation (8).

Since the geometry of exponential families is dualistic (cf. Efron, 1978; Amari, 1985), instead of the 'solution locus', well known in nonlinear regression (cf., e.g., Bates and Watts, 1980), we have to consider two surfaces: the expectation surface

$$\{\eta(\vartheta) : \vartheta \in \Theta\}$$

and the canonical surface

$$\{\gamma(\vartheta) : \vartheta \in \Theta\}.$$

Both are p-dimensional surfaces in $\mathbb{R}^k$, where k denotes the dimension of the sufficient statistic $t(y)$, and p is the dimension of ϑ. The projectors onto the tangent planes of these surfaces (at a given point ϑ) have the form

$$P^\vartheta = \dot\eta(\vartheta)M^{-1}(\vartheta)\dot\eta(\vartheta)^T\Sigma_\vartheta^{-1}$$

and

$$P_\vartheta = \dot\gamma(\vartheta)M^{-1}(\vartheta)\dot\gamma(\vartheta)^T\Sigma_\vartheta,$$

respectively. These projectors are orthogonal to the tangent planes of the respective surfaces, the expectation or the canonical surface, under the inner products $< a,b >^\vartheta = a^T\Sigma_\vartheta^{-1}b$ and $< a,b >_\vartheta = a^T\Sigma_\vartheta b$, respectively. We denote by $\|\cdot\|^\vartheta$ and $\|\cdot\|_\vartheta$ the corresponding norms. We have $P^\vartheta = (P_\vartheta)^T$. [For more arguments why the inner products $<,>^\vartheta$ and $<,>_\vartheta$ must be used, the reader is referred to Pázman (1993, Section 9.2).]

2.1. The Shift Vector

We define the 'shift vector' in $\mathbb{R}^k$,

$$u(\vartheta) = \dot\eta(\vartheta)M^{-1}(\vartheta)\dot l(\vartheta).$$

Evidently, $u(\vartheta)$ is invariant to any regular reparametrization of the statistical model. Further,

$$\|u(\vartheta)\|^\vartheta = \left([\dot l(\vartheta)]^T M^{-1}(\vartheta)\dot l(\vartheta)\right)^{\frac{1}{2}},$$

and we have $u(\vartheta) = 0$ when $\pi(\vartheta)$ is uniform. Let us use the notation

$$S(\vartheta, t(y)) = -\ln f(y|\vartheta) + l(\vartheta) = \psi(y) - t^T(y)\gamma(\vartheta) + \kappa[\gamma(\vartheta)] + l(\vartheta). \quad (9)$$

The normal equation for the estimator $\hat\vartheta(y)$ has the form

$$0 = \frac{\partial S(\vartheta, t(y))}{\partial\vartheta} = \dot\gamma(\vartheta)^T[\eta(\vartheta) - t(y)] + \dot l(\vartheta).$$

Multiplying this by $-\dot\eta(\vartheta)M^{-1}(\vartheta)$ we obtain other forms of the normal equation,

$$0 = P^\vartheta[t(y) - \eta(\vartheta)] - u(\vartheta) = P^\vartheta[t(y) - \eta(\vartheta) - u(\vartheta)]. \quad (10)$$

By $\mathcal{A}(\vartheta)$ we denote the 'ancillary space' of the estimator $\hat\vartheta$,

$$\mathcal{A}(\vartheta) = \{t(y) : P^\vartheta[t(y) - \eta(\vartheta) - u(\vartheta)] = 0\}.$$

This is the set of all 'samples' $t(y)$ giving the same solution of the normal equation (10).

The ancillary space of the ML estimator $\mathcal{A}_{ML}(\vartheta)$ is obtained from the set $\mathcal{A}(\vartheta)$ by puting $u(\vartheta) = 0$. We see that geometrically $\mathcal{A}_{ML}(\vartheta)$ can be obtained from the set $\mathcal{A}(\vartheta)$ by a parallel shift given by the shift vector $u(\vartheta)$. This is the geometrical meaning of $u(\vartheta)$.

Example 1. Consider the linear model (3), but with a non-normal prior. Since in this model $t(y) = y$, $\eta(\vartheta) = F\vartheta$, $\Sigma_\vartheta = \sigma^2 I$ and, hence, $\gamma(\vartheta) = \sigma^{-2}F\vartheta$, we have $M(\vartheta) = M = \sigma^{-2}F^TF$, and $u(\vartheta) = \sigma^2 F(F^TF)^{-1}\dot{l}(\vartheta)$. The ancillary spaces are

$$\mathcal{A}_{ML}(\vartheta) = \{y \in \mathbb{R}^N : (F^TF)\vartheta = F^Ty\} = \{y \in \mathbb{R}^N : F\vartheta = F(F^TF)^{-1}F^Ty\},$$

$$\begin{aligned}\mathcal{A}(\vartheta) &= \{y \in \mathbb{R}^N : (F^TF)\vartheta + \sigma^2\dot{l}(\vartheta) = F^Ty\}\\ &= \{y \in \mathbb{R}^N : F\vartheta + u(\vartheta) = F(F^TF)^{-1}F^Ty\}.\end{aligned}$$

2.2. The Information Matrix Modified by the Prior

We denote by $I(\vartheta)$ the information matrix modified by the prior,

$$I(\vartheta) = M(\vartheta) + G(\vartheta),$$

where

$$G(\vartheta) = \ddot{l}(\vartheta) - u^T(\vartheta)\ddot{\gamma}(\vartheta).$$

In the particular case of the linear model with normal errors (3) and with the normal prior (4), we have $G(\vartheta) = B$, hence $I(\vartheta) = M + B$, which is commonly accepted as an equivalent of the information matrix for such models. [To verify that $G(\vartheta) = B$, write first $l(\vartheta) = -\ln \pi(\vartheta) = \frac{1}{2}\vartheta^T B\vartheta + const.$, as it follows from (4). Hence, $\ddot{l}(\theta) = B$. Comparing (3) with (6) we obtain that $t(y) = y$, $\gamma(\vartheta) = \sigma^{-2}F\vartheta$, hence $\ddot{\gamma}(\vartheta) = 0$.]

The modified information matrix $I(\vartheta)$ will appear below in the expression for the curvature of the model, and in the approximate probability density of $\hat{\vartheta}(y)$. This is the pragmatic justification of $I(\vartheta)$.

Although the matrix $I(\vartheta)$ is positive definite in model (3) with prior (4), this may not be the case if the prior is not normal. Geometrically, the fact that $I(\vartheta)$ is not positive definite means that there are a vector $v \in \mathbb{R}^p$ and a small scalar $\alpha > O$ such that the parallel shift of the ancillary space $\mathcal{A}(\vartheta)$ given by the shift vector $u(\vartheta)$ is backward when moving forwards on the expectation surface from the point $\eta(\vartheta)$ to the point $\eta(\vartheta + \alpha v)$. This is formulated more exactly in the following proposition. [We remind that $\eta(\vartheta) + u(\vartheta) \in \mathcal{A}(\vartheta)$ and $\eta(\vartheta + \alpha v) + u(\vartheta + \alpha v) \in \mathcal{A}(\vartheta + \alpha v)$.]

Proposition 1. *The matrix $I(\vartheta)$ is positive definite if and only if for any $v \in \mathbb{R}^p, v \neq 0$, the limit*

$$\lim_{\alpha \to 0} < \frac{\eta(\vartheta + \alpha v) - \eta(\vartheta)}{\alpha}, \frac{[\eta(\vartheta + \alpha v) + u(\vartheta + \alpha v)] - [\eta(\vartheta) + u(\vartheta)]}{\alpha} >^\vartheta$$

is positive.

Proof. Obviously, the condition in the proposition is equivalent to the inequality

$$v^T[\dot{\gamma}(\vartheta)]^T[\dot{u}(\vartheta)+\dot{\eta}(\vartheta)]v > 0.$$

This can be reformulated using the definition of $u(\vartheta)$ and $M(\vartheta)$.

$$v^T\{[\dot{\gamma}(\vartheta)]^T\frac{\partial}{\partial\vartheta}[\dot{\eta}(\vartheta)M^{-1}(\vartheta)\dot{l}(\vartheta)] + M(\vartheta)\}v > 0.$$

After performing the indicated derivatives, we obtain that the left hand side of this inequality is equal to $v^T I(\vartheta)v$. □

We can say that priors giving a non-positive $I(\vartheta)$ for some $\vartheta \in supp\ \pi(.)$ should be avoided.

2.3. The Curvature

As is well known, measures of nonlinearity of a regression model can be obtained from some curvatures (cf. Bates and Watts, 1980, or Efron, 1978). In the presence of the prior density $\pi(\vartheta)$, we propose to use the following curvature,

$$C_\pi(\vartheta) = \sup_{v\in\mathbb{R}^p\setminus\{0\}} \frac{\|(I-P_\vartheta)\,v^T\ddot{\gamma}(\vartheta)v\|_\vartheta}{v^T I(\vartheta)v},$$

which is well defined when $I(\vartheta)$ is positive definite. Here $\ddot{\gamma}(\vartheta)$ is a 3-dimensional array with elements $\partial^2\gamma_r(\vartheta)/\partial\vartheta_i\partial\vartheta_j$; $(r = 1,...,k, i,j = 1,...,p)$, so the multiplications by vectors and matrices should be performed according to the dimensionality of the terms. In particular, $(I - P_\vartheta)v^T\ddot{\gamma}(\vartheta)v$ is a vector in $\mathbb{R}^k$. The same remark must be applied to the proof in Proposition 2.

The curvature $C_\pi(\vartheta)$ is invariant with respect to any regular reparametrization of the model. Hence, it is an intrinsic curvature, which is related to the canonical surface. In the particular case, when we have a nonlinear regression model with a uniform prior, $C_\pi(\vartheta)$ is equal to the intrinsic curvature of Bates and Watts (1980). In the case of an exponential family with a uniform prior, $C_\pi(\vartheta)$ is equal to the curvature of Efron [1978, equation (3.8.)], and in this case it is one of the 4 curvatures considered in Pázman (1993, Chapter 9).

We may call the expression

$$\rho_\pi(\vartheta) = [C_\pi(\vartheta)]^{-1}$$

the (intrinsic) radius of curvature (of the canonical surface). Its meaning [and hence a justification of $C_\pi(\vartheta)$] is given by the following proposition.

Proposition 2. *If*

(i) ϑ^* *is a solution of the normal equation* (10) [*i.e.,* ϑ^* *is a stationary point of* $S(\vartheta, t(y))$],

(ii) $I(\vartheta^*)$ *is positive definite, and*

(iii)

$$\|(I - P^{\vartheta^*})(\eta(\vartheta^*) - t(y))\|^{\vartheta^*} < \rho_\pi(\vartheta^*), \tag{11}$$

then ϑ^* is a (local) minimum of $S(\vartheta, t(y))$.

Proof. We will write ϑ instead of ϑ^* and t instead of $t(y)$ because of brevity. We have

$$\begin{aligned} -\ddot{S}(\vartheta,t) &= -\partial^2 S(\vartheta,t)/\partial\vartheta\partial\vartheta^T = \ddot{\gamma}(\vartheta)^T(t-\eta(\vartheta)) - \dot{\gamma}(\vartheta)^T\dot{\eta}(\vartheta) - \ddot{l}(\vartheta) \\ &= \ddot{\gamma}(\vartheta)^T(t-\eta(\vartheta)-u(\vartheta)) - I(\vartheta). \end{aligned}$$

With the use of (10) and of the equalities $P^\vartheta P^\vartheta = P^\vartheta$, $(P^\vartheta)^T = P_\vartheta$, we obtain

$$\begin{aligned} -\ddot{S}(\vartheta,t) &= \ddot{\gamma}(\vartheta)^T(I-P^\vartheta)(t-\eta(\vartheta)) - I(\vartheta) \\ &= [(I-P_\vartheta)\ddot{\gamma}(\vartheta)]^T(I-P^\vartheta)(t-\eta(\vartheta)) - I(\vartheta). \end{aligned}$$

Hence, for any $v \in \mathbb{R}^p$, $v \neq 0$, we obtain from the Schwarz inequality

$$\begin{aligned} -v^T[\ddot{S}(\vartheta,t)]v &\leq \|(I-P_\vartheta)(v^T\ddot{\gamma}(\vartheta)v)\|_\vartheta \|(I-P^\vartheta)(\eta(\vartheta)-t)\|^\vartheta - v^T I(\vartheta)v \\ &\leq [C_\pi(\vartheta)\,\|(I-P^\vartheta)(\eta(\vartheta)-t)\|^\vartheta - 1]\, v^T I(\vartheta)v \leq 0. \quad \square \end{aligned}$$

The meaning of Proposition 2 is in that $\rho_\pi(\vartheta)$ is a 'natural limit' for the distance of the sample point $t(y)$ from the expectation surface $\{\eta(\vartheta) : \vartheta \in \Theta\}$. This distance is measured in the 'sample space' of $t(y)$, in the direction orthogonal to the expectation surface. If the distance does not exceed the limit value $\rho_\pi(\vartheta)$, then the estimation is meaningful. We use this in the derivation of an approximate probability density of $\hat{\vartheta}$ as shown in the Appendix.

3. The Normal Approximation of the Posterior Density

The estimate $\hat{\vartheta}(y)$ is the modus of the posterior density $\pi(\vartheta|y)$. The shape of this density describes quite well the quality of the statistical inference, under the presence of prior. However, in nonlinear models or when the prior is not normal, we can not express $\pi(\vartheta|y)$ explicitly; nevertheless, approximations are still possible.

Let us consider the second order Taylor formula

$$\ln \pi(\vartheta|y) = \ln \pi(\hat{\vartheta}|y) + \frac{\partial \ln \pi(\vartheta|y)}{\partial\vartheta^T}|_{\hat{\vartheta}}(\vartheta-\hat{\vartheta})$$

$$+\frac{1}{2}(\vartheta-\hat{\vartheta})^T\frac{\partial^2 \ln \pi(\vartheta|y)}{\partial\vartheta\partial\vartheta^T}|_{\hat{\vartheta}}(\vartheta-\hat{\vartheta}) + o(\|\vartheta-\hat{\vartheta}\|^2).$$

We use here the notation $\hat{\vartheta} = \hat{\vartheta}(y)$, $t = t(y)$. From the normal equation we have that the term with the first derivative is zero. The second order derivative is evidently equal to the matrix $-\ddot{S}(\hat{\vartheta}, t)$. Hence, approximately we have

$$\pi(\vartheta|y) \doteq \pi(\hat{\vartheta}|y)\exp\{-\frac{1}{2}(\vartheta-\hat{\vartheta})^T[\ddot{S}(\hat{\vartheta},t)](\vartheta-\hat{\vartheta})\}. \tag{12}$$

Since $\hat{\vartheta}$ is also a local minimum of $S(\vartheta, t)$, the matrix $\ddot{S}(\hat{\vartheta}, t)$, is positive definite. Hence the pseudo-density (12) is normal and, after norming it to one, we obtain the final form of the normal approximation as

$$\pi(\vartheta|y) \doteq (2\pi)^{-p/2}\det{}^{-1/2}[\ddot{S}(\hat{\vartheta},t)]\exp\{-\frac{1}{2}(\vartheta-\hat{\vartheta})^T[\ddot{S}(\hat{\vartheta},t)](\vartheta-\hat{\vartheta})\}.$$

We note that this density is equal to the true, exact density when the model is linear, and the prior is normal. To see this, we compute first $S(\vartheta, t(y))$ [given by (9)] from (3) and (4). We obtain

$$S(\vartheta, t(y)) = \frac{1}{2\sigma^2}||y - F\vartheta||^2 + \frac{1}{2}\vartheta^T B\vartheta + d(y),$$

where $d(y)$ is a term not deppending on ϑ. It follows that

$$\ddot{S}(\vartheta, t(y)) = \sigma^{-2}F^T F + B,$$

hence the approximate posterior density is equal to

$$(2\pi)^{-p/2} \det{}^{-1/2}[M + B]\exp\{-\frac{1}{2}(\vartheta - \hat{\vartheta})^T[M + B](\vartheta - \hat{\vartheta})\},$$

with $M = \sigma^{-2}F^T F$ and $\hat{\vartheta}(y) = \sigma^{-2}(M+B)^{-1}F^T y$ [see the discussion after equation (4)]. But the same result can be obtained directly from (2) by taking the densities (3) and (4), which are normal, and hence, $\pi(\vartheta|y)$ is normal as well.

4. Accounting of the Boundaries of the Parameter Space, and the Probability Density of the Estimator $\hat{\vartheta}$

Let us suppose that we have to investigate the maximum likelihood estimator $\hat{\vartheta}_{ML}$, and that we are interested in the probability distribution on Θ. As long as Θ is open, the problem can be solved, at least approximately (cf. Pázman, 1993). However, in the cases where the parameter space Θ is closed, we usually have a non-negligible probability that $\hat{\vartheta}_{ML}$ is on the boundary of Θ. In general, especially for higher dimensions of ϑ, it is very difficult to obtain the probability distribution of $\hat{\vartheta}_{ML}$ on the boundary of Θ.

Example 2. Consider a direct observations of two parameters ϑ_1, ϑ_2:

$$y_i = \vartheta_1 + \varepsilon_i, \quad i = 1, 2,$$
$$y_i = \vartheta_2 + \varepsilon_i, \quad i = 3, 4,$$

with ε_i independent, $\varepsilon_i \sim \mathcal{N}(0, \sigma^2)$. Let us denote $\bar{y} = (y_1 + y_2)/2$, $\tilde{y} = (y_3 + y_4)/2$. When there is no constraint imposed on ϑ_1, ϑ_2, we have the ML estimator $\hat{\vartheta}_{ML} = (\bar{y}, \tilde{y})^T$. The density, moments, mean-square error, etc. can be obtained obviously for $\hat{\vartheta}_{ML}$.

Suppose now, that we have a constraint $\vartheta_1^2 + \vartheta_2^2 \leq 1$, i.e. the parameter space, $\Theta = \{(\vartheta_1, \vartheta_2)\colon \vartheta_1^2 + \vartheta_2^2 \leq 1\}$, is bounded. When $y_1, ..., y_4$ are such that $\bar{y}^2 + \tilde{y}^2 \leq 1$, then $\hat{\vartheta}_{ML}(y_1, ..., y_4) = (\bar{y}, \tilde{y})$, as before, otherwise $\hat{\vartheta}_{ML}(y_1, ..., y_4)$ is a point of the boundary of Θ. Hence there is a concentration of the distribution of $\hat{\vartheta}_{ML}$ on this boundary, which is singular, evidently non-normal, and the investigation of statistical properties of $\hat{\vartheta}_{ML}$ is difficult.

To solve such difficulties with the boundaries, we can make the approximation mentioned in Introduction. Instead of $\hat{\vartheta}_{ML}$ we consider the estimator $\hat{\vartheta}$ given in (5), but with a special choice of $l(\vartheta)$. First we take a closed set $\Theta_o \subseteq \text{int}(\Theta)$ such that the set $\text{int}(\Theta)\backslash\Theta_o$ is a 'narrow tube' along the boundary of Θ. In Example 2 we can take $\Theta_o = \{\vartheta : \vartheta_1^2 + \vartheta_2^2 \leq \frac{8}{9}\}$. Then we take $l(\vartheta)$ of the form

$$l(\vartheta) = \begin{cases} 0, & \vartheta \in \Theta_o, \\ \infty, & \vartheta \in \Theta\backslash\text{int}(\Theta), \end{cases}$$
$$l(\vartheta) > 0, \qquad \vartheta \in \Theta\backslash\Theta_o,$$

and such that $l(.)$ is continuous on Θ, and $\ddot{l}(\vartheta)$ exists and is continuous on $\text{int}(\Theta)$. When y is such that $\hat{\vartheta}_{ML}(y) \in \Theta_o$, then $\hat{\vartheta}_{ML}(y) = \hat{\vartheta}(y)$ since the shift vector $u[\hat{\vartheta}_{ML}(y)] = 0$. So, on the set Θ_o both estimators $\hat{\vartheta}$ and $\hat{\vartheta}_{ML}$ have the same probability distribution. The situation is different near the boundary of Θ. The estimate $\hat{\vartheta}(y)$ can never appear in the set $\Theta\backslash\text{int}(\Theta)$ since $l(.) = \infty$ on this set. But this estimate is not lost; simply the probability mass induced by the estimator $\hat{\vartheta}_{ML}$ on the set $\Theta\backslash\text{int}(\Theta)$ is shifted to $\text{int}(\Theta)\backslash\Theta_o$ by using $\hat{\vartheta}$ instead of $\hat{\vartheta}_{ML}$. How large is the shift for each individual point, depends on the shift vector u. But the individual shifts are not very important. What is important is that the whole mass of probability, which should lie on the boundary, is now on the "narrow tube" $\text{int}(\Theta)\backslash\Theta_o$. We can make the approximation as good as necessary by making the tube sufficiently narrow. This approach has been exploited in Pázman and Pronzato (1992).

So the whole problem can be reduced to the problem of obtaining the probability density of the estimator $\hat{\vartheta}$ on the set $\text{int}(\Theta)$, with an adequate choice of $l(\vartheta)$.

Now let us consider the probability density of the estimator $\hat{\vartheta}_{ML}$. In the general model (6), the computation of the approximate density in each point $\vartheta \in \text{int}(\Theta)$ requires solving a minimization problem (cf. Pázman, 1993). Of course, the difficulties can only increase when we use the estimator $\hat{\vartheta}$ instead of $\hat{\vartheta}_{ML}$. The situation is very different when we use the nonlinear regression model (7), where very precise approximations of the probability density of $\hat{\vartheta}_{ML}$ are available. By similar methods we can obtain the approximate probability density of the estimator also in the case when we have some prior. The exact probability density of $\hat{\vartheta}$ has then the form

$$p_\pi(\hat{\vartheta}|\bar{\vartheta}) = q_\pi(\hat{\vartheta}|\bar{\vartheta})Er(\hat{\vartheta}, \bar{\vartheta}).$$

Here $\bar{\vartheta}$ is the true value of ϑ, $Er(\hat{\vartheta}, \bar{\vartheta})$ is some error term, depending on the geometry of the model, and expressed explicitly in Pázman (1992). The approximate probability density of $\hat{\vartheta}$ is equal to (cf. Pázman and Pronzato, 1992)

$$q_\pi(\hat{\vartheta}|\bar{\vartheta}) = \frac{\det[I(\hat{\vartheta}) + (\eta(\hat{\vartheta}) - \eta(\bar{\vartheta}))^T \Sigma^{-1}(I - P^{\hat{\vartheta}}\ddot{\eta}(\hat{\vartheta})]}{(2\pi)^{\frac{p}{2}} \det^{1/2}[M(\hat{\vartheta})]}$$
$$\times \exp\{-\frac{1}{2}[||P^{\hat{\vartheta}}(\eta(\hat{\vartheta}) - \eta(\bar{\vartheta}) + u(\hat{\vartheta}))||^{\hat{\vartheta}}]^2\}. \tag{13}$$

Here again the modified information matrix $I(\vartheta)$ and the shift vector appear. We note that the sign before $u(\vartheta)$ is oposite to Pázman (1992), because $l(\vartheta)$, and hence $u(\theta)$, are defined with different signs.

In the Appendix we present the derivation of $q_\pi(\hat{\vartheta}|\bar{\vartheta})$ for the particular case of $\dim(\vartheta) = 1$.

Properties of $q_\pi(\hat{\vartheta}|\bar{\vartheta})$. When the prior is uniform, $q_\pi(\hat{\vartheta}|\bar{\vartheta})$ is equal to the 'saddle-point' or 'flat' approximation of the probability density of $\hat{\vartheta}_{ML}$ (cf. Pázman, 1993). When the model is linear and the prior is normal, then $q_\pi(\hat{\vartheta}|\bar{\vartheta})$ is the exact density of $\hat{\vartheta}$ (and it is normal). In general, $q_\pi(\hat{\vartheta}|\bar{\vartheta})$ is equivariant, i.e. when changing the parameters of the statistical model, say by $\beta = \beta(\vartheta)$, the density of the new parameters β is obtain from $q_\pi(\hat{\vartheta}|\bar{\vartheta})$, when multiplying by the Jacobian of the transformation of the parameters. That means, $q_\pi(\hat{\vartheta}|\bar{\vartheta})$ is transformed such as a probability density should be transformed. We note that the normal asymptotic approximation of the density of the estimator is not equivariant, although the exact density is.

Example 3. Consider the linear model (3). When the prior is normal, given by (4), then

$$\hat{\vartheta} \sim \mathcal{N}[(M+B)^{-1}M\vartheta,\ (M+B)^{-1}M(M+B)^{-1}].$$

However, when the prior is not normal, then we have to use the density (13). Since $\ddot{\eta}(\vartheta) = 0$, we obtain

$$q_\pi(\hat{\vartheta}|\bar{\vartheta}) = \frac{\det[I(\hat{\vartheta})]}{(2\pi)^{\frac{p}{2}}\det^{1/2}(M)}\exp\{-\frac{1}{2\sigma^2}||F(\hat{\vartheta}-\bar{\vartheta}) + u(\hat{\vartheta})||^2\},$$

where $u(\hat{\vartheta})$ is given in Example 1, the norm $||\cdot||$ is Euclidean, and

$$I(\hat{\vartheta}) = M + \ddot{l}(\hat{\vartheta}).$$

Putting the expressions for $I(\hat{\vartheta})$ and $u(\hat{\vartheta})$ we obtain

$$q_\pi(\hat{\vartheta}|\bar{\vartheta}) = \frac{\det[M + \ddot{\ell}(\hat{\vartheta})]}{(2\pi)^{\frac{p}{2}}\det^{1/2}(M)}\exp\{-\frac{1}{2\sigma^2}||F(\hat{\vartheta}-\bar{\vartheta}) + FM^{-1}\dot{l}(\hat{\vartheta})||^2\}.$$

Appendix

We now present a method used to derive the approximation $q_\pi(\hat{\vartheta}|\bar{\vartheta})$ of the probability density of the estimator $\hat{\vartheta}(y)$, when the nonlinear regression model (7) is considered, and when $\dim(\vartheta) = 1$. For a higher dimension the derivation should be different, as it is presented in Pázman and Pronzato (1992) and in Pázman (1992).

So we consider the density (7) where the parameter space is an interval, $\Theta = (a, b)$. Take a fixed point $\vartheta^* \in (a, b)$. A point $y \in \mathbb{R}^N$ lies in the ancillary space $\mathcal{A}(\vartheta^*)$ (see Section 2.1) if and only if it satisfies the normal equation

$$[y - \eta(\vartheta^*)]^T\Sigma^{-1}\dot{\eta}(\vartheta^*) = \dot{l}(\vartheta^*)$$

[see the formula following equation (9)]. Hence, y is a point of the set

$$\mathcal{S}(\vartheta^*) = \{y \in \mathbb{R}^N : [y - \eta(\vartheta^*)]^T\Sigma^{-1}\dot{\eta}(\vartheta^*) < \dot{l}(\vartheta^*)\}$$

if and only if the solution of the normal equation [i.e. $\hat{\vartheta}(y)$] is smaller than ϑ^*. [This depends on the orientation of the parameterization of the curve $\{\eta(\vartheta) : \vartheta \in (a, b)\}$;

when the orientation is opposite, we take the reverse inequality in the definition of $S(\vartheta^*)$.]

Denote by $\bar{\vartheta}$ the true value of ϑ, and denote by $F_{\hat{\vartheta}}(.)$ the distribution function of $\hat{\vartheta}$,

$$F_{\hat{\vartheta}}(\vartheta^*) = Pr_{\bar{\vartheta}}\{y : \hat{\vartheta}(y) < \vartheta^*\}.$$

We will assume that we can neglect the probability that y is such that

$$\|(I - P^{\hat{\vartheta}(y)})[\eta(\hat{\vartheta}(y)) - y]\| > \rho_\pi(\hat{\vartheta}(y)),$$

i.e. it is supposed that y is not 'too far' from the expectation 'surface' $\{\eta(\vartheta) : \vartheta \in (a,b)\}$. [Compare with (11).] Under this assumption we can write (approximately)

$$F_{\hat{\vartheta}}(\vartheta^*) \doteq Pr_{\bar{\vartheta}}[\mathcal{S}(\vartheta^*)].$$

To compute the probability of the set $\mathcal{S}(\vartheta^*)$ we define the random variable

$$z(y) = [y - \eta(\bar{\vartheta})]^T \Sigma^{-1} \dot{\eta}(\vartheta^*).$$

From $y - \eta(\bar{\vartheta}) \sim \mathcal{N}(0, \Sigma)$ it follows that

$$z(y) \sim \mathcal{N}(0, M(\vartheta^*)),$$

where $M(\vartheta^*) = \dot{\eta}^T(\vartheta^*)\Sigma^{-1}\eta(\vartheta^*)$ is the information matrix. The set $\mathcal{S}(\vartheta^*)$ can then be written in the form

$$\mathcal{S}(\vartheta^*) = \{y : z(y) < v(\vartheta^*) + \dot{l}(\vartheta^*)\},$$

where

$$v(\vartheta) = [\eta(\vartheta) \quad \eta(\bar{\vartheta})]^T \Sigma^{-1} \dot{\eta}(\vartheta).$$

Hence,

$$F(\vartheta^*) \doteq \int_{-\infty}^{v(\vartheta^*)+\dot{l}(\vartheta^*)} \frac{1}{\sqrt{2\pi M(\vartheta^*)}} \exp\left\{-\frac{t^2}{2M(\vartheta^*)}\right\} dt.$$

[Here $M(\vartheta^*)$ is a scalar, because $\dim(\vartheta) = 1$.] Thus,

$$F(\vartheta^*) \doteq \int_{-\infty}^{h(\vartheta^*)} \frac{1}{\sqrt{2\pi}} \exp\{-\frac{t^2}{2}\} dt,$$

where

$$h(\vartheta) = \frac{v(\vartheta) + \dot{l}(\vartheta)}{\sqrt{M(\vartheta)}}.$$

This allows to write

$$q_\pi(\vartheta|\bar{\vartheta}) = \frac{d}{d\vartheta} \int_{-\infty}^{h(\vartheta)} \frac{1}{\sqrt{2\pi}} \exp\{-\frac{t^2}{2}\} dt = \frac{1}{\sqrt{2\pi}} \dot{h}(\vartheta) \exp\{-\frac{1}{2}h^2(\vartheta)\}. \qquad (14)$$

We can write, according to the definition of $u(\vartheta)$ (Section 2.1),

$$v(\vartheta) + \dot{l}(\vartheta) = \{[\eta(\vartheta) - \eta(\bar{\vartheta})]^T \Sigma^{-1} + u^T(\vartheta)\Sigma^{-1}\}\dot{\eta}(\vartheta).$$

Hence,

$$h^2(\vartheta) = \{[\eta(\vartheta) - \eta(\bar{\vartheta})]^T + u^T(\vartheta)\}\Sigma^{-1}P^{\vartheta}\{[\eta(\vartheta) - \eta(\bar{\vartheta})] + u(\vartheta)\}$$
$$= \{\|P^{\vartheta}[\eta(\vartheta) - \eta(\bar{\vartheta}) + u(\vartheta)\|^{\vartheta}\}^2, \tag{15}$$

since in the nonlinear regression we have $\Sigma_{\vartheta} = \Sigma$, $P^{\vartheta} = \dot{\eta}(\vartheta)M^{-1}(\vartheta)\dot{\eta}^T(\vartheta)\Sigma^{-1}$, and since $P^{\vartheta}u(\vartheta) = u(\vartheta)$. Further we have

$$\dot{h}(\vartheta) = \frac{\dot{v}(\vartheta) + \ddot{l}(\vartheta)}{\sqrt{M(\vartheta)}} - \frac{1}{2}\frac{[v(\vartheta) + \dot{l}(\vartheta)]\dot{M}(\vartheta)}{M^{3/2}(\vartheta)},$$

where

$$\dot{v}(\vartheta) = M(\vartheta) + [\eta(\vartheta) - \eta(\vartheta)]^T\Sigma^{-1}\ddot{\eta}(\vartheta), \quad \dot{M}(\vartheta) = 2\dot{\eta}^T(\vartheta)\Sigma^{-1}\ddot{\eta}^T(\vartheta).$$

Hence,

$$\dot{h}(\vartheta) = [M(\vartheta)]^{-1/2}\{M(\vartheta) + \ddot{l}(\vartheta) + [\eta(\vartheta) - \eta(\bar{\vartheta})]^T\Sigma^{-1}\ddot{\eta}(\vartheta)$$
$$-[\eta(\vartheta) - \eta(\bar{\vartheta})]^T\Sigma^{-1}\dot{\eta}(\vartheta)M^{-1}(\vartheta)\dot{\eta}^T(\vartheta)\Sigma^{-1}\ddot{\eta}^T(\vartheta) - \dot{l}(\vartheta)M^{-1}(\vartheta)\dot{\eta}^T(\vartheta)\Sigma^{-1}\ddot{\eta}(\vartheta)\}.$$

We note that $\dot{\eta}(\vartheta)M^{-1}(\vartheta)\dot{\eta}^T(\vartheta)\Sigma^{-1} = P^{\vartheta}$ and that $\dot{l}(\vartheta)M^{-1}(\vartheta)\dot{\eta}^T(u) = u^T(\vartheta)$, $\ddot{l}(\vartheta) - u^T(\vartheta)\Sigma^{-1}\ddot{\eta}(\vartheta) = G(\vartheta)$, which allows to obtain

$$\dot{h}(\vartheta) = [M(\vartheta)]^{-1/2}\{M(\vartheta) + G(\vartheta) + [\eta(\vartheta) - \eta(\bar{\vartheta})]^T\Sigma^{-1}[I - P^{\vartheta}]\ddot{\eta}(\vartheta)\}. \tag{16}$$

Now it is sufficient to put (15) and (16) into (14) to obtain (13).

References

Amari, S.I. (1985). *Differential-Geometrical Methods in Statistics. Lecture Notes in Statistics* **28**. Springer-Verlag, Berlin.

Bates D.M. and Watts D.G. (1980). Relative curvature measure of nonlinearity. *Journal of the Royal Statistical Society B* **42**, 1–25.

Barndorff-Nielsen, O.E. (1979). *Information and Exponential Families in Statistical Theory.* Wiley Chichester.

Efron, B. (1978). The geometry of exponential families. *The Annals of Statistics* **6**, 362–376.

Firth, D. (1993). Bias reduction of maximum likelihood estimates. *Biometrika* **80**, 27–38.

Green, P.J. (1990). On the use of the EM algorithm for penalized likelihood estimation. *Journal of the Royal Statistical Society B* **52**, 443–452.

Lehmann, E.L. (1959). *Testing Statistical Hypotheses.* Wiley, New York.

Pázman, A. (1992). Geometry of the nonlinear regression with prior. *Acta Mathematica Universitatis Comenianae* **61**, 263–276.

Pázman A. (1993). *Nonlinear Statistical Models.* Kluwer, Dordrecht.

Pázman A. and Pronzato, L. (1992). Nonlinear experimental desing based on the distribution of estimates. *Journal of Statistical Planning and Inference* **33**, 385–402.

Pilz, J. (1983). *Bayesian Estimation and Experimental Design in Linear Regression Models.* Teubner, Leipzig.

GENERAL BALANCE: ARTIFICIAL THEORY OR PRACTICAL RELEVANCE?

R. A. BAILEY
Department of Mathematical Studies
Goldsmiths' College, University of London
New Cross
London SE14 6NW
United Kingdom

Abstract. If an experiment has orthogonal block structure then linear combinations of the data can be allocated to various strata, in such a way that combinations in different strata are uncorrelated and all normalized combinations in a single stratum have the same variance. The analysis of the experiment is computationally and conceptually simple if the design has the property of general balance, introduced by Nelder: that is, there is an orthogonal basis for the space of treatment contrasts, such that, in any stratum in which they are both estimable, the estimators of any pair of basis elements are uncorrelated.

More recent results show that general balance is related to commutativity of certain matrices. It is possible that neither computational simplicity nor abstract commutativity has anything to do with practical experiments in these days of high computer power. A contrary argument says that general balance aids interpretation and that the design which is generally balanced with respect to meaningful contrasts may be superior to a technically optimal design.

Key words: Basic contrasts, Efficiency, General balance, Incomplete-block design, Optimality, Orthogonal block structure, Strata.

1. What is General Balance?

1.1. Assumptions

We assume that there are N plots, each of which receives one of n treatments. The vector Y of responses on the N plots satisfies the equation

$$Y = X\tau + \varepsilon, \tag{1}$$

where X is the $N \times n$ design matrix, τ is the vector of n treatment parameters, and ε is the vector of random plot effects.

We assume that the real vector space $\mathbb{R}^N$ has an orthogonal direct-sum decomposition into known subspaces W_α, called *strata*, for α in some set A. Let S_α be the matrix of orthogonal projection onto W_α. Then

$$S_\alpha \text{ is symmetric,} \quad \textstyle\sum_{\alpha\in A} S_\alpha = I, \quad S_\alpha^2 = S_\alpha, \quad \text{and} \quad S_\alpha S_\beta = 0 \text{ if } \alpha \neq \beta. \tag{2}$$

These matrices are called the *stratum projection matrices*. Now we assume that $E(\varepsilon) = 0$ and

$$\mathrm{Cov}(\varepsilon) = \sum_{\alpha\in A} \xi_\alpha S_\alpha, \tag{3}$$

T. Caliński and R. Kala (eds.),
Proceedings of the International Conference on Linear Statistical Inference LINSTAT '93, 171–184.

so that the ξ_α are the (usually unknown) eigenvalues of $\mathrm{Cov}(\varepsilon)$.

Assumption (3) seems very strong, but it holds in many practical situations. For example, consider a proper incomplete-block design with b blocks of size k. If ε is the sum of the following independent random variables,

- a random variable on each plot, with variance σ^2,
- a random variable on each block, with variance σ_B^2,
- a single random variable on the entire experiment, with variance σ_0^2,

then

$$\mathrm{Cov}(\varepsilon) = \sigma^2 I + \sigma_B^2 J_B + \sigma_0^2 J, \tag{4}$$

where J is the all-1 matrix, and the (i,j)-entry of J_B is equal to 1 if plots i and j are in the same block, to 0 otherwise. Then

$$\mathrm{Cov}(\varepsilon) = \xi_{\text{plots}} S_{\text{plots}} + \xi_{\text{blocks}} S_{\text{blocks}} + \xi_0 S_0,$$

where $S_0 = N^{-1}J$, $S_{\text{blocks}} = k^{-1}J_B - N^{-1}J$, $S_{\text{plots}} = I - k^{-1}J_B$, $\xi_{\text{plots}} = \sigma^2$, $\xi_{\text{blocks}} = \sigma^2 + k\sigma_B^2$ and $\xi_0 = \sigma^2 + k\sigma_B^2 + N\sigma_0^2$. The S-matrices satisfy (2). A more general version of (4) is obtained by assuming that $\mathrm{Cov}(\varepsilon)$ has three different entries: one on the diagonal, one for pairs of distinct plots in the same block, and one for other pairs of plots. Then σ_B^2 and σ_0^2 may be either positive or negative, although, of course, each of ξ_{plots}, ξ_{blocks} and ξ_0 must be non-negative. See Speed and Bailey (1987).

Another example is the row-column design with a rows, b columns and 1 plot per row-column intersection. If we define J_R and J_C for rows and columns analogously to J_B for blocks, then a reasonable assumption is that

$$\mathrm{Cov}(\varepsilon) = \sigma^2 I + \sigma_R^2 J_R + \sigma_C^2 J_C + \sigma_0^2 J.$$

This can be rewritten as

$$\mathrm{Cov}(\varepsilon) = \xi_{\text{plots}} S_{\text{plots}} + \xi_{\text{rows}} S_{\text{rows}} + \xi_{\text{columns}} S_{\text{columns}} + \xi_0 S_0,$$

where now $S_0 = N^{-1}J$, $S_{\text{rows}} = b^{-1}J_R - N^{-1}J$, $S_{\text{columns}} = a^{-1}J_C - N^{-1}J$, and $S_{\text{plots}} = I - b^{-1}J_R - a^{-1}J_C + N^{-1}J$. Again, the S-matrices satisfy (2). Many further examples are given by Nelder (1965a) and Speed and Bailey (1987). Bailey (1991b) shows how many models of the form (3) can be justified by randomization.

Assumption (3) is crucial to our development. Without it, general balance cannot even be defined. It is common to assume also that ε has a multivariate normal distribution, but this assumption is not needed for all that follows. For this paper it will be assumed throughout that treatments are equally replicated with replication r. This assumption is not needed for general balance. The theory is described in the more general case by Nelder (1965b) and Houtman and Speed (1983), but the algebra is more straightforward in the equireplicate case, which gives us a sufficient richness of examples to discuss the pros and cons of general balance.

1.2. Estimation within each stratum

Projecting equation (1) into stratum W_α gives

$$S_\alpha Y = S_\alpha X\tau + S_\alpha \varepsilon$$

with $E(S_\alpha \varepsilon) = 0$ and

$$\mathrm{Cov}(S_\alpha \varepsilon) = S_\alpha \left(\sum_{\beta \in A} \xi_\beta S_\beta \right) S_\alpha = \xi_\alpha S_\alpha,$$

which is a scalar matrix on W_α (Bailey, 1981). Hence ordinary least squares gives the estimator $\hat{\tau}_{(\alpha)}$ of τ, or of some linear combinations of τ, from $S_\alpha Y$ by solving

$$(S_\alpha X)' S_\alpha Y = (S_\alpha X)' S_\alpha X \hat{\tau}_{(\alpha)} \text{ or } X' S_\alpha Y = L_\alpha \hat{\tau}_{(\alpha)},$$

where $L_\alpha = X' S_\alpha X$. The matrix L_α is called the *information matrix* in stratum W_α.

Denote by L_α^- a generalized inverse of L_α. As usual, denote the image of L_α by $\mathrm{Im}(L_\alpha)$. Let x_1, x_2 be contrasts in the treatment space $\mathbb{R}^n$. Then the standard least squares theory gives:

(i) if $x_i \in \mathrm{Im}(L_\alpha)$ then $x_i' \tau$ is estimable in W_α with variance $x_i' L_\alpha^- x_i \xi_\alpha$;

(ii) if x_1 and x_2 are in $\mathrm{Im}(L_\alpha)$ then the covariance of the estimators of $x_1' \tau$ and $x_2' \tau$ in stratum W_α is equal to $x_1' L_\alpha^- x_2 \xi_\alpha$.

In particular, if x_i is an eigenvector of L_α with eigenvalue $\lambda_{\alpha i}$ then

$$\mathrm{Var}(x_i' \hat{\tau}_{(\alpha)}) = \frac{x_i' x_i}{\lambda_{\alpha i}} \xi_\alpha,$$

and

$$\mathrm{Cov}(x_1' \hat{\tau}_{(\alpha)}, x_2' \hat{\tau}_{(\alpha)}) = \frac{x_1' x_2}{\lambda_{\alpha 2}} \xi_\alpha.$$

If $\mathrm{Cov}(\varepsilon) = \sigma^2 I$, then the variance of the ordinary least squares estimator of $x_i' \tau$ is $x_i' x_i \sigma^2 / r$. Thus the relative efficiency of estimation of $x_i' \tau$ in W_α is $\lambda_{\alpha i} \sigma^2 / r \xi_\alpha$. Since σ^2 and ξ_α are not known *a priori*, and are in any case assumed independent of X, the quantity $\lambda_{\alpha i}/r$ is called the *efficiency factor* for x_i in W_α, and $\lambda_{\alpha i}$ is the *effective replication* for x_i in W_α (Nelder, 1965b; James and Wilkinson, 1971; Caliński, Ceranka and Mejza, 1980).

Further, if x_1 and x_2 are orthogonal eigenvectors of L_α, then the estimators $x_1' \hat{\tau}_{(\alpha)}$ and $x_2' \hat{\tau}_{(\alpha)}$ are uncorrelated, and hence independent in the case that ε is multivariate normal.

The importance of the eigenvectors of the information matrices has been recognized by Morley Jones (1959), Martin and Zyskind (1966), James and Wilkinson (1971), Pearce, Caliński and Marshall (1974), who called them *basic contrasts*, and Corsten (1976).

1.3. Definition and Examples

In view of the importance of these eigenvectors, it is clearly of some interest if the eigenvectors of the information matrices are the same for each stratum. Nelder (1965b) defined a design to be *generally balanced with respect to* a given orthogonal direct-sum decomposition $\bigoplus_{\gamma \in \Gamma} T_\gamma$ of the treatment space $\mathbb{R}^n$ if each T_γ is an eigenspace (or contained in an eigenspace) of every information matrix L_α. Houtman and Speed (1983) generalized this definition, and called a design *generally balanced* if the information matrices $(L_\alpha)_{\alpha \in A}$ have a common eigenvector basis.

Example 1. Consider the 15 treatments of a half-diallel experiment with 6 parental lines; in other words, all unordered pairs from the 6-set $\{a, b, c, d, e, f\}$. The 15-dimensional treatment space is the orthogonal direct sum $T_0 \oplus T_1 \oplus T_2$, where T_0 is the 1-dimensional space for the general mean, T_1 is the 5-dimensional space of contrasts between parents, spanned by vectors representing comparisons like

$$ac + ad + ae + af - bc - bd - be - bf, \tag{5}$$

and $T_2 = (T_0 + T_1)^{\perp}$.

The treatments are allocated to the cells of a 6×10 rectangle as shown in Figure 1.

	$abc\|def$	$abd\|cef$	$abe\|cdf$	$abf\|cde$	$acd\|bef$	$ace\|bdf$	$acf\|bde$	$ade\|bcf$	$adf\|bce$	$aef\|bcd$
a	bc	bd	be	bf	cd	ce	cf	de	df	ef
b	ac	ad	ae	af	ef	df	de	cf	ce	cd
c	ab	ef	df	de	ad	ae	af	bf	be	bd
d	ef	ab	cf	ce	ac	bf	be	ae	af	bc
e	df	cf	ab	cd	bf	ac	bd	ad	bc	af
f	de	ce	cd	ab	be	bd	ac	bc	ad	ae

Fig. 1. Design in Example 1

The rows are labelled by the 6 parents, the columns by the 10 partitions of 6 of type 3^2. Consider the cell in the row labelled t and the column labelled $P|Q$. If $t \in P$ then that cell contains the treatment $P \setminus \{t\}$; otherwise it contains $Q \setminus \{t\}$. Thus the row labelled a contains all pairs which exclude a, and the column labelled by the partition $abc|def$ contains all pairs contained in either $\{a, b, c\}$ or $\{d, e, f\}$.

Now, each column contains each parent twice. Therefore, each of the six vectors representing sums like

$$ab + ac + ad + ae + af$$

has the same image under J_C. Hence vectors such as (5) are in the kernel of L_{columns} and so T_1 is orthogonal to columns. By projecting onto rows and then onto treatments, it can be checked that each contrast in T_1 is an eigenvector of L_{rows} with eigenvalue $2/5$ and an eigenvector of L_{plots} with eigenvalue $18/5$: each such contrast therefore has efficiency factors $1/10$ in rows and $9/10$ in plots.

The space T_2 is spanned by vectors representing comparisons like

$$ab + cd - ac - bd,$$

which sum to zero on each parent. Each such vector also sums to zero on each row, and so T_2 is orthogonal to rows. By projecting onto columns and then onto treatments, it can be checked that each contrast in T_2 is an eigenvector of L_{columns} and of L_{plots}, with efficiency factors $1/6$ and $5/6$ respectively. Hence the design is generally balanced with respect to the decomposition $T_0 \oplus T_1 \oplus T_2$. The efficiency factors are displayed in Table I.

TABLE I
Efficiency factors for the design in Figure 1

Strata	dim	T_0 1	T_1 5	T_2 9
mean	1	1	0	0
rows	5	0	1/10	0
columns	9	0	0	1/6
plots	45	0	9/10	5/6

Example 2. The plots in this design are arranged in 6 blocks, each of which has 2 rows and 2 columns. The 6 treatments have a factorial structure, consisting of all levels α, β, γ of a 3-level factor A with both levels 0, 1 of a 2-level factor B. The design in Figure 2 is generally balanced with respect to the treatment decomposition $T_0 \oplus T_A \oplus T_B \oplus T_{AB}$, where T_A and T_B are the main effects of A and B respectively and T_{AB} is the interaction. The efficiency factors are summarized in Table II.

<table>
<tr><td>γ 0</td><td>α 0</td><td></td><td>β 0</td><td>γ 0</td><td></td><td>α 0</td><td>β 0</td></tr>
<tr><td>β 1</td><td>β 0</td><td></td><td>α 1</td><td>α 0</td><td></td><td>γ 1</td><td>γ 0</td></tr>
<tr><td></td><td></td><td></td><td></td><td></td><td></td><td></td><td></td></tr>
<tr><td>γ 1</td><td>α 1</td><td></td><td>β 1</td><td>γ 1</td><td></td><td>α 1</td><td>β 1</td></tr>
<tr><td>β 0</td><td>β 1</td><td></td><td>α 0</td><td>α 1</td><td></td><td>γ 0</td><td>γ 1</td></tr>
</table>

Fig. 2. Design in Example 2

TABLE II
Efficiency factors for the design in Figure 2

Strata	dim	T_0 1	T_A 2	T_B 1	T_{AB} 2
mean	1	1	0	0	0
blocks	5	0	1/16	1/4	1/16
rows	6	0	9/16	1/4	1/16
columns	6	0	3/16	1/4	7/16
plots	6	0	3/16	1/4	7/16

Example 3. Lest the reader think that all designs are generally balanced, Figure 3 shows two designs for 3 treatments in a 3×3 square. Neither design is very

	Design (a)		Design (b)	
	a a c / a c b / c b b		a a b / c b c / c b a	
rows stratum	$a-b$	4/9	$a-c$	4/9
	$a+b-2c$	0	$a+c-2b$	0
columns stratum	$a-b$	4/9	$b-c$	4/9
	$a+b-2c$	0	$b+c-2a$	0
plots stratum	$a-b$	1/9	$a-b$	7/9
	$a+b-2c$	1	$a+b-2c$	1/3

Fig. 3. Two designs in Example 3

useful, but they suffice to make the point. Below each design are shown the eigenvectors (and their efficiency factors) on treatment contrasts in the three strata apart from the mean. The design in Figure 3(a) is generally balanced; that in Figure 3(b) is not.

1.4. Related Ideas

Many authors have defined concepts which are particular cases of general balance. Yates (1935) called a factorial design *factorially balanced* if it is generally balanced with respect to the usual factorial decomposition of the treatment space into main effects and interactions. Thus the design in Example 2 is factorially balanced.

John and Smith (1972) defined a proper incomplete-block design to have *orthogonal factorial structure* if, for every orthogonal pair of factorial effects, their estimators in the plots stratum are uncorrelated. Bailey (1985a) showed that orthogonal factorial structure is equivalent to general balance with respect to some refinement of the usual factorial decomposition.

Eccelston and Russell (1975) defined a design to have *adjusted orthogonality* if each treatment contrast is non-orthogonal to at most one of the nuisance factors, such as rows, columns etc. It follows that a design with adjusted orthogonality has general balance and that each common eigenvector has a non-zero efficiency factor in at most one stratum other than the bottom one. This is true whether or not the design is equireplicate, in spite of the doubt expressed by Lewis and Dean (1991). The design in Example 1 has adjusted orthogonality.

2. How to recognize General Balance

It is usually tedious to calculate all the information matrices and find their eigenvectors. Fortunately, there are some easier ways of recognizing general balance in the (assumed) equireplicate case.

Theorem 1 (Speed, 1983). *A design is generally balanced if and only if the information matrices commute, that is,* $L_\alpha L_\beta = L_\beta L_\alpha$ *for all* α, β *in* A.

Corollary 1 (Houtman and Speed, 1983). *All proper incomplete-block designs are generally balanced.*

Proof. In a proper incomplete-block design,

$$r^{-1}L_{\text{plots}} + r^{-1}L_{\text{blocks}} + n^{-1}J = I$$

and the $n \times n$ matrix J commutes with all information matrices. □

Similarly, every design which has all treatment contrasts orthogonal to all but two of the strata must be generally balanced. With three effective strata, as in row-column designs or split-plot designs, it is possible to have a design that is not generally balanced, as Example 3 shows. However, the information matrices sum to rI, and one of them ($rn^{-1}J$) always commutes with all of them, so general balance is equivalent to the commutativity of any of the other two. Several authors realised the importance of this commutativity condition for row-column designs without realising its equivalence to general balance: see Lewis and Dean (1991).

Many complicated structures whose natural covariance model is (3) are made by nuisance factors satisfying certain conditions described by Tjur (1984). For such a *Tjur block structure*, the stratum projection matrices are linear combinations of the matrices analogous to J_B, J_R and J_C in Section 1.1, and vice versa. Since $X'J_RX$ is the concurrence matrix in the row design, we have another corollary of Theorem 1.

Corollary 2 (Bailey and Rowley, 1990). *A design with Tjur block structure is generally balanced if and only if its concurrence matrices commute.*

In a *cyclic* incomplete-block design, the treatments are identified with a cyclic group G of order n. The blocks are all translates in G of one or more initial blocks. More generally, G may be taken to be any Abelian group of order n. More generally again, a design with Tjur block structure is called an *Abelian group design* if it has the above property with respect to every nuisance factor.

The *irreducible characters* of a finite Abelian group are straightforward but not well known to statisticians (Bailey, 1991a). The irreducible characters of the cyclic group of order n are the complex vectors

$$(1, \theta, \theta^2, \ldots, \theta^{n-1})$$

for complex numbers θ such that $\theta^n = 1$. For non-cyclic groups the irreducible characters are like the familiar AB^2 for a 3×3 factorial design.

Theorem 2 (Bailey and Rowley, 1990; Kobilinsky, 1990). *Every Abelian group design with treatment group* G *is generally balanced with respect to the irreducible* *characters of G.*

Example 2 is actually constructed from any one of its blocks by applying the 3-cycle $(\alpha\,\beta\,\gamma)$ and the 2-cycle $(0\,1)$. Thus Theorem 2 guarantees its general balance.

Theorem 2 may be regarded as a special case of two further results. Every association scheme defines a Bose-Mesner algebra and a set of common eigenspaces of the scheme; see Bose and Mesner (1959) and Bailey (1985b). Any design with a Tjur block structure which, for each nuisance factor, is partially balanced with respect to the *same* association scheme, is generally balanced with respect to the common eigenspaces of that scheme. Finally, the Abelian group design construction may be generalized to other permutation groups: those satisfying a technical condition called *proper constraint* (McLaren, 1963) always give general balance with respect to known eigenspaces.

3. What is Good about General Balance?

General balance is mathematically tidy. It is pleasing to be able to summarize efficiency factors by a table such as Tables I and II, and to estimate contrasts $x_i'\tau$ in each stratum for the same basis vector x_i. But is this mathematical tidiness worth anything statistically?

In a generally balanced design with common orthogonal eigenvector basis $(x_i)_{i=1}^n$, write $e_{\alpha i}$ for the efficiency factor of x_i in W_α (defined in Section 1.2). Then it can be shown that $\sum_\alpha e_{\alpha i} = 1$ for each i, as demonstrated in Tables I and II. Thus $e_{\alpha i}$ may be interpreted as the *proportion of information* on x_i in stratum W_α. Moreover, $e_{\alpha i}$ is also the square of the cosine of the angle between Xx_i and W_α. This double interpretation is not true for other treatment vectors. For example, in Figure 3(b) the contrast $a - b$ is not estimable in either the rows or the columns stratum, even though it makes an angle $\cos^{-1}(1/3)$ with each of them.

The estimator $x_i'\hat{\tau}_{(\alpha)}$ from data y is equal to

$$\frac{x_i'X'S_\alpha y}{e_{\alpha i} r},$$

a straightforward modification of the estimator when $\mathrm{Cov}(\varepsilon) = \sigma^2 I$. Two such estimators are uncorrelated unless $\alpha = \beta$ and $i = j$. Thus, under multivariate normality, estimators in different strata are independent, as are estimates of different $x_i'\tau$ in the same stratum. If the vectors x_i correspond to interesting treatment contrasts, this independence of estimation is useful.

The sum of squares for $x_i'\tau$ in W_α is equal to $\mathrm{SS}_{\alpha i}$, where

$$\mathrm{SS}_{\alpha i} = \frac{(x_i'X'S_\alpha y)^2}{x_i'x_i r e_{\alpha i}} = \frac{\left(x_i'\hat{\tau}_{(\alpha)}\right)^2}{x_i'x_i} r e_{\alpha i}.$$

Not only is this formula very similar to that for the sum of squares for $x_i'\hat{\tau}$ in a design with $\mathrm{Cov}(\varepsilon) = \sigma^2 I$. The $\mathrm{SS}_{\alpha i}$, for $e_{\alpha i}$ non-zero, partition the treatment sum of squares in W_α, and $\mathrm{SS}_{\alpha i}/\xi_\alpha$ has a χ^2 distribution on one degree of freedom if $x_i'\tau = 0$. The consultant statistician is very used to partitioning the treatment sum of squares into orthogonal meaningful components, giving mean squares which are independent χ^2 under the null hypothesis. For structured treatments, interpretation from such a partition of the sum of squares is more sensible, and quicker, than multiple comparison procedures. The statistician is denied this useful tool if the design is not generally balanced with respect to a relevant decomposition of the treatment space.

If the stratum variances ξ_α are known, then the best linear unbiased estimator of $x_i'\tau$ is

$$\frac{1}{v_i}\sum_\alpha \frac{e_{\alpha i}}{\xi_\alpha} x_i'\hat{\tau}_{(\alpha)}, \tag{6}$$

where $v_i = \sum_\alpha (e_{\alpha i}/\xi_\alpha)$ (Houtman and Speed, 1983). This property is called *simple combinability*. This leads to a simple iterative procedure for estimating τ when the ξ_α are not known (Nelder, 1968). At each step the current values of the ξ_α are used to estimate the $x_i'\tau$ from (6); the ensuing estimate of $X\tau$ is subtracted from y, to give revised mean-square estimates of the ξ_α.

4. What is Bad about General Balance?

If $\bigoplus T_\gamma$ is a meaningless decomposition of $\mathbb{R}^n$, then general balance with respect to $\bigoplus T_\gamma$ is worthless. There is no point in estimating $x_i'\tau$ if x_i is a meaningless contrast. Orthogonality between such estimators, and additivity of their sums of squares, is also useless if the contrasts themselves are not interesting. For these reasons, Pearce (1983) argues quite strongly that the concept of *general balance* by itself is worthless; only *general balance with respect to a named interesting decomposition of* $\mathbb{R}^n$ is useful.

Many of the arguments given above in favour of general balance refer to simplicity of formulae or of computation. However, with today's computers, such arguments carry little force. A mixed model can be analysed by first using restricted maximum likelihood (REML) (Patterson and Thompson, 1971) to estimate the relative values of the stratum variances, then using generalized least squares (GLS) to estimate τ. Houtman and Speed (1983) claim that this procedure gives the same results as Nelder's iterated procedure. Indeed, Patterson and Thompson, (1975) showed that the two procedures give solutions to the same equations.

General balance does not necessarily imply high efficiency factors, or any other measure of goodness. Indeed, disconnected and other poor designs can be generally balanced.

Finally, Pearce (1983) argues that Corollary 1 is a *disadvantage* of general balance. What is the point in naming a property that all incomplete-block designs have?

5. Discussion

5.1. Incomplete-Block Designs

For incomplete-block designs, I think that Pearce (1983) is partly right and partly wrong. Houtman and Speed's definition of general balance is perfectly valid, and makes no judgement on its desirability. It follows that all proper incomplete-block designs are generally balanced with respect to *some* decomposition of the treatment space: so be it. The definition is not vacuous for all structures, so it seems silly to argue about the words used.

However, it is clear from Section 3 that the main advantage of general balance comes when the common eigenvectors have meaning in terms of the treatment structure. Caliński (1993) argues this point in more detail. For a factorial design, factorial balance—or, at least, orthogonal factorial structure—is desirable. For a treatment set including one control, the contrast for 'control vs. the rest' should be a common eigenvector. General balance for the actual treatment structure is so useful for interpretation that it outweighs other considerations, such as optimality, or partial balance with few associate classes.

Among partially balanced designs, group divisible designs are generally balanced for both nested and crossed factorial designs with two treatment factors; triangular designs are generally balanced with respect to diallel treatment structure; rectangular designs are generally balanced for crossed factorial designs with two treatment factors. As Corsten (1976) points out, many other partially balanced incomplete-block designs with two associate classes have irrational eigenvalues in the Bose-Mesner algebra, so they are unlikely to be generally balanced with respect to any natural treatment structure.

Monod and Bailey (1992) contrast the use of general balance for analysis with its use for design, and point out that pseudofactorial decompositions (such as AB and AB^2 for a 3×3 factorial) are acceptable to many statisticians even if they have no direct meaning.

Table III shows the general balance decomposition of some common incomplete-block designs, including those just discussed, together with various other properties of those designs. Here 'optimality' refers to A-optimality, although many of the designs listed are optimal with respect to other criteria whenever they are A-optimal. For example, all group divisible designs are generally balanced with respect to a two-factor treatment structure, but they may or may not be efficient. Square lattice designs are optimal with respect to many criteria (Cheng and Bailey, 1991) but their general balance decomposition is not always meaningful. Rectangular lattices with three or more replicates have efficiency factors close to the upper bound (Williams 1977, Corsten 1985) but their general balance decomposition is quite hard to interpret (Bailey and Speed, 1986). The α-designs introduced by Patterson and Williams (1976) are the most extreme: many of them are very efficient indeed, but the eigenspaces are often low-dimensional and irrational, and so are hard to interpret. There is, thus, a clear conflict between meaningful general balance and optimality. For a completely unstructured treatment set, overall optimality may be desirable: for structured treatments I would sacrifice 10% on the overall harmonic mean efficiency factor to have a design more easily interpretable in terms of that treatment structure, especially if some chosen important contrasts have high efficiency factors.

5.2. Other Block Structures

With more complicated block structures, designs may no longer be generally balanced. Even for those that are, it is still important to know with respect to what

TABLE III
Properties of some incomplete-block designs

Design	Partial balance?	Efficient?	Optimal?	General balance decomposition
BIBD	✓	✓	✓	all contrasts
group divisible	✓	possibly	possibly	groups
triangular	✓	possibly	possibly	diallel, as in Example 1
cyclic	✓	possibly	possibly	the vectors $(1, \theta, \theta^2, \ldots, \theta^{n-1})$ where $\theta^n = 1$
factorial with classical confounding	✓	possibly	possibly	factorial, e.g. AB^2
square lattice	✓	✓	✓	pseudofactorial
rectangular lattice ($r = 2$)	×	✓	✓	mildly unpleasant, but rational[a]
rectangular lattice ($r \geq 3$)	not usually	✓	?	usually mildly unpleasant, but rational
α-designs[b]	×	possibly	possibly	usually irrational
resolvable 2-replicate designs from BIBDs[c]	×	✓	✓	mildly unpleasant
affine-resolvable	×	✓	✓[d]	sometimes factorial

[a] Bailey and Speed (1986).
[b] Patterson and Williams (1976).
[c] Williams, Patterson and John (1976, 1977).
[d] Bailey, Monod and Morgan (1993).

treatment decomposition they are generally balanced. For example, the design in Figure 1 is optimal among designs for 15 treatments in 6 rows by 10 columns. It is generally balanced with respect to the diallel structure with 6 parents, and so is suitable if the treatment structure is genuinely diallel. However, it is less suitable for a 3×5 factorial experiment. The best factorially balanced design is an Abelian group design with efficiency factors shown in Table IV: statisticians following the advice of Yates (1935) and Pearce (1983) will prefer this design to that in Figure 1, in spite of a 4% loss in overall efficiency. In fact, this is an excellent design if one is mainly interested in the interaction and the plots stratum has the smallest variance.

Another example is given by Bailey and Royle (1993). The optimal semi-Latin square for 12 treatments in blocks of size 2 is generally balanced with respect to an irrational decomposition of the treatment space; another semi-Latin square is generally balanced with respect to the 3×4 factorial structure, and only 1% less efficient. The second design would often be preferred in practice.

In fact, the whole issue of optimality for complicated block structures is not as simple as was once thought. Bailey (1993) shows that, whether or not information from different strata is combined, the current method of measuring optimality only

TABLE IV
Efficiency factors for the best factorially balanced Abelian group design for 3 × 5 treatments in 6 rows and 10 columns

Strata		T_0	T_A	T_B	T_{AB}
	dim	1	2	4	8
mean	1	1	0	0	0
rows	5	0	1/4	0	0
columns	9	0	0	3/8	0
plots	45	0	3/4	5/8	1

in the bottom stratum may not lead to the best designs.

For these structures, then, general balance may be more important than optimality. I agree with Mejza (1992) that it is very hard to understand either the structure of the design or the output from the analysis if the design is not generally balanced.

I am undecided on the computational issue. There does not seem to be universal agreement that REML plus GLS gives the same estimates as iterated Nelder. The estimates satisfy the same equations, but those equations may have more than one solution. Although ML and REML algorithms are notoriously slow to converge for non-orthogonal structures, they seem to perform well on structures like that in Figure 2. Variety trials in Australia and the U. K. are now being performed in that structure, using designs that may not be generally balanced. Does this matter?

My opinion is that, no matter what the block structure, general balance with respect to the pertinent treatment structure is always desirable. However, it would be absurd to insist on this if no such design exists, or if there are other designs for which the relevant contrasts can be estimated much more efficiently.

References

Bailey, R. A. (1981). A unified approach to design of experiments. *Journal of the Royal Statistical Society A* **144**, 214–223.

Bailey, R. A. (1985a). Balance, orthogonality and efficiency factors in factorial design. *Journal of the Royal Statistical Society B* **47**, 453–458.

Bailey, R. A. (1985b). Partially balanced designs. In: S. Kotz and N. L. Johnson, Eds., *Encyclopedia of Statistical Sciences, Vol. 6.* Wiley, New York, 593–610.

Bailey, R. A. (1991a). Cyclic designs and factorial designs. In: R. R. Bahadur, Ed., *Probability, Statistics and Design of Experiments.* Wiley Eastern, New Delhi, 51–74.

Bailey, R. A. (1991b). Strata for randomized experiments (with discussion). *Journal of the Royal Statistical Society B* **53**, 27–78.

Bailey, R. A. (1993). Recent advances in experimental design in agriculture. *Bulletin of the International Statistical Institute* **50**, 179–193.

Bailey, R. A., Monod, H. and Morgan, J. P. (1993). Construction and optimality of affine-resolvable designs. Submitted for publication.

Bailey, R. A. and Rowley, C. A. (1990). General balance and treatment permutations. *Linear Algebra and its Applications* **127**, 183–225.

Bailey, R. A. and Royle, G. (1993). Optimal semi-Latin squares with side six and block size two.

Submitted for publication.

Bailey, R. A. and Speed, T. P. (1986). Rectangular lattice designs: efficiency factors and analysis. *The Annals of Statistics* **14**, 874–895.

Bose, R. C. and Mesner, D. M. (1959). On linear associative algebras corresponding to association schemes of partially balanced designs. *The Annals of Mathematical Statistics* **30**, 21–38.

Caliński, T. (1993). The basic contrasts of a block experimental design with special reference to the notion of general balance. *Listy Biometryczne - Biometrical Letters* **30**, 13–38.

Caliński, T., Ceranka, B. and Mejza, S. (1980). On the notion of efficiency of a block design. In: W. Klonecki, A. Kozek and J. Rosiński, Eds., *Mathematical Statistics and Probability Theory. Lecture Notes in Statistics* **2**. Springer-Verlag, New York, 47–62.

Cheng, C.-S. and Bailey, R. A. (1991). Optimality of some two-associate-class partially balanced incomplete-block designs. *The Annals of Statistics* **19**, 1667–1671.

Corsten, L. C. A. (1976). Canonical correlation in incomplete blocks. In: S. Ikeda et al., Eds., *Essays in Probability and Statistics.* Shinko Tsusho, Tokyo, 125–154.

Corsten, L. C. A. (1985). Rectangular lattices revisited. In: T. Caliński and W. Klonecki, Eds., *Linear Statistical Inference. Lecture Notes in Statistics* **35**. Springer-Verlag, Berlin, 29–38.

Eccelston, J. A. and Russell, K. G. (1975). Connectedness and orthogonality in multi-factor designs. *Biometrika* **62**, 341–345.

Houtman, A. M. and Speed, T. P. (1983). Balance in designed experiments with orthogonal block structure. *The Annals of Statistics* **11**, 1069–1085.

James, A. T. and Wilkinson, G. N. (1971). Factorization of the residual operator and canonical decomposition of nonorthogonal factors in the analysis of variance. *Biometrika* **58**, 279–294.

John, J. A. and Smith, T. M. F. (1972). Two factor experiments in non-orthogonal designs. *Journal of the Royal Statistical Society B* **34**, 401–409.

Kobilinsky, A. (1990). Complex linear models and factorial designs. *Linear Algebra and its Applications* **127**, 227–282.

Lewis, S. M. and Dean, A. M. (1991). On general balance in row-column designs. *Biometrika* **78**, 595–600.

Martin, F. B. and Zyskind, G. (1966). On combinability of information from uncorrelated linear models by simple weighting. *The Annals of Mathematical Statistics* **37**, 1338–1347.

McLaren, A. D. (1963). On group representations and invariant stochastic processes. *Proceedings of the Cambridge Philosophical Society* **59**, 431–450.

Mejza, S. (1992). On some aspects of general balance in designed experiments. *Statistica* **52**, 263–278.

Monod, H. and Bailey, R. A. (1992). Pseudofactors: normal use to improve design and facilitate analysis. *Applied Statistics* **41**, 317–336.

Morley Jones, R. (1959). On a property of incomplete blocks. *Journal of the Royal Statistical Society B* **21**, 172–179.

Nelder, J. A. (1965a). The analysis of randomized experiments with orthogonal block structure. I. Block structure and the null analysis of variance. *Proceedings of the Royal Society of London A* **283**, 147–162.

Nelder, J. A. (1965b). The analysis of randomized experiments with orthogonal block structure. II. Treatment structure and the general analysis of variance. *Proceedings of the Royal Society of London A* **283**, 163–178.

Nelder, J. A. (1968). The combination of information in generally balanced designs. *Journal of the Royal Statistical Society B* **30**, 303–311.

Patterson, H. D. and Thompson, R. (1971). Recovery of inter-block information when block sizes are unequal. *Biometrika* **58**, 545–554.

Patterson, H. D. and Thompson, R. (1975). Maximum likelihood estimation of compenents of variance. In: L. C. A. Corsten and T. Postelnicu, Eds., *Proceedings of the 8th International Biometric Conference.* Editura Academiei Republicii Socialiste România, Bucureşti, 197–207.

Patterson, H. D. and Williams, E. R. (1976). A new class of resolvable incomplete block designs. *Biometrika* **63**, 83–92.

Pearce, S. C. (1983). *The Agricultural Field Experiment.* Wiley, Chichester.

Pearce, S. C., Caliński, T. and Marshall, T. F. de C. (1974). The basic contrasts of an experimental design with special reference to the analysis of data. *Biometrika* **61**, 449–460.

Speed, T. P. (1983). General balance. In: S. Kotz and N. L. Johnson, Eds., *Encyclopedia of Statistical Sciences, Vol. 3.* Wiley, New York, 320–326.

Speed, T. P. and Bailey, R. A. (1987). Factorial dispersion models. *International Statistical Review* **55**, 261–277.

Tjur, T. (1984). Analysis of variance models in orthogonal designs (with discussion). *International Statistical Review* **52**, 33–81.

Williams, E. R. (1977). A note on rectangular lattice designs. *Biometrics* **33**, 410–414.

Williams, E. R., Patterson, H. D. and John, J. A. (1976). Resolvable designs with two replications. *Journal of the Royal Statistical Society B* **38**, 296–301.

Williams, E. R., Patterson, H. D. and John, J. A. (1977). Efficient two-replicate resolvable designs. *Biometrics* **33**, 713–717.

Yates, F. (1935). Complex experiments. *Journal of the Royal Statistical Society, Supplement* **2**, 181–247.

OPTIMALITY OF GENERALLY BALANCED EXPERIMENTAL BLOCK DESIGNS

BARBARA BOGACKA and STANISŁAW MEJZA
Department of Mathematical and Statistical Methods
Agricultural University of Poznań
Wojska Polskiego 28
60-637 Poznań
Poland

Abstract. The subject of optimality of block designs under the mixed model has been undertaken in the eighties. Up to today there are not many papers considering the problem, contrary to the case of the fixed model. The papers of Bagchi (1987a,b), Mukhopdhyay (1981), Khatri and Shah (1984), Bhattacharya and Shah (1984) or Jacroux (1989) deal with the optimality of block designs under mixed model of a special simple kind.

Different ways of double randomization (depending on structure of experimental material) give more complicated mixed models. Many authors considered randomization models from the estimation and testing point of view, and they underlined the adequacy (by the nature of the problem) and applicability of the model to practical experiments (Fisher, 1926; Neyman, Iwaszkiewicz and Kołodziejczyk, 1935; Nelder, 1965; Ogawa and Ikeda, 1973; Caliński and Kageyama, 1991; Kala, 1991). We consider a class of block designs under such models having the property of general balance which makes considerations of optimality easier. We present conditions for a design to be optimal design in a general sense, i.e. optimal with respect to a wide class of criteria (considered formerly under the fixed model).

Key words: Block design, Mixed model, General balance, Optimality criteria.

1. The Model

Let $\mathcal{D}(v, b, k)$ denote a class of experimental designs such that v treatments are allocated to experimental units grouped into b blocks of size k. Vector of treatment replications $r = (r_1, \ldots, r_v)'$ fulfils the condition $bk = \sum_{i=1}^{v} r_i = n$, although the vector of replications can vary throughout the class $\mathcal{D}(v, b, k)$. We consider connected designs for which an n dimensional observable random vector y can be written as follows:

$$y = \Delta'\tau + D'\beta + \eta + e, \tag{1}$$

where τ denotes a v dimensional vector of treatment parameters, β stands for a b dimensional vector of random block effects, η is an n dimensional vector of random errors of experimental units and e is an n dimensional vector of random technical errors, Δ' and D' denote $(n \times v)$ and $(n \times b)$ dimensional design matrices for treatments and blocks, respectively, such that $\Delta D'$ is the incidence matrix, usually denoted by N. The expected value of the random vector y is $E(y) = \Delta'\tau$, and let us denote its dispersion matrix by V, i.e. $\mathrm{Cov}(y) = V$. We also assume that the vector y is normally distributed, i.e. $y \sim N_n(\Delta'\tau, V)$.

Such model with very general form of the matrix V has been lately derived by

T. Caliński and R. Kala (eds.),
Proceedings of the International Conference on Linear Statistical Inference LINSTAT '93, 185–194.

Caliński and Kageyama (1991) and by Kala (1991). This paper concerns some special cases of their matrix V. Different forms of the matrix V considered here are related to the different cases taken into account, depending on the size of the population of experimental units, on the way of grouping units, and on the randomization performed. However, all cases considered lead to the dispersion matrices which can be expressed in the same form (Bogacka, 1992),

$$V = \xi_1 S_1 + \xi_2 S_2 + \xi_3 S_3, \tag{2}$$

where the matrices S_i, $i = 1, 2, 3$, are orthogonal projection operators onto three subspaces of $\mathbb{R}^n$, so called strata. It means that $\oplus_i \mathcal{C}(S_i) = \mathbb{R}^n$, where $\oplus$ denotes a direct sum of subspaces, while $\mathcal{C}(A)$ is the column space of A. The matrices S_i are of the following forms:

$$S_1 = I_n - \frac{1}{k}D'D, \quad S_2 = \frac{1}{k}D'D - \frac{1}{n}1_n 1'_n, \quad S_3 = \frac{1}{n}1_n 1'_n,$$

where 1_n denotes the n dimensional vector of ones, and I_n is the identity matrix of order n. The coefficients ξ_i, $i = 1, 2, 3$, denote the strata variances, various in different cases of V. In the intrablock, interblock and total-area stratum, respectively, statistical analysis can be easily carried out (see Houtman and Speed, 1983; Bailey, 1991).

2. General Balance of the Designs

Let us decompose the treatment space $\mathcal{T} = \mathcal{C}(\Delta')$ into v orthogonal subspaces $\mathcal{T}_j = \mathcal{C}(\Delta' s_j)$, where s_j are eigenvectors of the matrix $C = \Delta S_1 \Delta'$ connected with the eigenvalues ϵ_j of C with respect to $r^\delta = \Delta\Delta' = \text{diag}(r_1, \ldots, r_v)$, i.e. vectors s_j, $j = 1, \ldots, v$, which fulfil the conditions:

$$C s_j = \epsilon_j r^\delta s_j, \quad s'_i r^\delta s_j = \delta_{ij}, \tag{3}$$

where δ_{ij} is the Kronecker delta. There is a strict connection between the structure of the dispersion matrix V in (2) and the treatment structure.

Lemma 1 (Caliński, 1993). *An experimental block design under model* (1) *with dispersion matrix* (2) *is generally balanced with respect to the decomposition of the treatment space* $\mathcal{T} = \oplus_j \mathcal{T}_j$.

Lemma 1 means that the following equalities hold:

$$TS_1T = \sum_{j=1}^{v-1} \epsilon_j T_j, \quad TS_2T = \sum_{j=1}^{v-1} (1 - \epsilon_j) T_j, \quad TS_3T = T_v, \tag{4}$$

where $T = \Delta' r^{-\delta} \Delta$, $T_j = \Delta' s_j s'_j \Delta$, $j = 1, \ldots, v-1$, $T_v = S_3$.

The property of general balance was introduced by Nelder (1965) and then considered by many authors, for instance: Bailey (1981, 1991), Houtman and Speed (1983), Payne and Tobias (1992), Mejza (1992). This property makes easier both the estimation of treatment parametric functions and of the corresponding variances, and the considerations on optimality of designs.

3. The Information Matrix of the Design

We are interested in optimality of designs from the efficiency of estimation of treatment parametric functions point of view. So, as the argument of optimality criteria we take the information matrix for estimating τ. For model (1) this matrix is given by $M = \Delta V^{-1}\Delta'$. Because of the simple form of the inverse of the matrix V, $V^{-1} = \sum_{i=1}^{3} \xi_i^{-1} S_i$, we can express matrix M as

$$M = \sum_{i=1}^{3} \xi_i^{-1} C_i, \tag{5}$$

where $C_i = \Delta S_i \Delta'$ $(C_1 = C)$ are the strata information matrices corresponding to submodels:

$$y_i = S_i y, \quad E(y_i) = S_i \Delta' \tau, \quad \mathrm{Cov}(y_i) = \xi_i S_i, \quad i = 1, 2, 3.$$

Vectors $y_i \in \mathcal{C}(S_i)$, $i = 1, 2, 3$, belong to the intrablock, interblock and total-area strata, respectively.

In considerations of optimality of designs the eigenvalues of information matrices play a principal role. Many of the optimality criteria are based on the eigenvalues.

Using equalities (4), (5), and the fact that $\Delta = \Delta T$, we obtain the spectral decomposition of M,

$$M = r^{\delta} \sum_{j=1}^{v} \nu_j s_j s_j' r^{\delta},$$

where the eigenvalues ν_j of M with respect to r^{δ} are of the form

$$\nu_j = \frac{\epsilon_j \xi_2 + (1 - \epsilon_j)\xi_1}{\xi_1 \xi_2}, \quad j = 1, \ldots, v - 1, \quad \nu_v = 1/\xi_3,$$

while $s_j, j = 1, \ldots, v$, are the eigenvectors defined in (3).

The eigenvalues ν_j, $j = 1, \ldots, v$, depend on the strata variances ξ_1, ξ_2 and ξ_3, and on the eigenvalues ϵ_j of the intrablock information matrix C with respect to r^{δ}. The strata variances ξ_1, ξ_2 and ξ_3 are, in fact, functions of the so called variance components which appear in the model because of randomization of blocks, of experimental units and because of technical errors. They are strictly connected with the structure of the experimental material and can depend on the number of blocks used in the experiment and on the block size (see Kala, 1991). The variance components do not depend on the way the treatments are assigned to units, that means, they do not depend on a design applied to the given experimental material provided the units are properly randomized and the unit-treatment additivity holds. Hence, in the class of designs $\mathcal{D}(b, k)$, with fixed parameters b and k, the variance components are also fixed, although they are unknown, and so are the strata variances $\xi_i, i = 1, 2, 3$. In the sequel, the coefficients ξ_i in (5) are considered as unknown positive constants.

4. Definitions of Design Opimality

In a mixed model case some optimality criteria can be defined analogously to a fixed model case. The main difference lies in the form of the information matrix and its eigenvalues taken into considerations. In the fixed model it is just the matrix C which does not depend on any unknown parameters, while in the mixed model considered here, the information matrix (5) is a linear combination of the strata information matrices with unknown coefficients ξ_i^{-1}. Moreover, in the fixed model case the ordinary eigenvalues of the information matrix are usually considered, while in the case of mixed generally balanced designs it is advantageous to use the eigenvalues of the information matrix with respect to the matrix r^δ. In both cases the general forms of the definitions of optimality criteria remain the same, although the meaning of the information matrix and its eigenvalues are different.

In this paragraph we do not distinguish between these cases, denoting the information matrix of a design $d \in \mathcal{D}(v, b, k)$ by Λ_d. The eigenvalues of Λ_d, denoted by $\lambda_j(\Lambda_d)$, can be the ordinary ones or can be taken with respect to some positive definite matrix, for instance r_d^δ. In next paragraphs we put C_d or M_d in place of Λ_d and $\epsilon_j(C_d)$ or $\nu_j(M_d)$ in place of $\lambda_j(\Lambda_d)$, respectively. Moreover, let $\mathcal{B} = \{\Lambda \geq 0\}$ denote a set of nonnegative definite or positive definite matrices, let $\mathcal{M} = \{\Lambda_d : d \in \mathcal{D}\}$ stand for a set of information matrices for designs d belonging to a class $\mathcal{D}$, such that $\mathcal{M} \subseteq \mathcal{B}$. Furthermore, let $\lambda(\Lambda) = (\lambda_1(\Lambda), \ldots, \lambda_m(\Lambda))'$, $0 < m \leq v$, be the vector of positive eigenvalues of matrix Λ, let $\lambda(\mathcal{B}) = \{\lambda(\Lambda) : \Lambda \in \mathcal{B}\}$, and let

$$\lambda^\delta(\Lambda) = \operatorname{diag}(\lambda_1(\Lambda), \ldots, \lambda_m(\Lambda)), \quad \lambda_\square^\delta(\Lambda) = \operatorname{diag}(\lambda_{[1]}(\Lambda), \ldots, \lambda_{[m]}(\Lambda)),$$

where $\lambda_{[j]}(\Lambda)$ is the j-th largest eigenvalue of Λ.

Given the function $\Phi : \mathcal{B} \mapsto \mathbb{R}$ (optimality criterion), a design $d^* \in \mathcal{D}$ is Φ-optimal in the class $\mathcal{D}$ if $\Phi(\Lambda_{d^*}) \leq \Phi(\Lambda_d)$ for all $d \in \mathcal{D}$.

We are interested in two general definitions of optimality comprising very wide classes of such criteria. The definitions were introduced in the fixed model case by Cheng (1978) and Bondar (1983).

Definition 1 (Cheng, 1978). A design $d^* \in \mathcal{D}$ is termed generally optimal in the class $\mathcal{D}$ if Λ_{d^*} minimizes in $\mathcal{M}$ every $\Phi : \mathcal{B} \mapsto \mathbb{R}$ of the form

$$\Phi(\Lambda) = \phi_f(\lambda(\Lambda)) = \sum_{j=1}^{m} f(\lambda_j(\Lambda)),$$

where the function $f : [0, x_0] \mapsto \mathbb{R}$, $x_0 = \max_{\Lambda \in \mathcal{M}} \operatorname{tr}\Lambda$, satisfies the following conditions:

(1) f is continuously differentiable on $(0, x_0)$,

(2) first, second and third derivatives of f satisfy, respectively: $f' < 0$, $f'' > 0$, $f''' < 0$ on $(0, x_0)$,

(3) $f(0) = \lim_{x \to 0+} f(x) = \infty$.

The class of functions Φ used by Cheng includes A- and D-optimality criteria.

Namely, taking $f_1(x) = x^{-1}$ and $f_2(x) = -\log x$ we obtain:

$$\phi_{f_1}(\lambda(\Lambda)) = \sum_{j=1}^{m} \lambda_j^{-1}(\Lambda), \qquad \phi_{f_2}(\lambda(\Lambda)) = \log \prod_{j=1}^{m} \lambda_j^{-1}(\Lambda).$$

These two functionals are also special cases of so called Φ_p criteria introduced by Kiefer (1974),

$$\Phi_p(\Lambda) = \left[\frac{1}{m}\sum_{j=1}^{m} \lambda_j^{-p}(\Lambda)\right]^{\frac{1}{p}}, \quad p \in (0, \infty).$$

Φ_p functionals are analogous to j_p-information functionals being the special cases of the j-information functional defined by Pukelsheim (1980) (see also Pukelsheim, 1983).

Bondar (1983) has also considered an interesting class of optimality criteria. He called the designs fulfilling conditions of his definition, the universally optimal designs. This name, however, was earlier used by Kiefer (1975) whose definition of optimality is not equivalent to that of Bondar. So, we propose to term the designs optimal with respect to Bondar's definition, the B-universally optimal designs (BU-optimal, for short).

Definition 2 (Bondar, 1983). A design $d^\star \in \mathcal{D}$ is termed BU-optimal in the class $\mathcal{D}$ if $\Lambda_{d^\star}$ minimizes in $\mathcal{M}$ every $\Phi : \mathcal{B} \mapsto \mathbb{R}$ satisfying the following conditions:

(a) $\Phi(\Lambda) = \phi(\lambda(\Lambda))$, i.e. Φ is a function defined on the eigenvalues of $\Lambda \in \mathcal{B}$,

(b) $\phi(\lambda(\Lambda))$ is a Schur-convex function on the set $\lambda(\mathcal{B})$, i.e. $\lambda(\Lambda_1) \prec \lambda(\Lambda_2) \Rightarrow \phi(\Lambda_1) \leq \phi(\Lambda_2)$, where the symbol $\prec$ denotes majorization of $\lambda(\Lambda_1)$ by $\lambda(\Lambda_2)$,

(c) $\lambda_{\Box}^{\delta}(\Lambda_1) \geq_L \lambda_{\Box}^{\delta}(\Lambda_2) \Rightarrow \Phi(\lambda_{\Box}^{\delta}(\Lambda_1)) \leq \Phi(\lambda_{\Box}^{\delta}(\Lambda_2))$, where $\geq_L$ denotes the Loewner ordering of matrices.

Denoting by $\mathcal{F}_C$ and $\mathcal{F}_B$ the classes of functionals fulfilling the conditions of Definition 1 and Definition 2, respectively, we have that $\mathcal{F}_C \subseteq \mathcal{F}_B$ (see Bogacka, 1992).

5. Results

Cheng's considerations were inspired by Kiefer's (1975) theorem saying that the design for which the information matrix is completely symmetric and its trace is maximal in $\mathcal{M}$, is universally optimal. Balanced Block Designs (BBD) satisfy such conditions, but not in every class $\mathcal{D}(v, b, k)$ do such designs exist. Cheng defined another class of criteria (Definition 1), included in the class defined by Kiefer (Bogacka, 1992), and proved a theorem presenting conditions the design has to fulfil to be optimal with respect to this class of criteria in case of no BBD existing in $\mathcal{D}(v, b, k)$.

The following theorem is based on the Cheng's result which was also examined with great care by Shah and Sinha (1989) in their book on optimality.

Theorem 1. *Let $\overline{\mathcal{D}}(v,b,k) \subseteq \mathcal{D}(v,b,k)$ be a class of designs d such that $p(C_d) = \left\{\sum_{j=1}^{v-1}\left[\epsilon_j(C_d) - \bar{\epsilon}\,(C_d)\right]^2\right\}^{\frac{1}{2}} > 0$. Let $d^\star \in \overline{\mathcal{D}}(v,b,k)$ be an experimental design under mixed model* (1) *for which the matrix $C_{d^\star}$ has two distinct positive eigenvalues with respect to $r_{d^\star}^{\delta}$, a greater one with multiplicity* 1 *and a smaller one with multiplicity $v-2$. Furthermore, let:*

$$\operatorname{tr}(r_{d^\star}^{-\delta} C_{d^\star}) = \max_{d\in\mathcal{D}} \operatorname{tr}(r_d^{-\delta} C_d), \tag{6}$$

$$\operatorname{tr}(r_{d^\star}^{-\delta} C_{d^\star}) - \gamma\, p(C_{d^\star}) \;\geq\; \operatorname{tr}(r_d^{-\delta} C_d) - \gamma\, p(C_d) \quad \textit{for all } d \in \mathcal{D}, \tag{7}$$

where $\gamma^2 = (v-1)/(v-2)$ and $\bar{\epsilon}\,(C_d)$ denotes the average of the eigenvalues $\epsilon_j(C_d)$. Then the design $d^\star$ minimizes every function

$$\phi_f(\nu(M_d)) = \sum_{j=1}^{v} f(\nu_j(M_d))$$

fulfilling the conditions of Cheng's definition.

Proof. Let us write the eigenvalues ν_j in the form

$$\nu_j = \epsilon_j(z_1 - z_2) + z_2, \tag{8}$$

$$\nu_v = z_3, \tag{9}$$

where $z_i = 1/\xi_i$, $i = 1, 2, 3$. Then, treating z_i as unknown constants, the proof of the theorem follows from Theorem 2.3.2 presented in the book of Shah and Sinha (1989). □

Corollary 1. *Let $\mathcal{D}_0(v,b,k) \subseteq \overline{\mathcal{D}}(v,b,k)$ denote a class of binary incomplete block designs under mixed model* (1). *Let $d^\star \in \mathcal{D}_0(v,b,k)$ be an experimental design for which the matrix $C_{d^\star}$ has two distinct positive eigenvalues with respect to $r_{d^\star}^{\delta}$, a greater one with multiplicity* 1 *and a smaller one with multiplicity $v-2$. Then the design $d^\star$ is generally optimal in $\mathcal{D}_0$ if it is MS-optimal.*

Proof. It suffices to check assumptions (6) and (7) of Theorem 1. Since $\operatorname{tr}(r_d^{-\delta} C_d)$ is a constant value for every $d \in \mathcal{D}_0$, the condition (6) is fulfilled. The condition (7) reduces now to the inequality $p(C_{d^\star}) \leq p(C_d)$ for all $d \in \mathcal{D}_0$, which, in the class $\mathcal{D}_0$, is equivalent to the MS-optimality criterion,

$$\sum_{j=1}^{v-1} \epsilon_j^2(C_{d^\star}) \leq \sum_{j=1}^{v-1} \epsilon_j^2(C_d), \quad \text{for all } d \in \mathcal{D}_0. \; \square$$

On the basis of Cheng's (1978) considerations and Theorem 1 we can say that some of Extreme Regular Graph Designs (ERGD) of type 1 and Most Balanced

Group Divisible PBB Designs (MBGDPBBD) of type 1 are generally optimal in the class $\mathcal{D}(v,b,k,r)$ in case of the mixed models considered here. Definitions of the designs can be found in Cheng (1978).

The conditions for the BU-optimality are given in the following

Theorem 2. *An experimental design $d^\star \in \mathcal{D}(v,b,k)$ under mixed model* (1) *is BU-optimal in the class $\mathcal{D}$ if and only if for all $d \in \mathcal{D}$*

$$\sum_{j=k}^{v-1} \epsilon_{[j]}(C_d) \leq \sum_{j=k}^{v-1} \epsilon_{[j]}(C_{d^\star}), \quad k = 1,\ldots,v-1, \tag{10}$$

where $\epsilon_{[j]}(C_d)$ is the j-th largest positive eigenvalue of the matrix C_d corresponding to the design d with respect to r_d^δ.

Proof. The proof follows from Bondar's (1983) Theorem 2.1 when we take into account the eigenvalues ν_j of M_d with respect to r_d^δ of the form (8) and (9), and we treat the unknown values z_i as fixed. □

Corollary 2. *If in a class $\mathcal{D}_1 = \{d \in \mathcal{D}(v,b,k) : \operatorname{tr}(r_d^{-\delta} M_d) = const.\}$ there exists a design $d^\star$ such that the eigenvalues of $C_{d^\star}$ with respect to $r_{d^\star}^\delta$ are all equal, then, under mixed model* (1), *the design $d^\star$ is BU-optimal in the class $\mathcal{D}_1$.*

Proof. Let us denote by $\epsilon(C_d)$ the vector of positive eigenvalues of C_d with respect to r_d^δ. Condition (10) of Theorem 2 can be written using the notion of weak majorization (Marshall and Olkin, 1979),

$$\epsilon(C_{d^\star}) \prec^w \epsilon(C_d), \quad \text{for all } \; d \in \mathcal{D}.$$

Furthermore, taking into account the form (8) and (9) of the eigenvalues $\nu_j, j = 1,\ldots,v$, we have

$$\operatorname{tr}(r_d^{-\delta} M_d) = \sum_{j=1}^{v} \nu_j(M_d) = (z_1 - z_2)\sum_{j=1}^{v-1} \epsilon_j(C_d) + (v-1)z_2 + z_3.$$

Thus, the sum $\sum_{j=1}^{v-1} \epsilon_j(C_d)$ is fixed for every $d \in \mathcal{D}_1$. Hence, the vector $\epsilon(C_{d^\star})$ is majorized by the vector $\epsilon(C_d)$, i.e. $\epsilon(C_{d^\star}) \prec \epsilon(C_d)$ for all $d \in \mathcal{D}_1$. Then, as a Schur-convex function attains its minimum at a point with equal coordinates (Marshall and Olkin, 1979), the vector $\epsilon(C_{d^\star})$ must be of that kind. □

Corollary 3. *If in the class $\mathcal{D}_1$ there exists an $r_d^{-\delta}$-balanced design* (*Caliński*, 1977) *then, under mixed model* (1), *it is BU-optimal in $\mathcal{D}_1$. Balanced Block Design* (BBD) *can be an example.*

Corollary 4. *If in the class $\mathcal{D}_0$ there exists Balanced Incomplete Block Design* (BIBD) *then, under mixed model* (1), *it is BU-optimal design in $\mathcal{D}_0$.*

TABLE I
Review of the literature

Optimality criterion	Design d^*	Class, in which d^* is optimal	AUTHORS
Universal Kiefer (1975)	BBD	$\mathcal{D}(v,b,k)$	Mukhopadhyay (1981) Khatri and Shah (1984)
	the design dual to BBD	$\mathcal{D}(v,b,k)$	Khatri and Shah (1984)
General Cheng (1978)	MBGDPBIBD(2) $\lambda_2 = \lambda_1 + 1$	$\mathcal{D}(v,b,k)$	Bagchi (1987b)
	ERGD of type 1	$\mathcal{D}_2(v,b,k,r)$ $\lambda_1 > 0$ or $\lambda_1 = 0, n_1 < v/2$	Khatri and Shah (1984)
		$\mathcal{D}_3 = \{d \in \mathcal{D} : \text{tr}RC$ $\geq \text{tr}R\,\text{tr}C/(v-1)\}$ $R = r^\delta - rr'/n$	Bhattacharya and Shah (1984)
	the design dual to BBD	$\mathcal{D}_2(v,b,k,r)$	Khatri and Shah (1984)
Φ_p, $p \in (0,\infty)$	the design dual to Φ_p -opt. design	$\mathcal{D}_2(v,b,k,r)$	Mukhopadhyay (1981)
E	(i) MBGDPBBD(2) $\lambda_2 = \lambda_1 + 1$, (ii) MBGDPBBD(2) $\lambda_2 = \lambda_1 - 1$, (iii) the design dual to BBD (iv) dual to (i), (ii)	$\mathcal{D}(v,b,k)$ $k < v$ or $k > v$	Bagchi (1987a)
	GDPBBD $\lambda_2 = \lambda_1 + 2$	$\mathcal{D}(v,b,k)$ group size $= v/2$	Jacroux (1989)

The advantage of the theorems and corollaries presented above is that the assumptions deal with the eigenvalues ϵ_j and the matrix C while the conclusions concern optimality of designs under mixed models. That means the conditions do not depend on the strata variances.

Investigation of some particular criteria, like A- or D-optimality, independently of strata variances is very difficult or even impossible. So, the results obtained for wide classes of criteria (including these particular ones) are especially valuable. Whereas, MS-optimality and E-optimality of block designs under the model (1) can be considered on the basis of the eigenvalues ϵ_j of the matrix C with respect to r^δ. It follows from the form of the functionals Φ_{MS} and Φ_E and from (8) and (9).

A few authors considered optimality of block designs under mixed models. We present the condensed review of some of their results in Table 1. They considered: universal optimality, general optimality, Φ_p-optimality, and E-optimality. All these authors considered the case of infinite population of units divided into infinite number of groups of infinite size, from which a random sample of b groups and a random sample of k units from every chosen group were taken to the experiment. That means, they considered a special case of the model (1).

References

Bagchi, S. (1987a). On the E-optimality of certain asymetrical designs under mixed effects model. *Metrika* **37**, 95–105.

Bagchi, S. (1987b). On the optimalities of the MBGDD's under mixed effects model. Report, Computer Science Unit, Indian Statistical Institute, Calcuta, India.

Bailey, R.A. (1981). A unified approach to design of experiments. *Journal of the Royal Statistical Society A* **144**, 214–223.

Bailey, R.A. (1991). Strata for randomized experiments (with discussion). *Journal of the Royal Statistical Society B* **53**, 27–78.

Bhattacharya, C. G. and Shah, K. R. (1984). On the optimality of block designs under a mixed effects model. *Utilitas Mathematica* **26**, 339–345.

Bogacka, B. (1992). Optimality of experimental block designs under mixed models. Ph.D. thesis. Agricultural University of Poznań, Poland, (in Polish).

Bondar, J. V. (1983). Universal optimality of experimental designs: definitions and a criterion. *The Canadian Journal of Statistics* **11**, 325–331.

Caliński, T. (1977). On the notion of balance in block designs. In: J. R. Barra, F. Brodeau, G. Romier and B. van Cutsem, Eds., *Recent Developments in Statistics*. North-Holland, Amsterdam, 365–374.

Caliński, T. (1993). The basic contrasts of a block experimental design with special references to the notion of general balance. *Listy Biometryczne - Biometrical Letters* **30**, 13–38.

Caliński, T. and Kageyama, S. (1991). On the randomization theory of intra-block and inter-block analysis. *Listy Biometryczne - Biometrical Letters* **28**, 97–122.

Cheng, C. S. (1978). Optimality of certain asymetrical experimental designs. *The Annals of Statistics* **6**, 1239–1261.

Fisher, R.A. (1926). The arrangement of field experiments. *Journal of Ministry of Agriculture* **23**, 503–513.

Houtman, A. M. and Speed, T. P. (1983). Balance in designed experiments with orthogonal block structure. *The Annals of Statistics* **11**, 1069–1085.

Jacroux, M. (1989). On the E-optimality of block designs under the assumption of random block effects. *Sankhyā B* **51**, 1–12.

Kala, R. (1991). Elements of the randomization theory. III. Randomization in block experiments. *Listy Biometryczne - Biometrical Letters* **28**, 3–23, (in Polish).

Khatri, C.G. and Shah, K.R. (1984). Optimality of block designs. In: *Proceedings of Indian Statistical Institute Golden Jubilee International Conference on Statistics: Applications and New Directions.* Calcutta, 326–332.

Kiefer, J. (1974). General equivalence theory for optimum designs (Approximate theory). *The Annals of Statistics* **2**, 849–879.

Kiefer, J. (1975). Construction and optimality of generalized Youden designs. In: J. N. Srivastava, Ed., *A Survey of Statistical Design and Linear Models.* North-Holland, Amsterdam, 333–353.

Marshall, A. W. and Olkin, I. (1979). *Inequalities: Theory of Majorization and Its Applications.* Academic Press, New York.

Mejza, S. (1992). On some aspects of general balance in designed experiments. *Statistica* **2**, 263–278.

Mukhopadhyay, S. (1981). On the optimality of block designs under mixed effects model. *Calcutta Statistical Association Bulletin* **30**, 171–185.

Nelder, J.A. (1965). The analysis of randomized experiments with orthogonal block structure. I. Block structure and the null analysis of variance. *Proceedings of the Royal Society of London A* **283**, 147–178.

Neyman, J., Iwaszkiewicz, K. and Kołodziejczyk, S. (1935). Statistical problems in agricultural experimentation (with discussion). *Journal of the Royal Statistical Society, Supplement* **2**, 107-180.

Ogawa, J. and Ikeda, S. (1973). On the randomization of block designs. In: J. N. Srivastava, Ed., *A Survey of Combinatorial Theory*. North-Holland, Amsterdam, 335–347.

Payne, R.W. and Tobias, R.D. (1992). General balance, combination of information and the analysis of covariance. *Scandinavian Journal of Statistics*

Pukelsheim, F. (1980). On linear regression designs which maximize information. *Journal of Statistical Planning and Inference* **4**, 339–364.

Pukelsheim, F. (1983). On optimality properties of simple block designs in the approximate design theory. *Journal of Statistical Planning and Inference* **8**, 193–208.

Shah, K. R. and Sinha, B. K. (1989). *Theory of Optimal Designs. Lecture Notes in Statistics* **54**. Springer-Verlag, New York.

OPTIMALITY OF THE ORTHOGONAL BLOCK DESIGN FOR ROBUST ESTIMATION UNDER MIXED MODELS

ROMAN ZMYŚLONY*
Institute of Mathematics of the Higher College of Engineering
Podgórna 50
65-246 Zielona Góra
Poland

and

STEFAN ZONTEK
Institute of Mathematics of the Polish Academy of Sciences
Kopernika 18
51-617 Wrocław
Poland

Abstract. In the paper a usual block design with treatment effects fixed and block effects random is considered. Comparison of experiments is based on the asymptotic covariance matrix of a robust estimator both for shift and scale parameters given by Fisher consistent and Fréchet differentiable functional (recently proposed by Bednarski and Zontek, 1994). In the class of equiblock-sized designs A- and D-optimality are discussed.

Key words: Two-way classification mixed model, Robust estimator, A-optimality, D-optimality.

1. Introduction

Consider an experiment with v treatments arranged in b blocks according to the $v \times b$ incidence matrix $N = (n_{ij})$. Thus the single observation y_{ijl} for $i = 1, \ldots, v$, $j = 1, \ldots, b$ and $l = 1, \ldots, n_{ij}$ is given by

$$y_{ijl} = \mu_i + \lambda_j + e_{ijl}, \tag{1}$$

where $\mu_1, \ldots, \mu_v$ are fixed effects (shift parameters), $\lambda_1, \ldots, \lambda_b$ are independent random effects normally distributed $N(0, \sigma_\lambda^2)$, while $e_{111}, \ldots, e_{vbn_{vb}}$ are independent random errors with $N(0, \sigma_e^2)$ distribution. Moreover, we assume that λ's and e's are independent.

For the model (1) there are different approaches to the problem of comparison of experiments. As one of them, the covariance matrix of the best linear unbiased estimator (BLUE) of shift parameters in an intra-block model is used (see Gaffke and Krafft, 1980; Christof and Pukelsheim, 1985). For the other approach only designs for which there exists BLUE of shift parameters are considered (see Kageyama and Zmyślony, 1993). For both cases the orthogonal designs are optimal among equiblock-sized designs.

* Also affiliated to the Institute of Mathematics of the Polish Academy of Sciences, Kopernika 18, 51-617 Wrocław, Poland

T. Caliński and R. Kala (eds.),
Proceedings of the International Conference on Linear Statistical Inference LINSTAT '93, 195–202.

In next sections we shall discuss A- and D-optimality of designs for estimation of shift and scale parameters on the base of the asymptotic covariance matrix of robust estimators proposed by Bednarski and Zontek (1994). The resulting estimators are consistent and asymptotically normal. It is shown that the orthogonal design is also asymptotically A- and D-optimal in the class of equiblock-sized designs both for shift and scale parameters. Since robust estimators are here considered, optimal properties of the orthogonal design are valid for small departure from the model distribution.

Throughout the paper w' stands for the transpose of a vector w and $\text{diag}(w)$ for a diagonal matrix with ith diagonal element equal to the ith coordinate of w. The a dimentional vector of 1's is denoted by $\mathbf{1}_a$. For any $n \times n$ matrix M, the symbol $M \geq 0$ means that M is non-negative definite, $\det M$ denotes the determinant of M and $\operatorname{tr} M$ the trace of M. A block diagonal matrix with blocks B_1 and B_2 is written as $\text{diag}(B_1, B_2)$. Finally, let $\mathbb{R}_+^2 = \{(\sigma_1, \sigma_2)' : \sigma_1 > 0, \ \sigma_2 > 0\}$.

2. Asymptotic Covariance Matrix of Robust Estimators

Optimal properties of experimental designs are derived by using an asymptotic covariance matrix (see Kurotschka and Müller, 1992) of a robust estimator for $\theta = (\mu', \sigma')'$, where $\mu = (\mu_1, \ldots, \mu_v)', \sigma = (\sigma_\lambda, \sigma_e)'$, proposed by Bednarski and Zontek (1994). The estimator is given by Fischer consistent and Fre̒chet differentiable functional. This minimizes a properly modified log-likelihood function. Proposed estimators are asymptotically normal under whole infinitesimal model given by the supremum norm. For details see Bednarski and Zontek (1994). To describe an asymptotic covariance matrix of the robust estimator we need some notations. Let $N_1, \ldots, N_p$ stand for different columns of the incidence matrix N, and let $b_i, i = 1, \ldots, p$, be the number of replications of N_i with $\sum_{i=1}^p b_i = b$. Then without loss of generality we can assume that the incidence matrix N has the following form

$$N = (N_1 \mathbf{1}'_{b_1}, \ldots, N_p \mathbf{1}'_{b_p}).$$

Under the assumption $\lim_{b\to\infty}(b_i/b) = q_i > 0$, for $i = 1, \ldots, p$, an asymptotic covariance matrix of the robust estimator is a function of σ and depends on $\widetilde{N} = (N_1, \ldots, N_p)$ and $q = (q_1, \ldots, q_p)'$, which is given by

$$\Sigma(\widetilde{N}, q, \sigma) = \text{diag}(w_1 V_1(\widetilde{N}, q, \sigma), \ w_2 V_2(\widetilde{N}, q, \sigma)), \tag{2}$$

where, for $j = 1, 2$,

$$V_j(\widetilde{N}, q, \sigma) = \left[\sum_{i=1}^p M_i^{(j)}(\sigma)\right]^{-1} \left[\sum_{i=1}^p \frac{1}{q_i} M_i^{(j)}(\sigma)\right] \left[\sum_{i=1}^p M_i^{(j)}(\sigma)\right]^{-1}, \tag{3}$$

while

$$M_i^{(1)}(\sigma) = \frac{1}{\sigma_e^2} \sum_{i=1}^p \left(\text{diag}(N_i) - \frac{\sigma_\lambda^2}{n_{\cdot i}\sigma_\lambda^2 + \sigma_e^2} N_i N_i'\right), \tag{4}$$

$$M_i^{(2)}(\sigma) = 2\begin{bmatrix} \frac{n_{\cdot i}^2\sigma_\lambda^2}{(n_{\cdot i}\sigma_\lambda^2+\sigma_e^2)^2} & \frac{n_{\cdot i}\sigma_\lambda\sigma_e}{(n_{\cdot i}\sigma_\lambda^2+\sigma_e^2)^2} \\ \frac{n_{\cdot i}\sigma_\lambda\sigma_e}{(n_{\cdot i}\sigma_\lambda^2+\sigma_e^2)^2} & \frac{\sigma_e^2}{(n_{\cdot i}\sigma_\lambda^2+\sigma_e^2)^2} + \frac{n_{\cdot i}-1}{\sigma_e^2} \end{bmatrix}. \tag{5}$$

Here $n_{\cdot i}$ are block sizes, i.e. $n_{\cdot i} = N_i'\mathbf{1}_v$, and w_1, w_2 in (2) are constants depending on a choosen functional defining the robust estimator.

Note that $\mathrm{diag}(V_1(\widetilde{N}, q, \sigma), V_2(\widetilde{N}, q, \sigma))$ is the asymptotic covariance matrix of a maximum likelihood estimator of θ.

3. Main Results

Throughout this section let $\widetilde{q} = (b_1/b, \ldots, b_p/b)'$ and assume that block sizes are constant, say k. For fixed b we will identify an experimental design with pair $(\widetilde{N}, \widetilde{q})$. The exact covariance matrix of a given robust estimator cannot be calculated. Therefore, we approximate it by $(1/b)\Sigma(\widetilde{N}, \widetilde{q}, \sigma)$ and we utilize this matrix to introduce two criteria of optimality of experimental designs for robust estimation of parameters μ and σ.

Definition 1. An experimental design $(\widetilde{N}, \widetilde{q})$ in a class $\mathcal{E}$ of experimental designs is called asymptotically A_σ (D_σ) optimal in $\mathcal{E}$, if $\mathrm{tr}\, V_2(\widetilde{N}, \widetilde{q}, \sigma)$ $(\det V_2(\widetilde{N}, \widetilde{q}, \sigma))$ is minimum in $\mathcal{E}$ for every $\sigma \in \mathbb{R}_+^2$.

Definition 2. An experimental design $(\widetilde{N}, \widetilde{q})$ in a class $\mathcal{E}$ of experimental designs is called asymptotically A_μ (D_μ) optimal in $\mathcal{E}$, if $\mathrm{tr}\, V_1(\widetilde{N}, \widetilde{q}, \sigma)$ $(\det V_1(\widetilde{N}, \widetilde{q}, \sigma))$ is minimum in $\mathcal{E}$ for every $\sigma \in \mathbb{R}_+^2$.

Let us define three classes of experimental designs, namely, $\mathcal{E}_o$ consisting of all equiblock-sized experiments with fixed v, b and k, $\mathcal{E}_1$ being the subset of $\mathcal{E}_o$ with $\widetilde{q} = (1/p)\mathbf{1}_p$, and finally $\mathcal{E}_2$ being the subset of $\mathcal{E}_1$ with $p = 1$.

First we find a class of experimental designs, which are asymptotically A_σ and D_σ optimal in $\mathcal{E}_o$. Next we indicate an experimental design, which is asymptotically A_μ and D_μ optimal in the class of experimental designs, which are asymptotically A_σ and D_σ optimal in $\mathcal{E}_o$. In this sense the resulting experimental design is optimal for estimation of shift and scale parameters.

Christof and Pukelsheim (1985) and Kageyama and Zmyślony (1993) considered optimality of designs only for estimation of shift parameters. Christof and Pukelsheim (1985) proposed simplification of the problem by considering the model induced by block totals insted of the original model. This in turn implies that the covariance matrix of the BLUE in the intra-block model corresponding to experimental designs from $\mathcal{E}_o$ is known up to an unknown factor. Kageyama and Zmyślony (1993) considered the original model but the optimality of the orthogonal design was obtained in the class $\mathcal{E}_2$, for which there exists a BLUE of μ. These are the main differences between their and our approaches.

Lemma 1. *For any experimental design $(\widetilde{N}, \widetilde{q})$ in $\mathcal{E}_o$ it holds that*

$$V_2(\widetilde{N}, \widetilde{q}, \sigma) - V_2(\widetilde{N}, (1/p)\mathbf{1}_p, \sigma) \geq 0$$

for every $\sigma \in \mathbb{R}^2_+$.

Proof. Note that since $n_{\cdot i} = k$ for $i = 1, \ldots, p$, then matrices $M_i^{(2)}(\sigma)$ defined as in (5) are all equal, say $M^{(2)}(\sigma)$. It then follows from (3) that

$$V_2(\widetilde{N}, \widetilde{q}, \sigma) = \frac{1}{p^2}\left(\sum_{i=1}^{p} \frac{1}{\widetilde{q}_i}\right)\left[M^{(2)}(\sigma)\right]^{-1},$$

where $\widetilde{q}_i = b_i/b$, $i = 1, \ldots, p$. Under conditions $\sum_{i=1}^{p} \widetilde{q}_i = 1$ and $\widetilde{q}_i > 0$ for $i = 1, \ldots, p$, it is obvious that the function $\sum_{i=1}^{p}(1/\widetilde{q}_i)$ attains the minimum if and only if $\widetilde{q}_i = 1/p$ for all i. □

If block sizes are not constant, then the optimal choice (for A_σ and D_σ criteria) of $\widetilde{q}$ depends on σ. Therefore, in the class of all experimental designs with fixed v, b and different block sizes $k_1, \ldots, k_p$, asymptotically A_σ and D_σ optimal designs in this class do not exist.

From Lemma 1, we immediately get the following

Theorem 1. *An experimental design $(\widetilde{N}, \widetilde{q})$ is asymptotically A_σ (D_σ) optimal in $\mathcal{E}_o$ if and only if $(\widetilde{N}, \widetilde{q}) \in \mathcal{E}_1$.*

From Theorem 1 it is clear that, if we are interested in effective estimation of σ only, then the number p of different columns in the incidence matrix N is not important. Only replications of vectors $N_1, \ldots, N_p$ in N have to be equal.

Now, in the class $\mathcal{E}_1$ of asymptotically A_σ and D_σ optimal designs in $\mathcal{E}_o$, we indicate an asymptotically A_μ and D_μ optimal designs.

Lemma 2. *For any experimental design $(\widetilde{N}, (1/p)\mathbf{1}_p)$ in $\mathcal{E}_1$*

$$V_1(\widetilde{N}, (1/p)\mathbf{1}_p, \sigma) - V_1\left((1/p)\textstyle\sum_{i=1}^{p} N_i, 1, \sigma\right) \geq 0$$

holds for every $\sigma \in \mathbb{R}^2_+$.

Proof. In our design $\widetilde{q} = (1/p)\mathbf{1}_p$. Then from (3) and (4) it follows that

$$V_1(\widetilde{N}, (1/p)\mathbf{1}_p, \sigma) = \sigma_e^2\left(\sum_{i=1}^{p} \frac{1}{p}\operatorname{diag}(N_i) - \frac{\sigma_\lambda^2}{k\sigma_\lambda^2 + \sigma_e^2}\frac{1}{p}\sum_{i=1}^{p} N_i N_i'\right)^{-1}.$$

Since $\sum_{i=1}^{p} N_i N_i' - (1/p)N_{\cdot} N_{\cdot}' \geq 0$, where $N_{\cdot} = \sum_{i=1}^{p} N_i$, it holds that

$$V_1(\widetilde{N}, (1/p)\mathbf{1}_p, \sigma) - \sigma_e^2\left(\frac{1}{p}\operatorname{diag}(N_{\cdot}) - \frac{\sigma_\lambda^2}{k\sigma_\lambda^2 + \sigma_e^2}\frac{1}{p}N_{\cdot}\frac{1}{p}N_{\cdot}'\right)^{-1} \geq 0.$$

Noting that

$$\sigma_e^2\left(\frac{1}{p}\text{diag}(N_.)-\frac{\sigma_\lambda^2}{k\sigma_\lambda^2+\sigma_e^2}\frac{1}{p}N_.\frac{1}{p}N_.'\right)^{-1}=V_1((1/p)N_.,1,\sigma),$$

the lemma follows. □

Lemma 2 shows that an asymptotically A_μ and D_μ optimal design in $\mathcal{E}_1$ belongs to $\mathcal{E}_2$.

Lemma 3. *For every $\sigma\in\mathbb{R}_+^2$ and every $(\widetilde{N},1)\in\mathcal{E}_2$ the following inequalities hold*

$$\text{tr }V_1((k/v)\mathbf{1}_v,1,\sigma)\leq\text{tr }V_1(\widetilde{N},1,\sigma)$$

and

$$\det V_1((k/v)\mathbf{1}_v,1,\sigma)\leq\det V_1(\widetilde{N},1,\sigma).$$

Proof. Note that for every design $(\widetilde{N},1)$ from $\mathcal{E}_2$, $\widetilde{N}$ is a vector, say $(\widetilde{n}_1,\ldots,\widetilde{n}_v)'$. In this case

$$V_1(\widetilde{N},1,\sigma)=\sigma_e^2\left(\text{diag}(\widetilde{N})-\frac{\sigma_\lambda^2}{k\sigma_\lambda^2+\sigma_e^2}\widetilde{N}\widetilde{N}'\right)^{-1}=\sigma_e^2\text{diag}^{-1}(\widetilde{N})+\sigma_\lambda^2\mathbf{1}_v\mathbf{1}_v'. \quad (6)$$

Taking the trace we have

$$\text{tr }V_1(\widetilde{N},1,\sigma)=\sigma_e^2\sum_{i=1}^{v}\frac{1}{\widetilde{n}_i}+v\sigma_\lambda^2.$$

Thus, the minimum of tr $V_1(\widetilde{N},1,\sigma)$ under condition $\sum_{i=1}^v\widetilde{n}_i=k$ is attained if and only if $\widetilde{n}_1,\ldots,\widetilde{n}_v$ are equal to k/v.

To prove the second inequality of the lemma note that from (6) we have

$$\det V_1(\widetilde{N},1,\sigma)=\sigma_e^{2(v-1)}(k\sigma_\lambda^2+\sigma_e^2)/\prod_{i=1}^{v}\widetilde{n}_i.$$

Since $\sum_{i=1}^v\widetilde{n}_i=k$, the minimum of $\det V_1(\widetilde{N},1,\sigma)$ is attained if and only if $\widetilde{n}_i=k/v$ for all i. □

As a consequence of the above lemmas we can obtain the following theorem.

Theorem 2. *An experimental design $(\widetilde{N},\widetilde{q})$ is asymptotically A_μ (D_μ) optimal in $\mathcal{E}_1$ if and only if $(\widetilde{N},\widetilde{q})=((k/v)\mathbf{1}_v,1)$, i.e. $(\widetilde{N},\widetilde{q})$ is the orthogonal design ($N=(k/v)\mathbf{1}_v'\mathbf{1}_b$).*

The proposed criteria of optimality are independent of factors w_1 and w_2 in (2). Therefore, optimal properties of the orthogonal designs are valid also for the maximum likelihood estimator.

4. Comparison of Experimental Designs

In this section we present some numerical comparisions of experimental designs. For $b = 20$, let $E_o = (\widetilde{N}, \widetilde{q})$ be an experimental design in $\mathcal{E}_o$ given by

$$\widetilde{N} = (N_1, \ldots, N_5) = \begin{bmatrix} 1 & 0 & 0 & 0 & 0 \\ 1 & 1 & 0 & 1 & 0 \\ 1 & 1 & 2 & 0 & 1 \\ 1 & 1 & 1 & 2 & 2 \\ 1 & 2 & 2 & 2 & 2 \end{bmatrix}$$

(in this case $v = k = p = 5$) and $\widetilde{q} = (.05, .15, .20, .25, .35)'$. Define the following experimental designs

$$E_1 = (\widetilde{N}, (1/5)\mathbf{1}_5) \text{ from } \mathcal{E}_1,$$

better than E_o with respect to A_σ and D_σ criteria by Theorem 1,

$$E_2 = \left((1/5) \textstyle\sum_{i=1}^{5} N_i, 1 \right) \text{ from } \mathcal{E}_2,$$

better than E_1 with respect to A_μ and D_μ criteria by Lemma 2, and finally

$$E_3 = (\mathbf{1}_5, 1) \text{ from } \mathcal{E}_2,$$

which is asymptotically A_μ and D_μ optimal orthogonal design by Theorem 2.

Note that in fact E_2 does not correspond to an experimental design, since components of $(1/5)\sum_{i=1}^{5} N_i$ are not integers. We consider it, however, to illustrate each step of improvement (lemmas 1, 2, 3) of a given experimental design, described in Section 3.

To compare the above experimental designs we introduce the following efficiency functions defined for $\tau \in \mathbb{R}_+^2$,

$$efA_\sigma(\tau) = \frac{\operatorname{tr} V_2(E_o, \tau) - \operatorname{tr} V_2(E_3, \tau)}{\operatorname{tr} V_2(E_o, \tau)} \times 100, \tag{7}$$

$$efD_\sigma(\tau) = \frac{\det V_2(E_o, \tau) - \det V_2(E_3, \tau)}{\det V_2(E_o, \tau)} \times 100, \tag{8}$$

for A_σ and D_σ criteria, respectively, and

$$efA_\mu^{(i)}(\tau) = \frac{\operatorname{tr} V_1(E_o, \tau) - \operatorname{tr} V_1(E_i, \tau)}{\operatorname{tr} V_1(E_o, \tau)} \times 100, \quad i = 1, 2, 3, \tag{9}$$

$$efD_\mu^{(i)}(\tau) = \frac{\det V_1(E_o, \tau) - \det V_1(E_i, \tau)}{\det V_1(E_o, \tau)} \times 100, \quad i = 1, 2, 3, \tag{10}$$

for A_μ and D_μ criteria, respectively. Since E_1 and E_2 are asymptotically A_σ and D_σ optimal in $\mathcal{E}_o$, the corresponding efficiency functions of E_1 and E_2 with respect to E_o for A_σ and D_σ criteria coincide with (7) and (8), respectively.

Note that the efficiency functions are invariant with respect to the multiplication τ by a constant. This implies that the efficiency function can be considered as a functions defined on a simplex

$$\{\tau = (\tau_1, 1 - \tau_1)' : 0 < \tau_1 < 1\}.$$

TABLE I
Efficiency functions for μ

τ_1	$efA_\mu^{(1)}(\tau)$	$efA_\mu^{(2)}(\tau)$	$efA_\mu^{(3)}(\tau)$	$efD_\mu^{(1)}(\tau)$	$efD_\mu^{(2)}(\tau)$	$efD_\mu^{(3)}(\tau)$
0.05	66.42	66.48	81.22	95.30	95.33	98.59
0.15	66.35	66.88	81.22	95.38	95.61	98.67
0.25	66.03	67.40	80.92	95.52	96.09	98.82
0.35	65.06	67.26	79.67	95.67	96.57	98.96
0.45	62.92	65.56	76.60	95.79	96.90	99.06
0.55	59.00	61.58	70.80	95.87	97.10	99.12
0.65	52.93	54.99	61.76	95.91	97.21	99.16
0.75	45.44	46.68	50.57	95.94	97.27	99.17
0.85	38.86	39.32	40.73	95.95	97.29	99.18
0.95	35.48	35.53	35.66	95.95	97.30	99.18

TABLE II
Efficiency functions for θ

τ_1	$efA_\theta^{(1)}(\tau)$	$efA_\theta^{(2)}(\tau)$	$efA_\theta^{(3)}(\tau)$	$efD_\theta^{(1)}(\tau)$	$efD_\theta^{(2)}(\tau)$	$efD_\theta^{(3)}(\tau)$
0.05	54.29	54.32	63.35	98.02	98.03	99.41
0.15	64.16	64.65	77.99	98.05	98.15	99.44
0.25	64.91	66.23	79.26	98.11	98.35	99.50
0.35	64.14	66.27	78.30	98.18	98.55	99.56
0.45	61.97	64.52	75.18	98.23	98.70	00.61
0.55	57.94	60.42	69.23	98.26	98.78	99.63
0.65	51.86	53.80	60.16	98.28	98.83	99.65
0.75	44.62	45.76	49.34	98.29	98.85	99.65
0.85	38.50	38.92	40.19	98.29	98.86	99.66
0.95	35.44	35.48	35.61	98.29	98.86	99.66

For the considered example efficiences (7) and (8) are constant and equal

$$efA_\sigma(\tau) = 35.11, \quad efD_\sigma(\tau) = 57.89.$$

The efficiency functions (9) and (10) are shown in Table I. Also, we define corresponding efficiency functions for simultaneous estimation of μ and σ by

$$efA_\theta^{(i)}(\tau) = \frac{\text{tr } \Sigma(E_o,\tau) - \text{tr } \Sigma(E_i,\tau)}{\text{tr } \Sigma(E_o,\tau)} \times 100, \quad i = 1,2,3, \tag{11}$$

$$efD_\theta^{(i)}(\tau) = \frac{\det \Sigma(E_o,\tau) - \det \Sigma(E_i,\tau)}{\det \Sigma(E_o,\sigma)} \times 100, \quad i = 1,2,3, \tag{12}$$

for A_θ and D_θ criteria, respectively.

We choose a robust estimator considered by Bednarski and Zontek (1994) for which $w_1 = 1.1$ and $w_2 = 1.29$. The efficiency functions (11) and (12) are shown in Table II.

From Table I and Table II it follows that an essential improvement of the asymptotic variance of the robust estimator we can get using designs with equal $\widetilde{q}_i$. Comparing the orthogonal design with designs form the class $\mathcal{E}_1$ we improve the asymptotic variance of the robust estimator but the increase is relatively small.

Acknowledgements

This paper was supported by Komitet Badań Naukowych, Grant No. 21052 9101. We are greatly indebted to Professor S. Kageyama for his suggestions while formulating the final version of this paper.

References

Bednarski, T., Zmyślony, R. and Zontek, S. (1992). On robust estimation of variance components via von Mises functionals. Preprint 504, Institute of Mathematics, Polish Academy of Sciences, Warszawa.

Bednarski, T. and Zontek, S. (1994). A note on robust estimation of parameters in mixed unbalanced models. In: T. Caliński and R. Kala, Eds., *Proceedings of the International Conference on Linear Statistical Inference LINSTAT'93*. Kluwer Academic Publishers, Dordrecht, 87-95.

Christof, K. and Pukelsheim, F. (1985). Approximate design theory for a simple block design with random block effects. In: T. Caliński and W. Klonecki, Eds., *Linear Statistical Inference. Lecture Notes in Statistics* **35**. Springer-Verlag, Berlin, 20–28.

Gaffke, N. and Krafft, O. (1980). D-optimum designs for the interblock model. In: W. Klonecki, A. Kozek and J. Rosiński, Eds., *Mathematical Statistics and Probability Theory. Lectures Notes in Statistics* **2**. Springer-Verlag, New York, 134–143.

Kurotschka, V. and Müller, Ch.H. (1992). Optimum robust estimation of linear aspects in conditionally contaminated linear models. *The Annals of Statistics* **20**, 331–350.

Kageyama, S. and Zmyślony, R. (1993). Characterization and optimality of block designs with estimation of parameters under mixed models. In: *Proceedings of the 3rd Pacific Area Statistical Conference*. To appear.

ON GENERALIZED BINARY PROPER EFFICIENCY-BALANCED BLOCK DESIGNS

ASHISH DAS
Statistical and Mathematical Division
Indian Statistical Institute
Calcutta 700 035,
India

and

SANPEI KAGEYAMA
Department of Mathematics
Hiroshima University
Shinonome, Hiroshima 734,
Japan

Abstract. Necessary conditions for a generalized binary proper block design with two different replication numbers to be efficiency-balanced (EB) are given. Certain characterizations of such designs have also been obtained. Various methods of construction with a list of 50 generalized binary proper EB designs are provided.

Key words: BIB, PBIB, EB, GB-EB designs, (r, λ)-design.

1. Introduction

We consider a block design with v treatments and b blocks of size k each, which is said to be *proper*. For it, let $R = \text{diag}\{r_1, ..., r_v\}$, $\mathbf{r} = (r_1, ..., r_v)'$ and $N = (n_{ij})$, the usual $v \times b$ incidence matrix, where n_{ij} is the number of times the ith treatment occurs in the jth block and r_i is the replication number of the ith treatment, for $i = 1, ..., v$ and $j = 1, ..., b$. The number of times the pair of treatments i, i' appears together in a block, over the whole design, is denoted by $\lambda_{ii'}$, i.e. $\lambda_{ii'} = \sum_{j=1}^{b} n_{ij}n_{i'j}$.

Under the usual additive homoscedastic linear model, the coefficient matrix of the reduced normal equation is given by $C = R - k^{-1}NN'$ with zero row sums. A design is said to be *connected* if $\text{rank}(C) = v - 1$. James and Wilkinson (1971) defined the *canonical efficiency factors* e_i, $i = 1, ..., v-1$, of a connected block design to be the non-zero eigenvalues of the matrix $R^{-1}C$. Let e denote their harmonic mean, called the *efficiency factor* of the design. A design is said to be *efficiency-balanced* (EB) if $e_i = e$ for every $i \in \{1, ..., v-1\}$ (cf. Caliński, 1971; Puri and Nigam, 1975). A necessary and sufficient condition, due to Williams (1975), for a proper connected design to be EB is that its incidence matrix satisfies the equality

$$NN' = k(1-e)R + (e/b)\mathbf{r}\mathbf{r}'. \tag{1.1}$$

A proper block design such that n_{ij}, $i = 1, ..., v$; $j = 1, ..., b$, can take one of two possible values, x and $x+1$, where $x = [k/v]$, i.e. the largest integer not exceeding

T. Caliński and R. Kala (eds.),
Proceedings of the International Conference on Linear Statistical Inference LINSTAT '93, 203–210.

k/v, is called *binary* when $x = 0$ (or $k < v$) and *generalized binary* when $x > 0$ (or $k \geq v$). Such designs have a significant role in theory of optimal designs since they maximize the trace of the C-matrix (see Shah and Sinha, 1989). Here the generalized binary proper EB design is denoted by a GB-EB design.

A binary and proper EB design is a balanced incomplete block (BIB) design. Kageyama and Das (1991) showed that a GB-EB design has at most two distinct replication numbers. There are a large number of equireplicate GB-EB designs. However, the existence of GB-EB designs with two different replication numbers has been an open problem but for a recent paper by Angelis, Kageyama and Moyssiadis (1993).

In this paper, we first note GB-EB designs when $e = 1$ or $v = 2$. Next, for GB-EB designs with $e < 1$ and $v \geq 3$ certain necessary conditions are obtained. Those conditions along with a sufficient condition on the existence of GB-EB designs give a complete list of possible GB-EB designs in the range of $v < k \leq 20$ and $b \leq 100$. Apart from various new construction methods, we also give an infinite series of GB-EB designs with two different replication numbers.

Here, we denote by $\otimes$ the Kronecker product of matrices, I_n the identity matrix of order n, and by $J_{n,m}$ the $n \times m$ matrix of ones. $\mathbf{1}_n = J_{n,1}$ and $J_n = J_{n,n}$. A BIB design with parameters (v, b, k) is denoted by BIB(v, b, k).

For definitions of BIB and 2-associate partially balanced incomplete block (PBIB) designs we refer to Shah and Sinha (1989).

2. Characterizations

We first give the following shown by Kageyama and Das (1991).

Lemma 2.1. *A* GB-EB *design with v treatments, b blocks of size k each has at most two distinct replication numbers.*

The above is due to the fact that the quadratic equation

$$(e/b)r_i^2 + (k(1-e) - 2x - 1)r_i + bx(x+1) = 0 \tag{2.1}$$

has at most two distinct solutions on r_i, independent of $i = 1, ..., v$.

We here focus our attention to the problem of constructing GB-EB designs with two different replication numbers. Consider $v = v_1 + v_2$ ($v_i > 0, i = 1, 2$) treatments in which v_1 treatments are replicated r_1 times and v_2 treatments are replicated r_2 times, $r_1 \neq r_2$. In this case, we have the following.

Lemma 2.2. *A necessary condition for the existence of a* GB-EB *design with v treatments, b blocks of size k each and replication numbers r_1 and r_2 is that*

$$(k - 2x - 1)r_1r_2 = bx(x+1)(bk - r_1 - r_2). \tag{2.2}$$

The efficiency factor of the design is given by $e = b^2x(x+1)/(r_1r_2)$.

Proof. From (2.1), it follows that $r_1 + r_2 = -(b/e)(k(1-e) - 2x - 1)$, $r_1r_2 = b^2x(x+1)/e$, which can yield the result. □

Lemma 2.3. *In a* GB-EB *design with* v $(= v_1 + v_2)$ *treatments,* b *blocks of size* k *each such that* v_i *treatments have replication number* r_i *for* $i = 1, 2$, *it holds that*

(i) *for every* $i, i' \in \{1, ..., v_1\}$, $i \neq i'$, $\lambda_{ii'} = bx(x+1)r_1/r_2 = \lambda_1$, *say,*

(ii) *for every* $i, i' \in \{v_1 + 1, ..., v\}$, $i \neq i'$, $\lambda_{ii'} = bx(x+1)r_2/r_1 = \lambda_2$, *say,*

(iii) *for every* i, i' *such that* $i \in \{1, ..., v_1\}$ *and* $i' \in \{v_1 + 1, ..., v\}$,

$$\lambda_{ii'} = bx(x+1) = \lambda_{12}, \textit{ say.}$$

Proof. It is seen from (1.1) and the expression of e. □

Without loss of generality, let $r_1 < r_2$. Note that since we are in a generalized binary set-up, $r_1 \geq bx$ and $r_2 \leq b(x+1)$. Let

$$r_1 = bx + s,\ r_2 = b(x+1) - t,\ s \geq 0,\ t \geq 0,\ s + t < b. \tag{2.3}$$

Then Lemma 2.2 and (2.3) can yield the following.

Corollary 2.1. *In a* GB-EB *design with parameters* v, b, k, $r_1 = bx + s$ *and* $r_2 = b(x+1) - t$,

$$bx(k - x)(t - s) = s(b - t)(k - 2x - 1). \tag{2.4}$$

The GB-EB designs can be classified into the three classes as (i) EB designs with $v \geq 2$ and $e = 1$; (ii) EB designs with $v = 2$ and $e < 1$; (iii) EB designs with $v \geq 3$ and $e < 1$. The EB designs of (i) are orthogonal designs. It is easy to show that a necessary and sufficient condition for the existence of such designs is $s = t = 0$, and that their incidence matrix is given by $(x\mathbf{1}'_{v_1}, (x+1)\mathbf{1}'_{v_2})'\mathbf{1}'_b$ with $v_1 = v(x+1) - k$ and $v_2 = k - vx$. On the other hand, when $v = 2$ and $e < 1$, (2.1), (2.3) and (2.4) give $k = 2x + 1$ and $0 < s = t < b/2$ which are shown to be necessary and sufficient conditions for the existence of GB-EB designs of (ii). The design is given by the incidence matrix as

$$\begin{pmatrix} x\mathbf{1}'_{b-s} & (x+1)\mathbf{1}'_s \\ (x+1)\mathbf{1}'_{b-s} & x\mathbf{1}'_s \end{pmatrix}$$

which is then EB with $e = b^2x(x+1)/(b^2x(x+1) + s(b-s))$.

3. Designs with $v \geq 3$ and $e < 1$

We first give a necessary condition for the existence of such designs. It is known that the Fisher inequality $b \geq v$ holds for all connected EB designs with $e < 1$.

Theorem 3.1. *A* GB-EB *design with* v (≥ 3) *treatments,* b $(\geq v)$ *blocks of size* k *each and* $e < 1$ *exists only if there are* $r_1 = bx + s$ *and* $r_2 = b(x+1) - t$ *such that*

(i) $1 < s < \min\{(b-1)/2,\ b(k/v - x),\ b(x + 1 - k/v) - 1\}$,

(ii) $s < t < \min\{2s,\ b - s,\ b(x + 1 - k/v)\}$,

(iii) $bx(k-x)(t-s) = s(b-t)(k-2x-1)$,
(iv) $(vr_2 - bk)/(r_2 - r_1) = v_1$, *an integer,*
(v) *for* $v_1 \neq 1$, $bx(x+1)r_1/r_2 = \lambda_1$, *an integer,*
(vi) *for* $v_1 \neq v-1$, $bx(x+1)r_2/r_1 = \lambda_2$, *an integer.*

Proof. From (2.4) and $v \geq 3$, it follows that $0 < (b-t)(k-2x-1)/(bx(k-x)) < 1$. Hence, $0 < (t-s)/s < 1$ which yields $s < t < 2s$. Let v_1 treatments have replication number r_1 and $v-v_1$ treatments have replication number r_2. Then $v_1r_1+(v-v_1)r_2 = bk$ which proves (iv). Now, by $0 < v_1 < v$, we have $0 < (vr_2 - bk)/(r_2 - r_1) < v$ which yields

$$t < b(x+1-k/v), \; s < b(k/v - x). \tag{3.1}$$

Thus, since $s < t < 2s$, (2.3) and (3.1) imply (ii), i.e.

$$s < t < \min\{2s, b-s, b(x+1-k/v)\}. \tag{3.2}$$

Here t satisfying (3.2) exists only if $s < \min\{2s-1, b-s-1, b(x+1-k/v)-1\}$, which, through (3.1), give (i). Finally, (iii) follows from Corollary 2.1, and (v) and (vi) follow from Lemma 2.3. □

Using Theorem 3.1 finds that for several sets (v, b, k), there does not exist a GB-EB design. An exhaustive search in the range of $2 < v < k \leq 20$ and $b \leq 100$ gave 121 design parameters satisfying the necessary conditions (i) – (vi) as in Theorem 3.1. Among them 50 designs are listed in Table I, by eliminating multiples of existing smaller designs. It is observed that majority of the designs have $v_1 = 1$ or $v-1$.

Stanton and Mullin (1966) generalized the concept of a BIB design to a (r, λ)-system in connection with the construction of BIB designs. A (r, λ)-system or (r, λ)-design is a binary block design with the $v \times b$ incidence matrix $N = (n_{ij})$ such that

$$\sum_{j=1}^{b} n_{ij} = r \text{ for } i = 1, ..., v, \; \sum_{j=1}^{b} n_{ij}n_{i'j} = \lambda \text{ for } i \neq i' = 1, ..., v.$$

Here (r, λ)-designs are utilized to construct our GB-EB designs.

Theorem 3.2. *Let* N *be the incidence matrix of a* GB-EB *design with parameters* $v_1 = 1$, $v_2 = v-1$, b, k, $r_1 = bx+s$, $r_2 = b(x+1)-t$. *Then*

$$N = \begin{pmatrix} x\mathbf{1}'_{b_1} & (x+1)\mathbf{1}'_{b_2} \\ N_1 + xJ & N_2 + xJ \end{pmatrix} \text{ with } b_1 = b-s \text{ and } b_2 = s$$

where N_i *is a* $(v-1) \times b_i$ *incidence matrix of a binary block design with replication number* $r_{(i)}$ *and constant block size* $k_{(i)}$, $i = 1, 2$. *Here* $r_{(1)} = b-t-x(t-s)$, $r_{(2)} = x(t-s)$, $k_{(1)} = k-vx$, $k_{(2)} = k-vx-1$, *and* $(N_1 : N_2)$ *is the incidence matrix of a* (r, λ)*-design with* $r = r_{(1)} + r_{(2)}$ *and* $\lambda = \lambda_2 - x(2b-2t+xb)$.

Proof. First note that s, $t > 0$. Since $xb_1 + (x+1)b_2 = r_1$ and $b_1 + b_2 = b$, we have $b_1 = b(x+1) - r_1 = b-s$ and $b_2 = r_1 - bx = s$. Also by $N'\mathbf{1}$

$= k\mathbf{1}$, it follows that $k_{(1)} = k - vx$ and $k_{(2)} = k - vx - 1$. It is clear that N_i is binary. Now, let $r_{(i)m}$ be the mth row sum of N_i, $m = 1, ..., v-1$, $i = 1,\ 2$. Then $(x+1)r_{(1)m} + x(b_1 - r_{(1)m}) + (x+1)r_{(2)m} + x(b_2 - r_{(2)m}) = r_2$, and using (iii) of Lemma 2.3,

$$x(x+1)r_{(1)m}+x^2(b_1-r_{(1)m})+(x+1)^2 r_{(2)m}+x(x+1)(b_2-r_{(2)m}) = bx(x+1). \quad (3.3)$$

Since $b_2 = s$, (3.3) on simplification can give $r_{(1)m} = b - t - x(t-s)$, $r_{(2)m} = x(t-s)$, $m = 1, ..., v-1$. Since $r_{(i)m}$ is independent of m, thus $r_{(1)} = b-t-x(t-s)$ and $r_{(2)} = x(t-s)$. Let $N_1N_1' + N_2N_2' = (\delta_{ii'})$, $i, i' \in \{2, ..., v\}$. Then using Lemma 2.3, we have $\lambda_2 = \delta_{ii'} + 2x(b-t) + x^2 b$ or $\delta_{ii'} = \lambda_2 - x(2b - 2t + xb) = \lambda$ for all $i \neq i'$. Also by $N_i\mathbf{1} = r_{(i)}\mathbf{1}$, $i = 1, 2$, it follows that $(N_1 : N_2)$ is the incidence matrix of a (r, λ)-design with $r = r_{(1)} + r_{(2)}$ and $\lambda = \lambda_2 - x(2b - 2t + xb)$. □

On a line similar to Theorem 3.2, we can get the following.

Theorem 3.3. *The incidence matrix of a* GB-EB *design with parameters* $v_1 = v-1$, $v_2 = 1$, b, k, $r_1 = bx + s$, $r_2 = b(x+1) - t$ *is given by*

$$\begin{pmatrix} N_1 + xJ & N_2 + xJ \\ x\mathbf{1}'_{b_1} & (x+1)\mathbf{1}'_{b_2} \end{pmatrix} \text{ with } b_1 = t \text{ and } b_2 = b - t,$$

where N_i *is the* $(v-1) \times b_i$ *incidence matrix of a binary block design with* $r_{(i)}$ *and* $k_{(i)}$, $i = 1,\ 2$. *Here* $r_{(1)} = s - x(t-s)$, $r_{(2)} = x(t-s)$, $k_{(1)} = k - vx$, $k_{(2)} = k - vx - 1$ *and* $(N_1 : N_2)$ *is a* (r, λ)*-design with* $r = r_{(1)} + r_{(2)}$ *and* $\lambda = \lambda_1 - x(xb + 2s)$.

Lemma 2.3 shows the necessary conditions to be satisfied by the incidence matrix of a GB-EB design. In fact, it can be shown that those conditions are sufficient as well for the incidence matrix to give a GB-EB design (see also Angelis, Kageyama and Moyssiadis, 1993). Hence,

Theorem 3.4. *A generalized binary block design with* v *treatments,* b *blocks of size* k *each such that the treatment* i, $i \in \{1, ..., v_1\}$, *has replication number* r_1 *and treatment* i', $i' \in \{v_1 + 1, ..., v\}$, *has replication number* r_2, $v_1 + v_2 = v$, *is a* GB-EB *design if and only if conditions* (i), (ii) *and* (iii) *of Lemma* 2.3 *hold.*

An appeal to Theorems 3.2 and 3.3 along with Theorem 3.4 gives some useful methods of constructing the required GB-EB designs. An application of Theorem 3.2 is only stated here.

Method A. Let $v(= v_1 + v_2)$, b, k, r_1 and r_2 satisfy the necessary conditions in Theorem 3.1. Then if $v_1 = 1$ $(v_2 = v - 1)$, the existence of BIB$(v-1, b-s, k-vx)$ and BIB$(v-1, s, k-vx-1)$ with incidence matrix N_1 and N_2, respectively, implies the existence of a GB-EB design with incidence matrix N, as in Theorem 3.2.

Let PBIB$(v, b, k, n_1, n_2, \lambda^{(1)}, \lambda^{(2)})$ denote a PBIB design with parameters v, b, k, n_1, n_2, $\lambda^{(1)}$ and $\lambda^{(2)}$ based on any association scheme.

TABLE I

GB-EB designs with $e < 1$ in the parametric range of $2 < v < k \leq 20$ and $b \leq 100$

EB#	v	b	k	v_1	v_2	r_1	r_2	e	Reference	Multiple
1	4	9	6	1	3	12	14	.964	2(3,2)+(3,1)	2 ~ 11
2	5	12	7	3	2	14	21	.980	A(1,1)	2 ~ 8
3	5	24	7	4	1	33	36	.970	2(4,2)+3(4,1)	2 ~ 4
4	5	30	8	1	4	36	51	.980	6(4,3)+(4,2)	2,3
5	6	25	8	5	1	32	40	.977	(5,2)+3(5,1)	2 ~ 4
6	6	30	9	1	5	40	46	.978	2(5,3)+(5,2)	2,3
7	7	45	9	6	1	55	75	.982	(6,2)+5(6,1)	2
8	4	15	10	1	3	36	38	.987	3(3,2)+2(3,1)	2 ~ 6
9	6	70	10	1	5	80	124	.988	12(5,4)+(5,3)	
10	7	40	11	1	6	50	65	.985	2(6,4)+BIB4	2
11	8	30	11	5	3	33	55	.992	A(1,2)	2,3
12	8	42	11	7	1	57	63	.982	3BIB1+(7,2)	2
13	5	14	12	4	1	33	36	.990	(4,2)+2(4,1)	2 ~ 7
14	8	21	12	1	7	28	32	.984	2BIB'1+BIB1	2 ~ 4
15	5	48	13	1	4	108	129	.992	9(4,3)+2(4,2)	2
16	10	60	13	9	1	76	96	.987	2BIB2+(9,2)	
17	4	21	14	1	3	72	74	.993	4(3,2)+3(3,1)	2 ~ 4
18	6	30	14	5	1	68	80	.993	(5,2)+4(5,1)	2,3
19	9	22	14	1	8	28	35	.988	R136+S6	2 ~ 4
20	6	50	15	1	5	120	126	.992	3(5,3)+2(5,2)	2
21	10	54	15	1	9	72	82	.988	2BIB'20+BIB20	
22	11	20	15	10	1	27	30	.988	T33+T9	2 ~ 5
23	11	56	15	7	4	60	105	.996	A(1,3)	
24	12	33	15	11	1	40	55	.990	–	
25	12	66	15	11	1	80	110	.990	–	
26	12	99	15	11	1	120	165	.990	–	
27	6	55	16	1	5	120	152	.995	9(5,4)+(5,3)	
28	10	39	16	1	9	48	64	.990	–	
29	10	78	16	1	9	96	128	.990	5BIB'2+BIB'20	
30	11	78	16	5	6	96	128	.990	–	
31	12	78	16	9	3	96	128	.990	–	
32	5	60	17	4	1	201	216	.995	4(4,2)+9(4,1)	
33	7	24	17	4	3	51	68	.997	A(2,1)	2 ~ 4
34	7	60	17	6	1	145	150	.993	3BIB4+2(6,2)	
35	10	84	17	1	9	96	148	.993	2(9,7)+BIB'2	
36	11	63	17	1	10	81	99	.990	3BIB'8+BIB27	
37	13	77	17	12	1	99	121	.990	BIB48+BIB47	
38	4	27	18	1	3	120	122	.996	5(3,2)+4(3,1)	2,3
39	5	22	18	1	4	72	81	.996	4(4,3)+(4,2)	2 ~ 4
40	7	65	18	1	6	150	170	.994	3(6,4)+(6,3)	
41	8	91	18	7	1	198	252	.996	(7,2)+10(7,1)	
42	11	25	18	1	10	30	42	.992	–	
43	11	50	18	1	10	60	84	.992	–	
44	11	75	18	1	10	90	126	.992	2BIB'26+BIB'8	
45	11	100	18	1	10	120	168	.992	–	
46	12	33	18	1	11	44	50	.990	2BIB'5+BIB5	2,3
47	14	78	19	13	1	105	117	.990	BIB93+3BIB3	
48	14	90	19	9	5	95	171	.997	A(1,4)	
49	6	65	20	5	1	212	240	.996	2(5,2)+9(5,1)	
50	8	35	20	1	7	84	88	.994	3BIB'1+2BIB1	2

Method B. Let $v(= v_1 + v_2)$, b, k, r_1 and r_2 satisfy the necessary conditions in Theorem 3.1. Then if $v_1 = 1$ $(v_2 = v - 1)$, the existence of PBIB$(v-1, b-s, k-vx, n_1, n_2, \lambda^{(1)}, \lambda^{(2)})$ and PBIB$(v-1, s, k-vx-1, n_1, n_2, \lambda_2-\lambda^{(1)}, \lambda_2-\lambda^{(2)})$ based on the same association scheme, for some integers n_1, n_2, $\lambda^{(1)}$ and $\lambda^{(2)}$ with incidence matrix N_1 and N_2, respectively, implies the existence of a GB-EB design with incidence matrix N, as in Theorem 3.2.

For the present investigation, our attention has been restricted to BIB designs, group divisible (GD), triangular, Latin square and cyclic PBIB designs. Individual plans or series of these designs are available in literature. Hence we can produce a number of GB-EB designs through the construction methods given here.

We will call GB-EB designs constructed through Method A as *type A designs*, and those constructed through Method B as *type B designs*.

Example 3.1. Let N_1 and N_2 be the incidence matrices of BIB(3,6,2) and BIB(3,3,1), respectively. Then Method A with $x = 1$ produces a type A design with $v_1 = 1$, $v_2 = 3$, $b = 9$, $k = 6$, $r_1 = 12$ and $r_2 = 28$.

Example 3.2. Let N_1 and N_2 be the incidence matrices of the GD designs R136 and S6, respectively, as listed in Clatworthy (1973). Then Method B with $x = 1$ produces a type B design with $v_1 = 1$, $v_2 = 8$, $b = 22$, $k = 14$, $r_1 = 28$ and $r_2 = 35$.

We also present an infinite series of designs for $v_1 \neq 1$ or $v - 1$.

Theorem 3.5. *For a pair of positive integers x and n, the following*

$$\begin{pmatrix} \mathbf{1}'_{v_2} \otimes (I_{v_1} + xJ_{v_1}) & xJ_{v_1, v_1 v_2} \\ \{(x+1)J_{v_2} - I_{v_2}\} \otimes \mathbf{1}'_{v_1} & (x+1)J_{v_2, v_1 v_2} \end{pmatrix}$$

is the incidence matrix of a GB-EB *design with parameters:*

$v_1 = (x+1)n + 1$, $v_2 = xn + 1$, $b = 2(xn+1)((x+1)n+1)$,
$k = 2x((x+1)n+1) + 1$, $r_1 = (xn+1)\{2x((x+1)n+1)+1\}$,
$r_2 = ((x+1)n+1)\{2x((x+1)n+1)+1\}$,
$e = 4x(x+1)(xn+1)((x+1)n+1)/\{2x((x+1)n+1)+1\}^2$.

Proof. It follows from Theorem 3.4 and the fact that $e = b^2x(x+1)/(r_1r_2)$. Here note that $x = [k/v]$. □

A GB-EB design belonging to the series, say *series A*, in Theorem 3.5, has two independent parameters x and n, denoted by $A(x, n)$.

4. Concluding Remarks and Tabulation

A list of possible GB-EB designs with $e < 1$ in the parametric range of $2 < v < k \leq 20$ and $b \leq 100$, which satisfies the necessary conditions in Theorem 3.1, originally includes 121 designs, where 86 designs are of type A and 9 of type B. Furthermore, 17 designs are of series A. Solutions to the remaining 9 designs are yet unknown. Here only 50 designs are listed among them, by eliminating multiples of existing smaller designs. Such information is given in Table I, where GB-EB designs are numbered in the ascending order of k. The number is followed by parameters $v, b, k, v_1, v_2, r_1, r_2$ and e. The entries under reference column refer to the solutions of designs. Here a BIB$(v, {}^vC_k, k)$ is denoted by (v, k). $p(v, k)$ refers to p copies of the BIB design. pBIBq refer to p (≥ 1) copies of the BIB design with serial number q listed in Hall (1986; Appendix). BIB' stands for the complement of a BIB design. Rq, Sq and Tq refer to PBIB designs in Clatworthy (1973). Finally, the figure a in the last column shows a copies of the design within the same scope of parameters. They have the same efficiency. A dash (–) shows that a solution is unknown.

Note that for given v, b, k if there exist two GB-EB designs (say d_i) with two different sets of replication numbers (r_{1d_i}, r_{2d_i}), $i = 1, 2$, then $e_{d_1} > e_{d_2}$ provided $r_{1d_1} + r_{2d_1} > r_{1d_2} + r_{2d_2}$.

References

Angelis, L., Kageyama, S. and Moyssiadis, C. (1993). Construction of generalized binary proper efficiency-balanced block designs with two different replication numbers. *Sankhyā B*. To appear.

Caliński, T. (1971). On some desirable patterns in block designs (with discussion). *Biometrics* **27**, 275–292.

Clatworthy, W. H. (1973). *Tables of Two-Associate-Class Partially Balanced Designs*. NBS Applied Mathematics Series 63, USA.

Hall, M., Jr. (1986). *Combinatorial Theory*, 2nd ed. Wiley, New York.

James, A. T. and Wilkinson, G. N. (1971). Factorisation of the residual operator and canonical decomposition of nonorthogonal factors in the analysis of variance. *Biometrika* **58**, 279–294.

Kageyama, S. and Das, A. (1991). A characterization of generalized binary proper efficiency-balanced designs. *Bulletin of Faculty of School Education*, Hiroshima University, Part II, **13**, 17–20.

Puri, P. D. and Nigam, A. K. (1975). On patterns of efficiency balanced designs. *Journal of the Royal Statistical Society B* **37**, 457–458.

Shah, K. R. and Sinha, B. K. (1989). *Theory of Optimal Designs. Lecture Notes in Statistics* **54**, Springer-Verlag, New York.

Stanton, R. G. and Mullin, R. C. (1966). Inductive methods for balanced incomplete block designs. *The Annals of Mathematical Statistics* **37**, 1348–1354.

Williams, E. R. (1975). Efficiency-balanced designs. *Biometrika* **62**, 686–688.

DESIGN OF EXPERIMENTS AND NEIGHBOUR METHODS

J.-M. AZAÏS
Université Paul Sabatier
Laboratoire de Statistique et Probabilités URA CNRS 745
118, route de Narbonne
31062 Toulouse Cedex
France

Abstract. This paper is a survey of different results concerning the design of experiments when neighbour phenomena occur. When these phenomena are neglected in the analysis, several results pertain to the optimality of neighbour balanced designs. However, when correlations are incorporated in the analysis, the nonlinearity of estimators gives no hope of proving such optimality properties. Moreover, it is never believed that the model is quite exact. For this reason two quantities are introduced: validity and efficiency, and new Monte Carlo results about properties of different designs are presented.

Key words: Design of experiments, Neighbour models, Efficiency, Validity, Universal optimality, Monte Carlo experiment.

1. Introduction

In a plot experiment, neighbour phenomena may cause departures from classical assumptions of the analysis of variance. They can be divided in two categories:

1. Neighbour effects: each plot is influenced not only by the actual treatment on the plot, but also by treatments present on neighbouring plots. Classical modelling is in the form of an additive model (Pearce, 1957).

2. Correlations: it is reasonable to assume that the nearer the plots are, the more correlated the fertilities are. Usually these correlations are broken by randomization (see Section 2.1), but when they are thought as a major effect, experimental precision can be improved by taking them into account.

In this paper, we investigate various consequences of these phenomena for the design of experiments. The situation is different depending on whether these phenomena are incorporated in the model on which the analysis is based or if they are still neglected and the ordinary analysis of variance (ANOVA) is used. The last case is studied first. Because of the linearity of estimators, many optimality results exist then. For the first case, most of the results have been obtained by Monte Carlo methods and we present some new findings in this paper.

2. Some Definitions

We consider a plot experiment with blocks.

We use the classical notation: n, v, b, k, r are the numbers of observations, treatments, blocks, plots within a block and replications, respectively.

T. Caliński and R. Kala (eds.),
Proceedings of the International Conference on Linear Statistical Inference LINSTAT '93, 211–221.

We suppose that the neighbours of a plot are well defined. Usually neighbours are taken as the nearest neighbours in one or both directions. In this paper we will often assume that neighbours are considered in one direction only and that they are taken within the block.

A design will be called neighbour balanced if each treament has each other treatment as a neighbour equally often.

A design will be called partially neighbour balanced if each treatment has each other treatment as a neighbour at most once.

A design will be called (according to Kiefer, 1975) universally optimal among a given class of designs if its information matrix for the estimation of treatment effects, C, minimizes every citerion $\Phi(C)$ such that:

(a) Φ is convex,

(b) Φ is invariant under permutations of treatment labels,

(c) $\Phi(bC)$ is a nonincreasing function of the scalar b ($b > 0$).

Such citeria are called Kiefer's criteria; among them we have: the trace, the determinant, the greatest eigenvalue, etc.

In the same framework, a design will be called (according to Kiefer and Wynn, 1981) universally weakly optimal among a given class if its covariance matrix for the estimators of treatment effects, D, minimizes any criterion $\Phi(D)$ such that:

(a) Φ is convex,

(b) Φ is invariant under permutations of treatment labels,

(c') $\Phi(bD)$ is a nondecreasing function of the scalar b ($b > 0$).

These two types of optimality differ by the convexity condition. None is included in the other, but the weak optimality is so-called because it does not contain the D-optimality.

2.1. Randomization

When estimators of treatment effects are linear functions of the observations, randomization theory can be applied. It shows that, in a certain sense, data can be analysed 'as if' they followed an analysis of variance model. The simplest notion is the weak validity that we describe briefly. We suppose that the observations are the sums of treatment effects and fertility plot values. The vector of fertility values is allowed to take any value in $\mathbb{R}^n$. Randomization theory shows that under the classical complete block randomization, the analysis of variance is weakly valid in the sense that, whathever the plot fertilities are, under the null hypothesis of the absence of treament effects, the treatment mean square and the residual mean square have the same expectation over the randomization. For any equireplicated experiment this criterion is equivalent to the equation

$$E(PAPV) = APV$$

that will be explained and used in Section 4.2.

2.2. Analysis with a Neighbour Model

In this paper we will consider only neighbour analysis in one direction. Let Y be the vector of yields, for which we assume that

$$E(Y) = X\tau, \quad \text{Cov}(Y) = V(\gamma).$$

Here $V(\gamma)$ depends on a vector γ containing a few scalar parameters. In many cases, observations are supposed to be independent from one block to another with the same intrablock banded covariance matrix. The analysis consists of two steps: first, estimate the unknown parameter γ by restricted maximum likelihood; then use the estimated $V(\hat{\gamma})$ to estimate the treatment values by generalized least squares.

3. Using Ordinary ANOVA for Analysing the Experiment

3.1. Neighbour Effects

In this section we suppose that data are analysed using the ordinary two-way (treaments, blocks) analysis of variance (analysis model), although in fact the data follow another model (true model) that includes additive neigbour effects. For simplicity we will suppose that there is only an influence of one neighbour, say left, and we state the following model

$$Y = X\tau + Z\lambda + B\beta + \varepsilon, \tag{1}$$

decomposing the yield Y into four components: treatment effect, left neighbour effect, block effect and error. Analysis using model (1) is possible, but the introduction of the term λ implies an important increase of the number of degrees of freedom. Thus, it is often preferable to use the classical model.

Neglecting λ in the analysis causes a bias for estimating the direct effect τ_i or the monoculture value $\tau_i + \lambda_i$ of treatment i (see, e.g., Azaïs, 1987 for details). Since this bias occurs on each treatment value, it can be viewed as a vector belonging to $\mathbb{R}^v$.

Consider the following two classes of designs: complete block designs and balanced incomplete block designs with blocks of size $(v-1)$, where v is the number of treaments.

It has been shown by Druilhet(1993) that neighbour balanced designs are optimal for bias reduction in the two above mentioned classes. Optimality is obtained for criteria that satisfy the same properties as Kiefer's criteria, namely:

(i) $\int_\lambda ||\text{bias due to } \lambda|| d\lambda$ is minimum,

(ii) $||\text{bias with the worst } \lambda||$ is minimum.

The choice of the norm does not matter and the domain of variation of λ is

$$\{\lambda \in \mathbb{R}^n : \sum \lambda_i = 0, ||\lambda|| = 1\}.$$

When using a randomized experiment, a part of the bias becomes random. It can be shown that the remaining bias is the same whether neighbour balanced is achieved or not. But the covariance matrix of a classical randomized experiment is greater with respect to the Löwner order (the difference is a nonnegative definite matrix) than the covariance matrix obtained with a neighbour balanced randomized experiment.

Generalizations of these results to other models than model (1), including two or four neighbours, are straightforward: bias due to these two or four kinds of neighbours are added.

3.2. Correlations

In this section, we suppose that the same analysis model is used and that this model is now false because errors are correlated. In case of constant lag-one correlations that are to be neglected, Kiefer and Wynn(1981) have shown that some kind of neighbour plus border balance gives optimality of the design. Namely, they consider two classes of designs: complete block designs and Latin square designs.

They suppose that correlations are in one direction only. Then designs satisfying the following condition (H) are weakly universally optimal in the above classes.

Condition H. For every pair of treatments (i, j),

$$\#(i \text{ and } j \text{ neighbours }) + \#(i \text{ or } j \text{ at one end of a block}) = \text{const.},$$

where # is the number of occurrences.

Of course border balance condition is met for latin squares. This result was generalized to lag-one and lag-two correlations by Morgan and Chakravarti(1988).

4. Using a Neighbour Model

In this part we will assume that the yield Y can be decomposed in three parts:

$$Y = X\tau + \xi + \varepsilon,$$

where $X\tau$ is the treatment effect part, ξ is the fertility variation of the field including block effects if any, and ε is an, eventually absent, vector of errors.

4.1. Adequate Models

Consider the case where the distribution of the trend is perfectly known and follows an autoregressive model with parameter λ. Let ξ_{ij} be the trend in plot j of block i, which is assumed as such that

$$\mathrm{Cov}(\xi_{ij}, \xi_{i'j'}) = \begin{cases} \lambda^{|j-j'|}, & \text{if } i = i', \\ 0, & \text{if } i \neq i', \end{cases} \tag{2}$$

where the parameter λ is known and the error term ε is absent. In such a simple case, treament estimates are the generalized least squares estimates and it can be proved that neighbour balanced designs are universally optimal in a certain class (see, e.g., Kunnert, 1987).

4.2. Efficiency and Validity of Inadequate Neighbour Methods

The result of Section 4.1 is for a very particular model. In most situations, the true model for ξ is unknown. Classical neighbour models involve only a few parameters for variance, because the estimation of such parameters is not easy. They are only rough approximations of the reality. In such situations, good designs are not only designs that correspond to small variances, but also designs which may help to fit the data. For example it is often written (see Besag and Kempton, 1986; Baird and Mead, 1991) that randomization may help for this last purpose.

Moreover, estimates are no longer linear, so that there exists no expression for their covariance matrix. Therefore, the hope of finding an optimal design must be abandoned and new properties of the design are to be considered.

To measure the performance of a neighbour model under inadequate conditions, we will consider the statistical analysis as aimed at two goals:

1. To estimate treatment comparisons;
2. To estimate the precision of the preceeding estimation.

How the first goal is achieved, can be measured by the average pairwise variance (Glesson and Eccleston, 1992),

$$APV = \frac{1}{n(n-1)} \sum_{i \neq j} \mathrm{Var}(\hat{\tau}_i - \hat{\tau}_j),$$

assuming that the treatment difference estimates are unbiased.

For comparing different designs, this quantity can be expressed as the inverse of the ratio with respect to the APV of the classical complete blocks analysis, APV_{class}. This quantity is called the efficiency,

$$Ef = \left(\frac{APV}{APV_{class}} \right)^{-1}.$$

Any statistical analysis provides an estimator of the APV that will be denoted by $PAPV$ (Predicted Average Pairwise Variance). The quantities APV and $PAPV$ are related to those called *Emp* and *Pre*, respectively, by Besag and Kempton (1986). For ordinary ANOVA, for example, the $PAPV$ is obtained through the residual mean square. When the $PAPV$ is unbiased, the second goal is achieved in the sense that we have a reliable estimator of the experiment precision. In this particular case the pair 'design' and 'analysis model' is said to be valid. This can be appreciated by comparing the efficiency with the predicted efficiency,

$$PEf = \left(\frac{E(PAPV)}{APV_{class}} \right)^{-1}.$$

Note that the two quantities Ef and PEf are related to the pair: 'design' and 'analysis model'. This pair will be called 'the method' in the sequel. Regardless whether or not the model is adequate, the method may or may not be valid and/or efficient. There is of course a hierarchy between the two properties in the sense that invalid methods may lead to completely false conclusions. On the other hand, valid but inefficient methods lead to the correct decision that such methods should no longer be used.

A Monte Carlo experiment has been conducted by Baird and Mead (1991) to appreciate the validity and efficiency of two neighbour models under classical complete or incomplete block randomization. The models are:

1. The first difference (FD) analysis of Besag and Kempton (1986) defined under the assumptions that

(i) the variables $\eta_{ij} = \xi_{ij} - \xi_{i,j-1}$, $i = 1 \ldots b$, $j = 2 \ldots k$, are independent with some common variance,

(ii) no ε errors are present.

2. The error in variable (EV) model of Besag and Kempton (1986), also called linear variance model by Williams (1986), which includes ε errors in the model above.

True models are several and may include linear, sinusoidal trends, autoregressive models, etc. The authors have shown that the FD model is approximatively valid and more efficient than the incomplete block analysis. Error in variable model, EV, is more efficient but is only valid for large designs.

4.3. Some New Results

Azaïs, Bailey and Monod have conducted Monte Carlo experiments to compare several designs when using inadequate neighbour models. The designs are:

1. Classical complete block randomization (CLASS).
2. Partially neighbour balanced restricted randomization (RNB) (Azaïs, Bailey and Monod, 1993).
3. Very systematic partially neighbour balanced design (SNB) based on arithmetic progressions. For example, with $v = 7, r = 3$,

$$\begin{matrix} 1&2&3&4&5&6&7\\ 2&4&6&1&3&5&7\\ 4&7&3&6&2&5&1 \end{matrix}\ .$$

4. The silly design in which treatments are always in the same order. For example, with $v = 7, r = 2$,

$$\begin{matrix} 1&2&3&4&5&6&7\\ 1&2&3&4&5&6&7 \end{matrix}\ .$$

4.3.1. Analysis Models

Six analysis models were considered.

1. The classical two-way analysis of variance (ANOVA).
2. The classical inter- and intra-block analysis of incomplete block designs of size 5(IB5).
3. Neighbour analysis by an autoregressive model as defined by formula (2) with λ unknown and ε errors added (AR).
4. Neighbour analysis by the FD model.
5. Neighbour analysis by the EV model.
6. Neighbour analysis by the Least Squares Smoothing model (LSS), which is a generalisation of the EV model to second order differences (Green, Jennison and Seheult, 1985).

General presentations of the tables. The results are presented in the tables. Each table relates to only one true model, indicated in the heading. Rows indicate the design, columns, the analysis models (if several), and cells, the Ef and PEf of the pair: 'design' and 'analysis model'. For reading convenience, these quantities have been multiplied by 100. There is a star when Ef and PEf differ significantly with respect to simulation variability.

TABLE I
Behavior of the silly design. Case of 20 treatments, 3 replicates. True model: nonmonotonous intrablock correlation + independent errors (300 simulations). Ef and PEf multiplied by 100 are shown in the table

	ANOVA		AR		EV		FD	
Class Design	100	102	187	189	175	189*	187	223*
Silly design	103	102	103	118*	103	65*	103	37*

TABLE II
Behaviour of the autoregressive analysis model with various designs. True model: 20 treatments, 3 replicates, very correlated random trend (1000 simulations)

Design	Ef	PEf
Class	672	370*
RNB	762	391*
SNB	829	391*
IB5	179	175

4.3.2. The Silly Design Gives Very Bad Results

All neighbour methods assume that the fertility trend is independent from one block to another with the same intrablock banded covariance matrix. Associated with the silly design this gives a case where the generalized least squares are equivalent to the ordinary least squares (Kruskall, 1968). So the efficiency is always one, but the predicted efficiency can be very different (Table I).

It can be seen, in this special case, that some situations exist where neighbour models are efficient ($Ef > 1$) and almost valid, when associated with the classical randomization. When associated with silly design, the advantage in efficiency is destroyed and the predicted efficiency is completely biased, sometimes in the direction of unconservativeness (AR) sometimes in the direction of conservativeness (EV,FD).

4.3.3. Systematic Neighbour Balanced Design is More Conservative and More Efficient

In almost all the cases we have considered, neighbour methods have greater efficiency than classical methods. Sometimes the gain is very small, but sometimes it is rather important. For very correlated true models and rather large design sizes (20×3) the gain is important and the methods show conservative bias. This happens even more with the systematic neighbour balanced design (Table II).

Results for incomplete blocks of size 5 (IB5) correspond to the classical interblock and intrablock analysis. They are only given for comparison. In such a situation, neighbour methods are not valid, associated with any design but this does not matter

TABLE III
Unconservative behaviour of SNB design associated with LSS model. The same situation as in Table II except that a systematic sinusoidal trend has been added to the true model

Design	AR model		EV model		LSS model	
Class	298	223*	339	275*	335	355*
RNB	325	229*	379	285*	373	362
SNB	286	225*	277	312*	311	780*

because of their very high efficiency. Unvalidity causes only the statistician to loose some part of the efficiency because he will use too large confidence intervals. One puzzling result is that the systematic SNB design is better, though less valid. Since the PEf is the same as for other designs, it will give the same confidence interval size, but since the true Ef is greater, the α error rate of such intervals will be smaller.

4.3.4. The Only Case of Bad Behavior of Systematic Neighbour Balanced Design
Similar results to those of Table II have happened with various shapes and sizes. Neighbour methods are rather valid or conservative and in that case conservativeness is compensated by high efficiency; the SNB design seems to be the best choice since it has greater efficiency with the same predicted efficiency. Nevertheless, among all that cases we have found one case (Table III) where the behavior under SNB design clearly differs from other designs. It is for the Least Squares Smoothing model(LSS).

In such a situation, a statistician using LSS model with SNB design will considerably overestimate the precision of the experiment and will draw completly false conclusions. Table III illustrates the fact that using systematic designs is always dangerous.

4.3.5. Models with an Important Independent Errors Part
When the variance of independent ε errors is substantial relatively to the variance of the increments of ξ (say when it is 5 times bigger), there are of course less differences between the designs, because design and randomization does not change the distribution of the ε's. In fact, only the silly design can be then distinguished from the others. In these cases the gain of efficiency due to neighbour methods is very small with respect to incomplete block design and analysis. It is very diminished by the small unconservative bias shown in Table IV.

In this case the efficiency of the neighbour method will be overestimated.

4.3.6. Small Sizes
It is sometimes believed that for very small design sizes, neighbour methods cannot bring an improvement with respect to ANOVA because there are not enough data to estimate variance parameters. We found that when using a randomized design, and

TABLE IV
Models with an independant error part. 20 treat., 3 rep. True model : nearly autoregressive model with independent errors. Autoregressive method used for the analysis, except for IB5 (300 simulations)

Design	*Ef*	*PEf*
Class	108	117*
RNB	108	117*
SNB	108	117*
IB5	105	107

TABLE V
Small designs with classical randomization (500 simulations). In rows: size of the design, in columns: type of analysis, cells: *Ef* and *PEf*

	ANOVA		IB5		AR		EV	
7 treat. 2 rep.	100	98			127	150*	129	156*
10 treat. 2 rep.	100	96	131	137	193	190	197	205

when correlation between plots are essential (for example lag-one 0.9 correlation), neighbour methods bring an improvement with respect to the classical analysis.

Note that the incomplete block analysis is not possible with 7 treatments.

5. Conclusion

When using ordinary ANOVA, neglecting neighbour effects and correlations, some theoretical results exist concerning the kind of design which should be used. Neighbour balanced designs seem to be a good choice. When using a neighbour model, conclusions are more difficult to draw. In many cases, both neighbour balanced and randomized design bring an improvement of efficiency with approximate validity or conservative bias. It remains that:

1. Conservativeness causes a loss of power of tests.

2. Some cases of small unconservativeness with small efficiency, as in Table IV, are very dangerous: the improvement due to neighbour method is very small and is widely overestimated by the statistical analysis. In such situations classical incomplete block analysis may be preferable.

Finding some valid randomization theory for neighbour models would prevent from such situations.

5.1. Towards a Randomization Theory for Neighbour Models

The simplest neighbour model is the FD model which is an ordinary linear model based on the differences between adjacent plots. As shown by the simulations, the classical way of estimating the variance is not valid, so we must find another estimator.

Since block effects disappear by taking differences, the analysis model can be written as

$$Y = X\tau + \varepsilon,$$

where:

(a) Y is the vector of yield differences; it is formed by substracting from each yield value the yield of the preceding plot in the block,

(b) $X\tau$ is the vector of treatment effect differences; the matrix X is obtained by calculating relevant differences between rows of the original design matrix,

(c) ε satisfies the usual assumptions of the analysis of variance.

Estimators of variance are wanted that are

(i) quadratic for simplicity,

(ii) invariant with respect to treatment effects and

(iii) weakly valid.

Let $P = X(X'X)^{-}X'$ be the orthogonal projector on the treatment space, and $Q = I_d - P$, where $d = b(k-1)$. Relations (i) and (ii) imply that there exists a symmetric matrix A such that

$$\hat{\sigma}^2 = Y'QAQY.$$

To check the weak validity, we suppose that the true model is

$$Y = X\tau + \xi,$$

where ξ is the vector formed with the differences of fertilities between adjacent plots within blocks. It can take any value in $\mathbb{R}^{b(k-1)}$. In that model, condition of weak validity is equivalent to

$$\xi'\overline{QAQ}\xi = \frac{1}{t-1}\xi'\bar{P}\xi \text{ for any } \xi,$$

where the bar means the mean over randomization. Thus (iii) is equivalent to

$$(t-1)\overline{QAQ} = \bar{P} \tag{3}$$

(see Bailey and Rowley, 1987, for details).

For some particular restricted randomizations, it can be shown that the solution of equation (3) exists. Works by Azaïs, Bailey and Monod are in progress on this subject.

References

Azaïs, J.-M. (1987). Design of experiments for studying intergenotypic competition. *Journal of the Royal Statistical Society B* **49**, 334–345.

Azaïs, J.-M., Bailey, R.A. and Monod, H. (1993). A catalogue of efficient neighbour-designs with border plots. *Biometrics* **49**, 1252-1261.

Bailey, R. and Rowley, C.A. (1987). Valid randomization. *Proceedings the Royal Society A* **410**, 105–124.

Baird, P. and Mead, R. (1991). The empirical efficiency and validity of two neighbour models. *Biometrics* **47**, 1473–1487.

Besag, J. and Kempton, R. (1986). Statistical analysis of field experiments using neighbouring plots. *Biometrics* **42**, 231–251.

Druilhet, P. (1993). Optimality of neighbour balanced designs. Unpublished manuscript.

Gleeson, A.C. and Eccleston, J.A. (1992). On the design of field experiments under a one-dimensional neighbour model. *The Australian Journal of Statistics* **34**, 91–97.

Green, P.J., Jennison, C. and Seheult, A.M. (1985). Analysis of field experiments by least squares smoothing. *Journal of the Royal Statistical Society B* **47**, 299–315.

Kiefer, J. (1975). Construction and optimality of generalized Youden designs. In: J.N. Srivastava, Ed., *A Survey of Statistical Design and Linear Models.* North-Holland, Amsterdam, 333–353.

Kiefer, J. and Wynn, H.P. (1981). Optimum balanced block and Latin square designs for correlated observations. *The Annals of Statistics* **9**, 737–757.

Kruskall, W. (1968). When are Gauss-Markov and least squares estimators identical? A coordinate-free approach. *The Annals of Mathematical Statistics* **39**, 70–75.

Kunert, J. (1987). Neighbour balanced block designs for correlated errors. *Biometrika* **74**, 717–724.

Morgan, J.P. and Chakravarti, I.M. (1988). Block designs for first and second order neighbour correlations. *The Annals of Statistics* **16**, 1206–1224.

Pearce, S.C. (1957). Experimenting with organisms as blocks. *Biometrika* **44**, 141–149.

Williams, E.R. (1986). A neighbour model for field experiments. *Biometrika* **73**, 279–287.

A NEW LOOK INTO COMPOSITE DESIGNS

SUBIR GHOSH and WALID S. AL-SABAH
Department of Statistics
University of California, Riverside
Riverside, CA 92521-0138
USA

Abstract. This paper is concerned with the selection of factorial points (FP's) in composite designs. A characterization of orthogonal FP's is given under a submodel of the second order response surface model. Nonorthogonal FP's are considered and their relationship to orthogonal FP's is demonstrated for a special case. The rotatability of composite designs with these FP's is discussed.

Key words: Factorial points, Orthogonality, Rotatability, Response surface, Second order model, Submodel.

1. Introduction

Consider a second order response surface model with the response variable y and k explanatory variables coded to $x_1, \ldots, x_k$. The N runs (or points) in the design are $(x_{u1}, \ldots, x_{uk})$ and the observations are $y(x_{u1}, \ldots, x_{uk})$, $u = 1, \ldots, N$. The expectation of $y(x_{u1}, \ldots, x_{uk})$ is

$$E[y(x_{u1}, \ldots, x_{uk})] = \beta_0 + \sum_{i=1}^{k} \beta_i x_{ui} + \sum_{i=1}^{k} \beta_{ii} x_{ui}^2 + \sum_{\substack{i=1 \\ i<j}}^{k} \sum_{j=1}^{k} \beta_{ij} x_{ui} x_{uj}, \tag{1}$$

where the intercept β_0, the linear coefficients β_i, the pure quadratic coefficients β_{ii} and the interaction coefficients β_{ij} are unknown constants. The model is

$$E(\underline{y}) = X\underline{\beta}, \quad V(\underline{y}) = \sigma^2 I, \tag{2}$$

where $\underline{y}(N \times 1)$ is the vector of observations, $\underline{\beta}(p \times 1), p = 1 + 2k + \binom{k}{2}, p \leq N$, is the vector of β's in (1), $X(N \times p)$ is the design matrix and σ^2 is an unknown constant. For a second order design with N runs $(x_{u1}, \ldots, x_{uk}), u = 1, \ldots, N$, rank $X = p$. A special second order design, called composite design (CD), consists of F factorial points (FP's) which are a fraction of 2^k points $(\pm 1, \ldots, \pm 1)$, $2k$ axial points (AP's) $(\pm\alpha, \ldots, 0), \ldots, (0, \ldots, \pm\alpha)$, where α is a given constant, and $n_0 (\geq 0)$ center points (CP's) $(0, \ldots, 0)$. The total number of runs is $N = F + 2k + n_0$. Box and Wilson (1951) introduced such designs. Box and Hunter (1957), Hartley (1959), Westlake (1965), Draper (1985), Draper and Lin (1990) and others constructed such designs. In their research, a particular effort was being made for requiring small number of runs in the choice of FP's.

T. Caliński and R. Kala (eds.),
Proceedings of the International Conference on Linear Statistical Inference LINSTAT '93, 223–228.

This paper presents a fresh new look into the problem of constructing such designs, particularly in choosing the factorial points. A submodel of (1) is introduced, which is new in this context. A new characterization of orthogonal fractional factorial plans under the submodel is given. Several series of orthogonal FP's (OFP's) are given in Table I. Series IV is in fact different from earlier known plans as explained in Example 1. Some nonorthogonal FP's available in the literature are listed in Table II. The interrelation between orthogonal and nonorthogonal FP's is studied for $k = 5$. The rotatability of CD's is also studied for both orthogonal and nonorthogonal FP's.

2. Factorial Points

Observations at AP's permit the unbiased estimation (UE) of β_i and $\beta_0+\alpha^2\beta_{ii}$. The CP observations provide the UE of β_0. For a second order CD, observations at FP's must at least allow the UE of $\beta_0+\sum_{i=1}^{k}\beta_{ii}$ and β_{ij}, $i<j$, $i,j=1,\ldots,k$, given that the estimators of $\beta_1,\ldots,\beta_k$ are available from AP's. (In case $n_0 = 0$, $\alpha \neq k^{1/2}$.) In view of this, the following submodel of (1) is introduced for the choice of FP's.

$$E[y(x_{u1},\ldots,x_{uk})] = \beta_0 + \sum_{\substack{i=1\\ i<j}}^{k}\sum_{j=1}^{k}\beta_{ij}x_{ui}x_{uj}, \quad u=1,\ldots,F. \tag{3}$$

In matrix notation,

$$E(\underline{y}_s) = X_s\underline{\beta}_s, \quad V(\underline{y}_s) = \sigma^2 I, \tag{4}$$

where $\underline{y}_s(F\times 1)$ is the vector of observations at FP's, $\underline{\beta}_s(p_s\times 1)$, $p_s = 1+\binom{k}{2}$, is the vector of β's in (3) and $X_s(F\times p_s)$ is the design matrix. Note that if one FP is the negative of another FP then the corresponding rows of X_s are identical. A CD with $\alpha > 0$ for $n_0 \geq 0$, and in addition $\alpha \neq k^{1/2}$ for $n_0 = 0$, is of second order if and only if rank $X_s = p_s$. For OFP's, $X_s'X_s$ is a diagonal matrix. Let x_{i_1}, x_{i_2}, x_{i_3} and x_{i_4} be 4 distinct explanatory variables; $i_1,i_2,i_3,i_4 \,\epsilon\{1,\ldots,k\}$. Denote

$$\begin{aligned} N(a,b) &= \text{Number of points from } (x_{u1},\ldots,x_{uk}), u=1,\ldots,F, \\ &\quad\ \text{satisfying } x_{ui_1}x_{ui_2}=a \text{ and } x_{ui_3}x_{ui_4}=b, \\ N(a,\cdot) &= \text{Number of points from } (x_{u1},\ldots,x_{uk}), u=1,\ldots,F, \\ &\quad\ \text{satisfying } x_{ui_1}x_{ui_2}=a, \\ N(\cdot,b) &= \text{Number of points from } (x_{u1},\ldots,x_{uk}), u=1,\ldots,F, \\ &\quad\ \text{satisfying } x_{ui_3}x_{ui_4}=b. \end{aligned} \tag{5}$$

The $N(a,b)$ defined above is in fact $N_{i_1i_2i_3i_4}(a,b)$, and the subscripts i_1,i_2,i_3 and i_4 are omitted for brevity.

Note that

$$\begin{aligned} N(a,\cdot) &= N(a,1)+N(a,-1), \\ N(\cdot,b) &= N(1,b)+N(-1,b). \end{aligned} \tag{6}$$

Theorem 1. *A necessary and sufficient condition for F points $(x_{u1}, \ldots, x_{uk}), u = 1, \ldots, F$, to be OFP's under (4) is that for every (x_{i_1}, x_{i_2}) and (x_{i_3}, x_{i_4}), where i_1, i_2, i_3 and i_4 are 4 distinct members of $\{1, \ldots, k\}$,*

$$N(1,1) = N(-1,1) = N(1,-1) = N(-1,-1) = F/4.$$

Proof. The columns of X_s for $x_{i_1}x_{i_2}$ and $x_{i_3}x_{i_4}$ are orthogonal if and only if $N(-1,-1) + N(1,1) = N(-1,1) + N(1,-1) = F/2$. The first column and the column for $x_{i_1}x_{i_2}$ are orthogonal if and only if $N(1,\cdot) = N(-1,\cdot) = F/2$. The same is true for columns corresponding to $x_{i_1}x_{i_3}$ and $x_{i_2}x_{i_3}$. The rest follows from (6). This completes the proof. □

Table I presents four series of OFP's for various values of k. Hartley (1959) pointed out that FP's need not be of resolution V. Let it be recalled that FP's are of resolution V if the UE of β_0, β_i's and β_{ij}'s is possible in (1) assuming β_{ii}'s are known. This could be as low as of resolution III [i.e., the UE of β_0 and β_i's is possible in (1) from FP's assuming β_{ii}'s and β_{ij}'s are known], with an additional condition that no two β_{ij}'s are aliased with each other in (1) [i.e., the UE of β_{ij}'s is possible in (1) assuming β_i's are known]. Draper and Lin (1990) named such plans as resolution III*. The following example shows that OFP's may not even be of resolution III.

Example 1. Consider $k = 4$ and the following 8 FP's (i.e., $F = 8$) given in a $k \times F$ matrix. The points satisfy $x_4 = 1$ (Series IV in Table I for $k = 4$).

$$\begin{bmatrix} 1 & 1 & -1 & -1 & 1 & 1 & -1 & -1 \\ 1 & 1 & 1 & 1 & -1 & -1 & -1 & -1 \\ 1 & -1 & 1 & -1 & 1 & -1 & 1 & -1 \\ 1 & 1 & 1 & 1 & 1 & 1 & 1 & 1 \end{bmatrix}. \tag{7}$$

The points also satisfy the condition of Theorem 1 and hence they are OFP's. However, the points are not of resolution III.

Table II presents a list of nonorthogonal FP's that are available in the literature. FP's given in Westlake (1965) may or may not be of Resolution III* but satisfy rank $X_s = p_s$. FP's given in Draper (1985) and Draper and Lin (1990) are however of Resolution III*. Two sets of 12 FP's given in Table II on page 178 of Draper (1985) are isomorphic under (4) in the sense that they give the same X_s matrix. Moreover, the first two rows of X_s are identical for both sets.

The OFP's for $k = 5$ are given by $x_1x_2x_3x_4x_5 = 1$ or -1 (Series III in Table I). There are 16 points. This is a regular fraction and also a Type A Orthogonal Array (OA) of strength 4. [See Rao (1947), Srivastava and Chopra (1973).] The nonorthogonal FP's given in Westlake (1965) mentioned in Table II for $k = 5$ consist of 12 points; 4 of which satisfy $x_1x_3x_4 = 1$, $x_2 = 1$, $x_5 = -1$, another 4 satisfy $x_1x_3x_4 = 1$, $x_2 = -1$, $x_5 = 1$ and the remaining 4 satisfy $x_1x_3x_4 = -1$, $x_2 = 1$, $x_5 = 1$. This is an irregular or parallel flat fraction obtained from the OFP's satisfying $x_1x_2x_3x_4x_5 = -1$ by deleting the 4 points satisfying $x_1x_3x_4 = -1, x_2 = -1$ and $x_5 = -1$. The nonorthogonal FP's in Plan a of Draper (1985) mentioned

TABLE I
Four series of orthogonal FP's

Series	k	Orthogonal FP's	F	Reference	Property
I	$k = 3t$, $t = 1, 2, \ldots$	$x_1x_2x_3 = 1$ or -1, $x_4x_5x_6 = 1$ or -1, $\vdots$ $x_{k-2}x_{k-1}x_k = 1$ or -1	2^{2t}	Hartley (1959)	Res III*
II	$k = 3t + 1$, $t = 1, 2, \ldots$	$x_1x_2x_3 = 1$ or -1, $x_4x_5x_6 = 1$ or -1, $\vdots$ $x_{k-2}x_{k-1}x_k = 1$ or -1	2^{2t+1}	Hartley (1959)	Res III*
III	$k = 3t + 2$, $t = 1, 2, \ldots$	$x_1x_2x_3x_4x_5 = 1$ or -1, $x_4x_5x_6x_7x_8 = 1$ or -1, $\vdots$ $x_{k-4}x_{k-3}x_{k-2}x_{k-1}x_k = 1$ or -1	2^{2t+2}	Hartley (1959) Box and Hunter (1957)	Res V
IV	All k	$x_k = 1$ or -1	2^{k-1}		Not Res III

TABLE II
List of Nonorthogonal FP's

k	F	Source
5	12	Westlake (1965), pages 329, 333-335.
	12, 11	Draper (1985), Plan b and Plan a (deleting the repeat run), Table 2, page 175.
7	24	Westlake (1965), page 335 (deleting two repeat runs).
	24, 22	Draper and Lin (1990), pages 191, 192.
8	36, 30	Draper and Lin (1990), pages 191, 192.
9	38, 40	Draper and Lin (1990), pages 191, 192.
10	48, 46	Draper and Lin (1990), pages 191, 192.

in Table II for $k = 5$ are obtained from 12 run Plackett and Burman design. They are given below.

$$\begin{bmatrix} -1 & -1 & -1 & -1 & -1 & -1 & 1 & 1 & 1 & 1 & 1 & 1 \\ -1 & -1 & -1 & 1 & 1 & 1 & -1 & -1 & -1 & 1 & 1 & 1 \\ -1 & -1 & 1 & -1 & 1 & 1 & -1 & 1 & 1 & -1 & -1 & 1 \\ -1 & -1 & 1 & 1 & -1 & 1 & 1 & -1 & 1 & -1 & 1 & -1 \\ -1 & -1 & 1 & 1 & 1 & -1 & 1 & 1 & -1 & 1 & -1 & -1 \end{bmatrix}.$$

This is a Type C OA of strength 2 and also a Balanced Array of strength 5. [See Srivastava and Chopra (1973).] The last 10 points appear in the OFP's satisfying $x_1x_2x_3x_4x_5 = 1$. The points $(-1,-1,-1,-1,-1)$ and $(1,1,1,1,1)$ give identical rows in X_s. Plan a is therefore a subset of the OFP's satisfying $x_1x_2x_3x_4x_5 = 1$.

3. Rotatability

Let $\hat{y}(x_1,\ldots,x_k)$ be the predicted response at the point $(x_1,\ldots,x_k)$. Then

$$\hat{y}(x_1,\ldots,x_k) = b_0 + \sum_{i=1}^{k} b_i x_i + \sum_{i=1}^{k} b_{ii} x_i^2 + \sum_{\substack{i=1 \\ i<j}}^{k} \sum_{j=1}^{k} b_{ij} x_i x_j, \tag{8}$$

where b's are the least squares estimators of β's under (2). Denote

$$\underline{x}*' = (1; x_1,\ldots,x_k; x_1^2,\ldots,x_k^2; x_1x_2,\ldots,x_{k-1}x_k).$$

Now $\sigma^{-2}V[\hat{y}(x_1,\ldots,x_k)] = \underline{x}*'(X'X)^{-1}\underline{x}*$. A second order CD is rotatable if $\sigma^{-2}V[\hat{y}(x_1,\ldots,x_k)]$ depends on $x_1,\ldots,x_k$ only through $\rho^2 = x_1^2+\ldots+x_k^2$. [See Box and Draper (1987), Khuri and Cornell (1987) and Draper, Gaffke and Pukelsheim (1993).] If a CD consists of OFP's, then the matrix $X'X$ takes the following form:

$$X'X = \begin{bmatrix} N & \underline{a}_1' & F+2\alpha^2 \ldots F+2\alpha^2 & \underline{0}' \\ \underline{a}_1 & (F+2\alpha^2)I & \underline{a}_1 \cdots \underline{a}_1 & A \\ F+2\alpha^2 & \underline{a}_1' & & \\ \vdots & \vdots & 2\alpha^4 I + FJ & 0' \\ F+2\alpha^2 & \underline{a}_1' & & \\ \underline{0} & A' & 0 & FI \end{bmatrix}, \tag{9}$$

where I is the identity matrix, J is a matrix with all elements being unity, $\underline{a}_1'(1\times k)$ is a vector, $A(k\times\binom{k}{2})$ is a matrix. It can be seen that

$$\begin{aligned} \underline{a}_1 &= \underline{0}, A = 0, \text{ for OFP's of Series III in Table I,} \\ \underline{a}_1 &= \underline{0}, A \neq 0, \text{ for OFP's of Series I and II in Table I,} \\ \underline{a}_1 &\neq \underline{0}, A \neq 0, \text{ for OFP's of Series IV in Table I.} \end{aligned} \tag{10}$$

For Series I and II in Table I, only $3t$ elements of A are in fact nonzero. It is now clear from (10) that a CD with OFP's of Series III in Table I is a rotatable design. However, CD's with OFP's of Series I, II and IV in Table I are not rotatable. The following example explains this clearly.

Example 2. Consider $k = 3$ and the 4 OFP's satisfying $x_1x_2x_3 = -1$ (Series I in Table 1 for $k = 3$), $n_0 = 1$ and a CD with 11 points. Clearly $F = 4$, $\underline{a}_1 = \underline{0}$ and

$$A = \begin{bmatrix} 0 & 0 & -4 \\ 0 & -4 & 0 \\ -4 & 0 & 0 \end{bmatrix}.$$

Define

$$\begin{aligned} \omega &= 84 + 10\alpha^4 - 48\alpha^2, \\ \omega a &= 12 + 2\alpha^4, \quad \omega b = -(4 + 2\alpha^2), \\ 2\omega c\alpha^4 &= 56 + 14\alpha^4 - 32\alpha^2. \end{aligned} \tag{11}$$

It can be checked that

$$\sigma^{-2}V[\hat{y}(x_1, x_2, x_3)] =$$

$$a + (2b + \frac{4}{8\alpha^2})\rho^2 + c\rho^4 + \frac{2(\alpha^2 - 2)}{8\alpha^4}(x_1^2x_2^2 + x_1^2x_3^2 + x_2^2x_3^2) + \frac{24}{8\alpha^2}x_1x_2x_3, \tag{12}$$

where $\rho^2 = x_1^2 + x_2^2 + x_3^2$. The last two terms on the right hand side in (12) are contributing to the departure from rotatability. One of these two terms becomes zero when $\alpha^2 = 2$. A CD with nonorthogonal FP's is not rotatable.

4. Conclusion

The submodel introduced in this paper [formula(3)] is instrumental in the selection of FP's in composite designs. It is demonstrated that nonorthogonal FP's in Westlake (1965) and Draper (1985) for $k = 5$ are in fact embedded in OFP's given in Series III (Table I). Composite designs with OFP's are not necessarily rotatable.

References

Box, G.E.P. and Wilson, K.B. (1951). On the experimental attainment of optimum conditions. *Journal of the Royal Statistical Society B* **13**, 1–45.

Box G.E.P. and Hunter, J.S. (1957). Multi-factor experimental designs for exploring response surfaces. *The Annals of Mathematical Statistics* **28**, 195–241.

Box, G.E.P. and Draper, N.R. (1987). *Empirical Model-Building and Response Surfaces.* Wiley, New York.

Draper, N.R. (1985). Small composite designs. *Technometrics* **27**, 173–180.

Draper, N.R. and Lin, D.K.J. (1990). Small response-surface designs. *Technometrics* **32**, 187–194.

Draper, N. R., Gaffke, N. and Pukelsheim, F. (1993). Rotatability of variance surfaces and moment matrices. *Journal of Statistical Planning and Inference* **36**, 347–355.

Hartley, H.O. (1959). Smallest composite designs for quadratic response surfaces. *Biometrics* **15**, 611–624.

Khuri, A.I. and Cornell, J.A. (1987). *Response Surfaces, Designs and Analyses.* Marcel Dekker, New York.

Rao, C.R. (1947). Factorial experiments derivable from combinatorial arrangement of arrays. *Journal of the Royal Statistical Society, Supplement* **9**, 128–139.

Srivastava, J. N. and Chopra, D.V. (1973). Balanced arrays and orthogonal arrays. In: J.N. Srivastava, Ed., *A Survey of Combinatorial Theory.* North-Holland, Amsterdam, 411–428.

Westlake, W.J. (1965). Composite designs based on irregular fractions of factorials. *Biometrics* **21**, 324–336.

USING THE COMPLEX LINEAR MODEL TO SEARCH FOR AN OPTIMAL JUXTAPOSITION OF REGULAR FRACTIONS

H. MONOD and A. KOBILINSKY
INRA-Versailles
Laboratoire de Biométrie
Route de Saint-Cyr
F-78026 Versailles Cedex
France

Abstract. The use of a linear model with complex parameters to study factorial designs, has been proposed and detailed by Bailey (1982, 1990) and Kobilinsky (1985, 1990). The complex parameterisation arises quite naturally from the representation theory of abelian groups, which had been used earlier by Foata (1961) and Chakravarti (1976) in the context of factorial designs. This approach has been shown to lead to much simplified calculations to study the efficiency and optimality properties of factorial designs (Collombier, 1989; Kobilinsky and Monod, 1991).

Despite these advantages, the complex linear model is not yet well known in the statistical community. The aim of this paper is to illustrate its usefulness. After a short presentation, we do this by concentrating on a specific problem: the search for an optimal half-fraction of a $2\times2\times2\times p\times q$ factorial design of resolution V.

Key words: Linear model, Factorial designs, Information matrix, Optimality.

1. The Complex Linear Model

1.1. Complex Parameterization of Factorial Effects

Consider an experiment with s factors F_i at m_i levels, for $i = 1, \ldots, s$. In the sequel, the levels of factor F_i are identified with the elements of the cyclic additive group $\mathbf{Z}_{m_i}$ of integers modulo m_i. The treatments are denoted by $(f_1, \ldots, f_s) \in T$, where $T = \mathbf{Z}_{m_1} \times \ldots \times \mathbf{Z}_{m_s}$ denotes the set of treatments.

Let $\tau(f_1, \ldots, f_s)$ denote the effect of treatment $(f_1, \ldots, f_s)$. We define exactly $m_1 \times \ldots \times m_s$ complex parameters $e(F_1^{\phi_1} \ldots F_s^{\phi_s})$, for $(\phi_1 \ldots \phi_s) \in \mathbf{Z}_{m_1} \times \ldots \times \mathbf{Z}_{m_s}$, by

$$e(F_1^{\phi_1} \ldots F_s^{\phi_s}) = \frac{1}{m_1 \times \ldots \times m_s} \sum_{f_1=0}^{m_1-1} \cdots \sum_{f_s=0}^{m_s-1} \eta_{m_1}^{-\phi_1 f_1} \ldots \eta_{m_s}^{-\phi_s f_s} \tau(f_1, \ldots, f_s),$$

where η_m denotes the primitive m^{th} root of unity.

We use the notational convention that, for instance, $e(1) = e(F_1^0 \ldots F_s^0)$ and $e(F_1^\phi) = e(F_1^\phi F_2^0 \ldots F_s^0)$. It is clear that $e(1)$ represents the general mean of all treatment effects. Let $\tau_\bullet(F_i = f)$ denote the average treatment effect over all treatments with level f of F_i, then $e(F_i^\phi)$ is equal to

$$e(F_i^\phi) = \frac{1}{m_i} \sum_{f=0}^{m_i-1} \eta_{m_i}^{-\phi f}\ \tau_\bullet(F_i = f),$$

T. Caliński and R. Kala (eds.),
Proceedings of the International Conference on Linear Statistical Inference LINSTAT '93, 229–237.

so it represents a contrast associated with the main effect of F_i. In particular, if $m_i = 2$, we get

$$e(F_i) = \frac{1}{2}\{\tau_\bullet(F_i = 0) - \tau_\bullet(F_i = 1)\}.$$

More generally, the parameters $e(F_1^{\phi_1} \dots F_s^{\phi_s})$ associated with the main effect of F_i are those for which only ϕ_i is different from 0. Those associated with the interaction between any given factors are those for which the corresponding ϕ's, and those only, are different from 0.

The definitions above express the complex parameters associated with factorial effects as linear combinations of the treatment effects. We thus have the matrix relation

$$e = \frac{1}{t}K^*\tau,$$

where e is the vector of factorial effects parameters, K is a square matrix with complex entries, K^* is its conjugate transpose, and $t = | T | = m_1 \times \dots \times m_s$. It can be verified that the columns of K form an orthonormal basis of $\mathbf{C}^T$, for the usual inner product on a complex field defined by $\langle u, w\rangle = \sum u_i \overline{w}_i$. Therefore, K^*K is equal to t times the identity matrix of order t, and

$$\tau = Ke. \tag{1}$$

There is a direct relationship between the complex parameterization described above and the representation theory of abelian groups. We refer to Kobilinsky (1990) or Bailey (1990) for more details.

1.2. Linear Model

The models we consider are given by

$$y = X\tau + \varepsilon, \tag{2}$$

where y is the response vector, X is the 0-1 design matrix, and ε is a vector of independent and identically distributed error terms.

It is commonly assumed that some factorial effects, usually high-degree interactions, are negligible. We can then use (1) to write model (2) as

$$y = Z\underline{e} + \varepsilon, \tag{3}$$

where $\underline{e}$ is the vector e restricted to the non-negligible factorial effects, and $Z = X\underline{K}$, with $\underline{K}$ being the matrix K restricted to the columns associated with non-negligible factorial effects.

It is easy to construct Z. Let $(F_1^{\phi_1} \dots F_s^{\phi_s})$ denote its column corresponding to the factorial effect $e(F_1^{\phi_1} \dots F_s^{\phi_s})$. Then column (1) is filled by 1's only; in column (F_i^{ϕ}), the entry in the row associated with any unit receiving a treatment with level f_i of factor F_i is equal to $\eta_{m_i}^{\phi f_i}$; the column $(F_1^{\phi_1} \dots F_s^{\phi_s})$ associated with an interaction is then obtained by componentwise multiplication of the columns $(F_1^{\phi_1}), \dots, (F_s^{\phi_s})$.

Remark. In the sequel, we occasionally use the notation $(F_1^{\phi_1} \dots F_s^{\phi_s})$ even if $e(F_1^{\phi_1} \dots F_s^{\phi_s})$ is not included in the model. It then denotes the column of Z which would be associated with $e(F_1^{\phi_1} \dots F_s^{\phi_s})$ if it were included in the model.

1.3. Information Matrix

What we are most interested in, here, is the complex information matrix Z^*Z. Indeed, Kobilinsky (1990) shows that the efficiency and optimality properties of a factorial design can be studied through Z^*Z in much the same way as through a real information matrix.

The entries of Z^*Z are the inner products between the columns of Z. Note that all the entries of Z have their modulus equal to 1, so that the diagonal of Z^*Z is entirely filled by the number N of experimental units. Two important properties are given by the propositions below.

Proposition 1. *The inner product between columns $(F_1^{\phi_1} \ldots F_s^{\phi_s})$ and $(F_1^{\psi_1} \ldots F_s^{\psi_s})$ of Z is equal to the inner product between $(F_1^{\phi_1-\psi_1} \ldots F_s^{\phi_s-\psi_s})$ and* (1).

Proposition 2. *If all the combinations of levels of factors $F_1, \ldots, F_r$ are equally replicated, then the inner product between any two columns $(F_1^{\phi_1} \ldots F_r^{\phi_r})$ and $(F_1^{\psi_1} \ldots F_r^{\psi_r})$ of Z is equal to N if $\phi_i = \psi_i$ for $i = 1, \ldots, r$, and equal to zero otherwise.*

1.4. Optimality

We suppose that all the treatment effects which are not negligible are of equal interest. We are looking for a design which makes it possible to estimate all these effects with maximal efficiency. To compare designs, we consider criteria based on the canonical efficiencies, which are equal to the eigenvalues of $(1/N)Z^*Z$.

In the present setting, the trace of Z^*Z is equal to N times the number g of parameters in $\underline{e}$, so the sum of the canonical efficiencies of any design is equal to g. The notion of Schur-optimality (Giovagnoli and Wynn, 1981; Shah and Sinha,1989; Bailey, Monod and Morgan, 1993) is well adapted to this setting.

Definition 1. Let $\lambda_1 \leq \ldots \leq \lambda_g$ and $\mu_1 \leq \ldots \leq \mu_g$ denote the canonical efficiency factors of designs d_L and d_M, respectively, in non-decreasing order. Then d_L is said to be Schur-better than d_M if

$$\lambda_1 + \ldots + \lambda_k \geq \mu_1 + \ldots + \mu_k \text{ for } k = 1, \ldots, g,$$

with equality when $k = g$. A design is said to be Schur-optimal in a given class if it is Schur-better than any other design in the class.

A Schur-optimal design minimizes all Schur-convex functions of the canonical efficiency factors (see Marshall and Olkin, 1979). In particular, it is A-, D- and E-optimal.

2. Application

2.1. The Problem

We now consider an experiment to study five factors A, B, C, D and E at 2, 2, 2, p and q levels, respectively. Only $N = 4pq$ units are available. It is assumed that all interactions between three or more factors are negligible, and we want to estimate all main effects and all two-factor interactions. So we are looking for a half-fraction of resolution V.

This problem is very specific, but it arises quite often in practice with values of p and q small but larger than 2. Some methods in the literature provide efficient solutions, but the conditions for such fractions to be optimal are not yet known.

2.2. Solutions from the Literature

When $p = q = 2$, it is well-known that the half-fraction consisting of all the treatments (a, b, c, d, e) satisfying $a + b + c + d + e = 0 \bmod 2$ is orthogonal and optimal for the estimation of each main effect or two-factor interaction (Kobilinsky, 1990).

When $p = q = 3$, Fry (1961) proposes a method which is equivalent to taking all the treatments satisfying $a + b + c + n(d) + n(e) = 0 \bmod 2$, where, for instance, $n(0) = n(1) = 0$ and $n(2) = 1$. Connor and Young (1961) propose to take the juxtaposition of the three regular 1/6 fractions defined by:

$$\begin{array}{ccc} (1) & (2) & (3) \\ a+b+c=0 \bmod 2 & a+b+c=1 \bmod 2 & a+b+c=0 \bmod 2 \\ d+e=0 \bmod 3 & d+e=1 \bmod 3 & d+e=2 \bmod 3 \end{array}$$

The DSIGN key method (Patterson, 1976) is a more general method, which gives the two examples above as special cases.

2.3. A Class of Fractions

Both examples above consist of a half-fraction defined by the equation

$$a + b + c + n(d, e) = 0 \bmod 2, \tag{4}$$

where n is a 0-1 function of the levels of D and E (see Table I).

TABLE I
Values of $n(d, e)$ for Fry's design (a) and for Connor and Young's design (b)

(a)

		level of E		
	$n(d,e)$	0	1	2
level	0	0	0	1
of	1	0	0	1
D	2	1	1	0

(b)

		level of E		
	$n(d,e)$	0	1	2
level	0	0	1	0
of	1	1	0	0
D	2	0	0	1

For given values of p and q, we will denote by $C_{p,q}$ the class of all half-fractions satisfying equation (4) for some function $n(d, e)$. A particular fraction in $C_{p,q}$ is then entirely determined by the function $n(d, e)$.

In the sequel, we are going to look for an optimal fraction within the class $\mathcal{C}_{p,q}$. There will be no guarantee that such a fraction is optimal over all possible fractions. It is particularly interesting, however, to note that the D-optimal fractions found by the DetMax algorithm implemented in SAS (SAS Institute Inc., 1989), for $p = q = 3$ and $p = 3, q = 5$ belong to the class $\mathcal{C}_{p,q}$ (up to permutations of the levels of each factor). For $p = q = 3$, the D-optimal fraction corresponds to the method of Connor and Young (Table Ib). For $p = 3, q = 5$, the D-optimal fraction corresponds to the function $n(d, e)$ given in Table II.

TABLE II
Values of $n(d, e)$ for a D-optimal fraction

		level of E				
	$n(d,e)$	0	1	2	3	4
level	0	0	0	0	1	1
of	1	0	1	1	0	0
D	2	1	1	0	0	0

2.4. Information Matrix for the Fractions of $\mathcal{C}_{p,q}$

We start from the following results.

Proposition 3. *For any fraction in $\mathcal{C}_{p,q}$, the combinations of levels of A, B, D, E, those of A, C, D, E, and those of B, C, D, E, are replicated exactly once.*

Proof. For a given combination (a, b, d, e) of levels of A, B, D, E, there is exactly one level c of C satisfying $a + b + c + n(d, e) = 0 \bmod 2$. □

The model we consider contains the general mean, the main effects, and the two-factor interactions. Let the columns of Z be ordered in the following way:

$$\begin{array}{l}(1),\\ (BC),(A),(AD)\ldots(AD^{p-1}),(AE)\ldots(AE^{q-1}),\\ (AC),(B),(BD)\ldots(BD^{p-1}),(BE)\ldots(BE^{q-1}),\\ (AB),(C),(CD)\ldots(CD^{p-1}),(CE)\ldots(CE^{q-1}),\\ (D)\ldots(D^{p-1}),(E)\ldots(E^{q-1}),(DE)\ldots(D^{p-1}E^{q-1}).\end{array}$$

As a consequence of Propositions 1, 2 and 3, any two distinct columns $(A^{\alpha_1}B^{\beta_1}C^{\gamma_1}D^{\delta_1}E^{\varepsilon_1})$ and $(A^{\alpha_2}B^{\beta_2}C^{\gamma_2}D^{\delta_2}E^{\varepsilon_2})$ are orthonormal unless $\alpha_1 + \alpha_2 = \beta_1 + \beta_2 = \gamma_1 + \gamma_2 = 1 \bmod 2$. So, matrix $(1/N)Z^*Z$ is a block-diagonal matrix, with the division into blocks corresponding to the five groups of parameters given in separate rows just above. The first diagonal block is just the scalar 1, and the last diagonal block is the identity matrix of order $pq-1$. The three intermediate blocks, which we denote by M_A, M_B, M_C, have the following form

$$M_l = \begin{pmatrix} 1 & u_l^* \\ u_l & I_{p+q-1} \end{pmatrix},$$

where u_l, for $l \in \{A,B,C\}$, is a vector of length $p+q-1$, and I_{p+q-1} is the identity matrix of order $p+q-1$.

The eigenvalues of M_l are 1 with multiplicity $p+q-2$, $1-||u_l||$ with multiplicity 1, and $1+||u_l||$ with multiplicity 1. So the eigenvalues of $(1/N)Z^*Z$ are:

$$\begin{aligned}
&1-||u_A||, 1-||u_B||, 1-||u_C|| \text{ each with multiplicity 1,}\\
&1 \text{ with multiplicity } 3(p+q-2)+pq,\\
&1+||u_A||, 1+||u_B||, 1+||u_C|| \text{ each with multiplicity 1.}
\end{aligned}$$

2.5. CALCULATIONS

The results of the previous section show that a fraction of $\mathcal{C}_{p,q}$ is Schur-optimal if it minimizes the norms of u_A, u_B, u_C simultaneously. In this section, we detail the relation between these quantities and the function $n(d,e)$.

Let $n_{d+}(\alpha)$, for $\alpha \in \mathbf{Z}_2$, denote the number of levels e of factor E such that $n(d,e)=\alpha$, and let $\Delta_{d+} = n_{d+}(0) - n_{d+}(1)$. Similarly, let $n_{+e}(\alpha)$, for $\alpha \in \mathbf{Z}_2$, denote the number of levels d of factor D such that $n(d,e)=\alpha$, and let $\Delta_{+e} = n_{+e}(0) - n_{+e}(1)$. Finally, let Δ_{++} be equal to $\sum_d \Delta_{d+} = \sum_e \Delta_{+e}$.

Proposition 4. *For any fraction in* $\mathcal{C}_{p,q}$,

$$||u_A||^2 = ||u_B||^2 = ||u_C||^2 = \frac{2^4}{N^2}\left\{p\sum_{d=0}^{p-1}\Delta_{d+}^2 + q\sum_{e=0}^{q-1}\Delta_{+e}^2 - \Delta_{++}^2\right\}.$$

Proof. The entries of u_A are equal to

$$\begin{aligned}
u_A[A] &= (1/N)\langle A, BC\rangle = (2^2/N)\Delta_{++},\\
u_A[AD^\delta] &= (1/N)\langle AD^\delta, BC\rangle = (2^2/N)\textstyle\sum_d \eta_p^{\delta d}\Delta_{d+}, \text{ for } \delta = 1,\dots,p-1,\\
u_A[AE^\varepsilon] &= (1/N)\langle AE^\varepsilon, BC\rangle = (2^2/N)\textstyle\sum_e \eta_q^{\varepsilon e}\Delta_{+e}, \text{ for } \varepsilon = 1,\dots,q-1.
\end{aligned}$$

Hence,

$$||u_A||^2 = \frac{2^4}{N^2}\left\{\sum_{\delta=1}^{p-1}\left|\sum_d \eta_p^{\delta d}\Delta_{d+}\right|^2 + \sum_{\varepsilon=1}^{q-1}\left|\sum_e \eta_q^{\varepsilon e}\Delta_{+e}\right|^2 + \Delta_{++}^2\right\}.$$

Developing this expression and using

$$\sum_{\delta=0}^{p-1}\eta_p^{\delta(d_1-d_2)} = \begin{cases} 0, & \text{if } d_1 \neq d_2 \bmod p,\\ p & \text{otherwise,}\end{cases}$$

and similar property with η_q, gives the result. □

2.6. Optimal Fractions in $C_{p,q}$

Gathering the results of the previous sections, we arrive at the following result.

Proposition 5. *A fraction of $C_{p,q}$ is Schur-optimal if and only if it minimizes*

$$\Gamma = p\sum_{d=0}^{p-1} \Delta_{d+}^2 + q\sum_{e=0}^{q-1} \Delta_{+e}^2 - \Delta_{++}^2.$$

To minimize Γ, we note that

$$\Gamma = p\sum_{d=0}^{p-1}(\Delta_{d+} - \frac{1}{p}\Delta_{++})^2 + q\sum_{e=0}^{q-1} \Delta_{+e}^2.$$

Thus, Γ is necessarily positive. In addition, each Δ_{+e} is necessarily odd if p is odd. Consequently, Γ is greater than or equal to q^2 if p is odd. These remarks lead us to the following solutions.

If p and q are even.

The fraction is Schur-optimal in $C_{p,q}$ if and only if $\Delta_{d+} = 0$ for all d's and $\Delta_{+e} = 0$ for all e's. A particular solution is given by setting $n(d,e) = n_1(d) + n_2(e) \bmod 2$, where

$$n_1(d) = 0 \text{ if } 0 \leq d \leq (p/2) - 1, \quad \text{and} \quad n_1(d) = 1 \text{ if } (p/2) \leq d \leq p - 1,$$
$$n_2(e) = 0 \text{ if } 0 \leq e \leq (q/2) - 1, \quad \text{and} \quad n_2(e) = 1 \text{ if } (q/2) \leq e \leq q - 1.$$

Note that this solution can be obtained by using pseudofactors for D and E (Kobilinsky and Monod, 1991). The fraction is completely orthogonal for the general mean, the main effects, and the two-factor interactions.

If p is odd and q is even.

The fraction is Schur-optimal in $C_{p,q}$ if $\Delta_{d+} = 0$ for all d's and $\Delta_{+e} = \pm 1$ for all e's. A particular solution is given by setting $n(d,e) = n_1(d) + n_2(e) \bmod 2$, where

$$n_1(d) = 0 \text{ if } 0 \leq d \leq (p+1)/2, \quad \text{and} \quad n_1(d) = 1 \text{ if } (p+1)/2 \leq d \leq p - 1,$$
$$n_2(e) = 0 \text{ if } 0 \leq e \leq (q/2) - 1, \quad \text{and} \quad n_2(e) = 1 \text{ if } (q/2) \leq e \leq q - 1.$$

If p and q are odd.

Without loss of generality, suppose $p \leq q$. The fraction is Schur-optimal in $C_{p,q}$ if $\Delta_{d+} = 1$ for all d's and $\Delta_{+e} = \pm 1$ for all e's. Note that Δ_{++} is then equal to p, so there must be $(q+p)/2$ Δ_{+e}'s equal to $+1$ and $(q-p)/2$ equal to -1.

It is more difficult than in the previous cases, to construct a table giving values of $n(d,e)$ consistent with the optimality requirements. We propose the following solution:

• the rows of the table are identified with the levels of D, that is, with the elements of $\mathbf{Z}_p$, and similarly, the columns of the table are identified with the elements of $\mathbf{Z}_q$;

• for $0 \leq d \leq (p-1)/2$, row d is made of zeros in columns d to $(q-1)/2+d$ mod q, and of ones in the other columns;

• for $(p+1)/2 \leq d \leq p-1$, row d is made of zeros in columns $(q-p)/2+d$ to $-(p+1)/2+d$ mod q, and of ones in the other columns.

This method ensures of course that each row has $(q+1)/2$ zeros and $(q-1)/2$ ones. In addition, Δ_{+e} is equal to +1 for the first $(q+p)/2$ columns and to −1 for the last $(q-p)/2$ columns. An example is given in Table III for $p=5$ and $q=9$.

TABLE III
Values of $n(d,e)$ for an optimal fraction with $p=5$ and $q=9$

		level of E								
	$n(d,e)$	0	1	2	3	4	5	6	7	8
	0	0	0	0	0	0	1	1	1	1
level	1	1	0	0	0	0	0	1	1	1
of	2	1	1	0	0	0	0	0	1	1
D	3	0	1	1	1	1	0	0	0	0
	4	0	0	1	1	1	1	0	0	0

References

Bailey, R. A. (1982). Dual Abelian groups in the design of experiments. In: P. Schultz, C. E. Praeger and R. P. Sullivan, Eds., *Algebraic Structures and Applications.* Marcel Dekker, New York, 45–54.

Bailey, R. A.(1990). Cyclic designs and factorial designs. In: R.R. Bahadur and K. Sen, Eds., *Proceedings of the R. C. Bose Symposium on Probability, Statistics and Design of Experiments.* Eastern Wiley, New Dehli, 51–74.

Bailey, R. A., Monod, H. and Morgan, J. P. (1993). Construction and optimality of affine-resolvable designs. Manuscript, Goldsmiths' College, London. Submitted for publication.

Chakravarti, I. M. (1976). Optimal linear mapping of a Burnside group and its applications. *Atti dei Convegni Lincei* **17**, 171–181.

Collombier, D. (1989). Optimality of some fractional factorial designs. In: Y. Dodge, V.V. Fedorov and H.P. Wynn, Eds., *Optimal Design and Analysis of Experiments.* North-Holland, Amsterdam, 39–45.

Connor, W. S. and Young, S. (1961). Fractional factorial designs with factors at two and three levels. *National Bureau of Standards, Applied Mathematics Series* **58**.

Foata, C. (1961). Sur la construction des plans factoriels et certains codes correcteurs à l'aide des caractères des groupes abéliens. In: *Le Plan d'Expériences, Colloques Internationaux du Centre National de la Recherche Scientifique No 110.* CNRS, Paris, 137–146.

Fry, R. E. (1961). Finding new fractions of factorial experimental designs. *Technometrics* **3**, 359–370.

Giovagnoli, A. and Wynn, H. P. (1981). Optimum continuous block designs. *Proceedings of the Royal Society London A* **377**, 405–416.

Kobilinsky, A. (1985). Confounding in relation to duality of finite Abelian groups. *Linear Algebra and its Applications* **70**, 321–347.

Kobilinsky, A. (1990). Complex linear models and factorial designs. *Linear Algebra and its Applications* **127**, 227–282.

Kobilinsky A. and Monod H. (1991). Experimental designs and group morphisms: an introduction. *Scandinavian Journal of Statistics* **18**, 119–134.

Marshall, A. W. and Olkin, I. (1979). *Inequalities: Theory of Majorization and its Applications.* Academic Press, New York.

Patterson, H. D. (1976). Generation of factorial designs. *Journal of the Royal Statistical Society B* **38**, 175–179.

SAS Institute Inc. (1989). *SAS/QC® Software: Reference, Version 6.* SAS Institute Inc., Cary. North Carolina.

Shah, K. R. and Sinha, B. K. (1989). *Theory of Optimal Designs. Lecture Notes in Statistics* **54**. Springer-Verlag, New York.

SOME DIRECTIONS IN COMPARISON OF LINEAR EXPERIMENTS: A REVIEW

CZESLAW STĘPNIAK and ZDZISLAW OTACHEL
Institute of Applied Mathematics
Agricultural University of Lublin
Akademicka 13
20-950 Lublin
Poland.

Abstract. In this expository work basic ideas and recent results in comparing linear experiments are systematized and discussed.

Key words: Linear experiment, Normal linear experiment, Comparison of experiments, Sufficient transformation.

1. Introduction

Any statistical experiment can be perceived as a random machinery induced by various factors. Some of the factors called unknown parameters are beyond our control, while the others, such as configuration of regression points or allocation of experimental units remain in our hand and define *a design of experiment.*

A question may arise how to compare possible designs with respect to potential statistical information. Further specifications of the task may lead to different concepts of comparison of statistical experiments.

The first idea, due to Blackwell (1951, 1953) is closely related to the well known notion of sufficiency and, in consequence, involves all statistical decision problems. It was developed, among others, by LeCam (1964), Torgersen (1972, 1991), Goel and DeGroot (1979) and Lehmann (1988). For comprehensive information in the subject we refer to a book by Torgersen (1991). The reader should be aware that LeCam (1964) developed a more general theory of approximate comparisons. Applications to normal linear experiments may be found in LeCam (1986), Swensen (1980) and Torgersen (1991). We shall here restrict ourselves to the comparison by sufficiency. This tool has been used to compare *normal linear experiments*, among others, by Kiefer (1959), DeGroot (1966), Sinha (1973), Hansen and Torgersen (1974) and Stępniak (1982).

Independently of the Blackwell's concept which refers to *general statistical experiments*, Ehrenfeld (1955) originated a simple and efficient idea for comparing the *usual linear experiments*, taking into consideration only classical problems of linear estimation. After restoring by Stępniak (1976), the idea was developed in the past decade by Baksalary and Kala (1981), Stępniak and Torgersen (1981), Drygas (1983), Baksalary (1984), Stępniak (1983, 1985, 1987a), Stępniak, Wang and Wu (1984) and Torgersen (1984). In the past years, in addition to linear estimation,

T. Caliński and R. Kala (eds.),
Proceedings of the International Conference on Linear Statistical Inference LINSTAT '93, 239–244.

some problems of quadratic estimation and F-testing have also been included in the problem by Mueller (1987) and Oktaba, Kornacki and Wawrzosek (1988).

Because of an extensive dispersion of the results in the literature, the actual information on comparisons of linear experiments is difficult to obtain. Our goal is to systematize both the ideas and the recent results in this subject.

2. Preliminaries

Let X and Y be random vectors with corresponding distributions P_θ and Q_θ, where θ is an unknown parameter. In this context we consider a statistical decision problem involving a loss function $L(a,\theta)$, where a is an action. Referring the problem to the experiment induced by X we deal with a decision rule $\delta = \delta(X)$ and the risk function $R(\delta,\theta) = E_\theta L(\delta(X),\theta)$. For a corresponding decision rule in the experiment induced by Y we use the symbol $\gamma = \gamma(Y)$.

Any idea of a comparison of experiments reduces to a goodness-type partial ordering. Intuitively, the random vector X is at least as good as the random vector Y if for any decision rule $\gamma(Y)$ there exists a rule $\delta(X)$ such that $R(\delta,\theta) \le R(\gamma,\theta)$ for all θ. In order to make this precise we need to specify:

(a) the decision problems, and

(b) the classes of decision rules δ and γ.

In this paper we shall restrict our attention to linear experiments and to linear normal experiments. The main options considered here are the following.

1. Total ordering $\succ^t$ [all possible decision rules δ and γ and all possible loss functions].
2. Ordering $\succ^l$ for linear estimation [decision rules of the type $a'X$ and $b'Y$ under the quadratic loss].
3. Ordering $\succ^q$ for quadratic estimation [decision rules of the type $X'QX$ and $Y'SY$ under the quadratic loss].

Let X be a random vector with expectation $EX = A\beta$ and covariance matrix of the form $\mathrm{Cov}X = \sum_{i=1}^q \sigma_i V_i$, where A is a known $n_A \times p$ matrix, $V_1,\ldots,V_q$ are known symmetric nonnegative definite matrices of order n_A, while $\beta \in \mathbb{R}^p$ and $\sigma_1,\ldots,\sigma_q, \sigma_i^2 \ge 0, i = 1,\ldots,q$, are unknown parameters. An experiment which may be expressed in terms of such an observation vector X will be called a *linear experiment* and will be denoted by $\mathcal{L}(A\beta, \sum \sigma_i V_i)$. For a corresponding *normal linear experiment* we use the symbol $\mathcal{N}(A\beta, \sum \sigma_i V_i)$.

Special attention in the statistical literature is given to experiments of the form $\mathcal{N}(A\beta, \sigma I_{n_A})$, where σ may be known or not. The basic result in the *total* comparison of such experiments is given in the following

Theorem 1 (Hansen and Torgersen, 1974). *According as σ is known or not,*

(i) *if σ is known, then $\mathcal{N}(A\beta, \sigma I_{n_A})$ is totally at least as good as $\mathcal{N}(B\beta, \sigma I_{n_B})$ if and only if $A'A - B'B$ is n.n.d.*

(ii) *if σ is unknown, then $\mathcal{N}(A\beta, \sigma I_{n_A})$ is totally at least as good as $\mathcal{N}(B\beta, \sigma I_{n_B})$ if and only if $A'A - B'B$ is n.n.d. and $n_A \ge n_B + \mathrm{rank}(A'A - B'B)$.*

Sinha (1973) and Stępniak (1982) have studied the utility of the ordering $\succ^t$ to the comparison of allocations of experimental units in the one-way classification with fixed and random effects, respectively. In both cases only equivalent experiments appeard to be comparable.

3. Comparison for Linear Estimation

The idea was originated by Ehrenfeld (1955) for experiments of the form $\mathcal{L}(A\beta, \sigma I_{n_A})$. In this case, for any estimable parametric function $k'\beta$ there exists the Best Linear Unbiased Estimator and it coincides with the Least Squares Estimator. Thus the problem of the comparison of such experiments with respect to linear estimation reduces to the comparison of variances of the underling LSE's. Considering the experiments $X \sim \mathcal{L}(A\beta, \sigma I_{n_A})$ and $Y \sim \mathcal{L}(B\beta, \sigma I_{n_B})$, Ehrenfeld (1955) has proved that if $A'A - B'B$ is n.n.d., then

(i) any parametric function $\Psi = k'\beta$ being estimable by Y is also estimable by X,

(ii) the variance of the LSE $\hat{\Psi}$ based on X is not greater than the variance of the LSE $\tilde{\Psi}$ based on Y.

The idea has been used by Stępniak (1976) to define the ordering $\succ^l$ for general linear experiments.

Formally, $X \succ^l Y$ if for any parametric function $\Psi = k'\beta$ being estimable by Y and for any unbiased estimator $b'Y$ of it there exists an unbiased estimator $a'X$ such that $\text{var}(a'X) \leq \text{var}(b'Y)$. If $X \sim \mathcal{L}(A\beta, \sum \sigma_i V_i)$ and $Y \sim \mathcal{L}(B\beta, \sum \sigma_i W_i)$ then, instead of $X \succ^l Y$, we will also use the symbol $\mathcal{L}(A\beta, \sum \sigma_i V_i) \succ^l \mathcal{L}(B\beta, \sum \sigma_i W_i)$.

The ordering $\succ^l$ may be also expressed in terms of quadratic risk for linear estimation without unbiasedness (cf. Stępniak, 1983) and by stochastic ordering of linear estimators (Stępniak, 1989). Moreover, as proved in Torgersen (1984), for the normal Gauss-Markov models with a known variance the ordering $\succ^l$ coincides with the total ordering $\succ^t$.

Now let us state some basic results in the subject.

Theorem 2 (Stępniak and Torgersen, 1981). *$\mathcal{L}(A\beta, \sum \sigma_i V_i) \succ^l \mathcal{L}(B\beta, \sum \sigma_i W_i)$ if and only if $\mathcal{L}(A\beta, V_\sigma) \succ^l \mathcal{L}(B\beta, W_\sigma)$ for all $\sigma = (\sigma_1, \ldots, \sigma_q) \in [0, \infty)^q$, where $V_\sigma = \sum \sigma_i V_i$ and $W_\sigma = \sum \sigma_i W_i$.*

Theorem 3 (Stępniak, Wang, and Wu, 1984). *$\mathcal{L}(A\beta, V) \succ^l \mathcal{L}(B\beta, W)$ if and only if $A'(V + AA')^- A - B'(W + BB')^- B$ is n.n.d., where $^-$ denotes a generalized inverse.*

It is worth to emphasize that Theorem 3 reduces the comparison of general linear experiments to the point-by-point comparisons of corresponding subexperiments with known covariances. Here are some applications of the result.

Example 1. *Allocations of T experimental units in k subclasses.* To each allocation $n = (n_1, \ldots, n_k)$ corresponds a one-way linear experiment $\mathcal{E}(n) = \mathcal{L}(\mu J_T, \sigma_a D + \sigma_e I_T)$ with random effects, where $D = \text{diag}(J_{n_1} J'_{n_1}, \ldots, J_{n_k} J'_{n_k})$ and J_s means a column of s ones. Consider also an alternative allocation $m = (m_1, \ldots, m_k)$. Then $\mathcal{E}(n) \succ^l \mathcal{E}(m)$ if and only if

$$\sum_{i=1}^{k} \frac{n_i}{1 + n_i \varrho} \geq \sum_{i=1}^{k} \frac{m_i}{1 + m_i \varrho} \quad \text{for all } \varrho \geq 0. \tag{1}$$

(cf. Stępniak, 1982). In particular if m majorizes n, then the condition (1) is satisfied. However, the condition (1) does not imply the majorization (see Stępniak, 1989, for a counterexample).

Example 2. *Allocations of T experimental units in a q-way cross classification with k_i subclasses for the i-th classification, $i = 1, \ldots, q$.* To each allocation $\mathcal{A}$ there corresponds a linear experiment $\mathcal{E}(\mathcal{A})$ with random effects. Denote by $\mathcal{C} = \mathcal{C}(T; k_1, \ldots, k_q)$ the class of possible allocations of the experimental units in the subclasses and assume that an orthogonal allocation, say $\mathcal{A}_0$, is a member of $\mathcal{C}$. Then the allocation $\mathcal{A}_0$ is optimal in $\mathcal{C}$ (cf. Stępniak, 1983). A similar result, for a q-way hierarchical classification with random effects, has also been proved (ibid.).

Example 3. *Allocations of r treatments with replications $t_1, \ldots, t_r$ in k blocks of sizes $b_1, \ldots, b_k$.* Each allocation of treatments may be indentified with a $T \times r$ matrix $D = (d_{ij})$, where $T = \sum t_i = \sum b_i$, defined by

$$d_{ij} = \begin{cases} 1, & \text{if the } i\text{th observation refers to } j\text{th treatment,} \\ 0, & \text{otherwise.} \end{cases}$$

To each D corresponds a linear experiment

$$\mathcal{L}(\mu J_T + D\alpha, \sigma_a B + \sigma_e I_T), \tag{2}$$

called a block design with random block effects, where $B = \mathrm{diag}(J_{b_1}, \ldots, J_{b_k})$. The essential information on the block design is contained in the incidence matrix $N = D'B$.

It was shown in Stępniak (1987b) that an allocation D_1 is at least as good as an allocation D_2, in context of model (2), if and only if $H_{D_1}(\varrho) - H_{D_2}(\varrho)$ is n.n.d. for all $\varrho \geq 0$, where $H_D(\varrho) = N \mathrm{diag}^{-1}((1 + b_1\varrho), \ldots, (1 + b_k\varrho))N'$. In particular, any orthogonal design with equal block sizes is optimal, however, the assumption $b_1 = \ldots = b_k$ is essential in this case. A corresponding result for block designs with fixed block effects can also be found in Stępniak (1987b).

4. Miscellaneous Results

In experiment $\mathcal{N}(A\beta, \sigma V)$ we are mainly interested in the following problems:

(a) linear unbiased estimation of the parametric functions $k'\beta$,

(b) quadratic unbiased estimation of σ, and

(c) testing linear hypotheses of the form $H\beta = 0$.

Some aspects of a comparison of experiments for such problems has been considered by Baksalary and Kala (1981), Drygas (1983), Baksalary (1984), Mueller

(1987), Stępniak (1987a) and Oktaba, Kornacki and Wawrzosek (1988) in the context of sufficient transformations. It can be easily deduced from Theorem 3 that the experiment $X \sim \mathcal{N}(A\beta, \sigma V)$ is at least as good as the experiment $Y \sim \mathcal{N}(B\beta, \sigma W)$ with respect to the problems (a) and (b) [notation: $X \succ^m Y$] if and only if

$$A'(V + AA')^- A - B'(W + BB')^- B \text{ is n.n.d.}$$

and

$$\operatorname{rank}(V + AA') - \operatorname{rank}(A) \geq \operatorname{rank}(W + BB') - \operatorname{rank}(B).$$

Moreover, the relation $X \succ^m Y$ implies that any Y-testable hypothesis $H\beta = 0$ is also X-testable and the Uniformly Most Powerful Invariant test $F(X)$ is not weaker than the corresponding UMPI test $F(Y)$.

Now let us list some necessary and sufficient conditions for the relations $\succ^t$ and $\succ^l$ in terms of linear transformations.

Theorem 4 (Torgersen, 1984, 1991).

(i) *Under assumptions that* $X \sim \mathcal{N}(A\beta, \sigma V)$ *and* $Y \sim \mathcal{N}(B\beta, \sigma W)$, $X \succ^t Y$ *if and only if there exists a linear transformation* F *such that* FX *has the same distribution as* Y.

(ii) *Under assumptions that* $X \sim \mathcal{L}(A\beta, \sigma V)$ *and* $Y \sim \mathcal{L}(B\beta, \sigma W)$, $X \succ^l Y$ *if and only if there exists a linear transformation* F *and a centered random vector* Z *uncorrelated with* X *such that* $FX + Z$ *has the same first two moments as* Y.

Remark. The random vector Z appearing in Theorem 4 may be considered as a *white noise.*

Acknowledgements

This paper was supported by Grant KBN No 212558 9101. The authors wish to express their gratitude to one of the reviewers for his valuable suggestions and comments. Thanks to them it was possible to avoid some misreadings and to improve the presentation of the paper.

References

Baksalary, J.K. (1984). A study of the equivalence between a Gauss-Markov model and its augmentation by nuisance parameters. *Mathematische Operationsforschung und Statistik, Series Statistics* **15**, 3–35.

Baksalary, J.K. and Kala, R. (1981). Linear transformations preserving best linear unbiased estimators in a general Gauss-Markoff model. *The Annals of Statistics* **9**, 913–916.

Blackwell, D. (1951). Comparison of experiments. In: J. Neyman, Ed., *Proceedings of the Second Berkeley Symposium on Mathematical Statistics and Probability.* University of California Press, Berkeley, 93–102.

Blackwell, D. (1953). Equivalent comparison of experiments. *The Annals of Mathematical Statistics* **24**, 265–272.

DeGroot, M.H. (1966). Optimal allocation of observations. *Annals of the Institute of Statistical Mathematics* **18**, 13–28.

Drygas, H. (1983). Sufficiency and completeness in the general Gauss-Markov model. *Sankhyā A* **45**, 88–98.

Ehrenfeld, S. (1955). Complete class theorem in experimental design. In: J. Neyman, Ed., *Proceedings of the Third Berkeley Symposium on Mathematical Statistics and Probability, Vol. 1.* University of California Press, Berkeley, 69–75.

Goel, P.K. and DeGroot, M.H. (1979). Comparison of experiments and information measures. *The Annals of Statistics* **7**, 1055–1077.

Hansen, O.H. and Torgersen, E. (1974). Comparison of linear normal experiments. *The Annals of Statistics* **2**, 365–373.

Kiefer, J. (1959). Optimum experimental designs. *Journal of the Royal Statistical Society B* **21**, 272–304.

LeCam, L. (1964). Sufficiency and approximate sufficiency. *The Annals of Mathematical Statistics* **35**, 1419–1455.

LeCam, L. (1986). *Asymptotic Methods in Statistical Decision Theory.* Springer-Verlag, New York.

Lehmann, E.L. (1988). Comparing location experiments. *The Annals of Statistics* **16**, 521–533.

Mueller, J. (1987). Sufficiency and completeness in a linear model. *Journal of Multivariate Analysis* **21**, 312–323.

Oktaba, W., Kornacki, A. and Wawrzosek, J. (1988). Invariant linearly sufficient transformations of the general Gauss-Markoff model. Estimation and testing. *Scandinavian Journal of Statistics* **15**, 117–124.

Sinha, B.K. (1973). Comparison of some experiments from sufficiency consideretion. *Annals of the Institute of Statistical Mathematics A* **25**, 501–520.

Stępniak, C. (1976). Optimal planning in hierarchical linear experiments with random effects. Ph. D. thesis, Polish Academy of Sciences, Warsaw (in Polish).

Stępniak, C. (1982). Optimal allocation of observations in one-way random normal model. *Annals of the Institute of Statistical Mathematics A* **34**, 175–180.

Stępniak, C. (1983). Optimal allocation of units in experimental designs with hierarchical and cross classification. *Annals of the Institute of Statistical Mathematics A* **35**, 461–473.

Stępniak, C. (1985). Ordering of nonnegative definite matrices with applications to comparison of linear models. *Linear Algebra and its Applications* **70**, 67–71.

Stępniak, C. (1987a). Reduction problems in comparison of linear models. *Metrika* **34**, 211–216.

Stępniak, C. (1987b). Optimal allocation of treatments in block design. *Studia Scientiarum Mathematicarum Hungarica* **22**, 341–345.

Stępniak, C. (1989). Stochastic ordering and Schur-convex functions in comparison of linear experiments. *Metrika* **36**, 291–298.

Stępniak, C. and Torgersen, E. (1981). Comparison of linear models with partially known covariances with respect to unbiased estimation. *Scandinavian Journal of Statistics* **8**, 183–184.

Stępniak, C., Wang, S.G. and Wu, C.F.J. (1984). Comparison of linear experiments with known covariances. *The Annals of Statistics* **12**, 358–365.

Swensen, A.R. (1980). Deficiencies between linear normal experiments. *The Annals of Statistics* **8**, 1142–1155.

Torgersen, E. (1972). Comparison of translation experiments. *The Annals of Mathematical Statistics* **43**, 1383–1399.

Torgersen, E. (1984). Ordering of linear models. *Journal of Statistical Planning and Inference* **9**, 1–17.

Torgersen, E. (1991). *Comparison of Statistical Experiments.* Cambridge University Press, Cambridge.

PROPERTIES OF COMPARISON CRITERIA OF NORMAL EXPERIMENTS

JAN HAUKE* and AUGUSTYN MARKIEWICZ
Department of Mathematical and Statistical Methods
Agricultural University of Poznań
Wojska Polskiego 28
60-637 Poznań
Poland

Abstract. The class of linear models induced by possible allocations of units in an experimental design with hierarchical or cross classification is investigated. Properties of comparison criteria of experiments are given. The sufficient condition for a model to be *better* than another one, given by Shaked and Tong (1992), is generalized.

Key words: Comparison of experiments, Majorization, Group majorization.

1. Introduction

Shaked and Tong (1992) compare some linear normal experiments. Their results can be presented equivalently (in a reduced form) in the context of comparison of one–way random models (see Torgersen, 1991, p.422; Stępniak, 1987, Theorem 1).

Let $\varepsilon(\mathbf{k})$ be an experiment corresponding to a vector $\mathbf{k}' = (k_1, \ldots, k_a)$ of allocation of $n = \sum_{i=1}^{a} k_i$ individuals in a groups, and realized by a normal random vector $\mathbf{y}$ with the expectation $E(\mathbf{y}) = \theta \mathbf{1}_n$ and the dispersion matrix

$$\mathrm{Cov}(\mathbf{y}) = \sigma_0^2 \mathbf{I}_n + \sigma_1^2 \mathrm{diag}(\mathbf{1}_{k_1}\mathbf{1}'_{k_1}, \ldots, \mathbf{1}_{k_a}\mathbf{1}'_{k_a}),$$

where θ is an unknown real parameter, while $\sigma_0^2 > 0$, $\sigma_1^2 \geq 0$ are arbitrary but fixed variance components. Shaked and Tong (1992) have shown that if $\mathbf{k}^*$ majorizes $\mathbf{k}$, in symbols $\mathbf{k} \leq_m \mathbf{k}^*$, then the experiment $\varepsilon(\mathbf{k})$ is at least as informative as the experiment $\varepsilon(\mathbf{k}^*)$.

In the present paper more complex experiments, namely two–stage hierarchical and two–way cross experiments, are compared. In both cases fixed, mixed and random models, with possibly unknown variance components, are considered. The properties of comparison criteria and the possibility of replacing majorization by group majorization are examined. Finally, counterparts of Shaked and Tong's condition are given.

It is worth noticing that several more authors, e.g., Bailey (1981) and Dawid (1988), have applied group theory to random and mixed models, and that group majorization was used for fixed effects models by Giovagnoli, Pukelsheim and Wynn (1987). We also emphasize that problems of comparison of experiments are recently

* Also affiliated to the Institute of Socio-Economic Geography and Spatial Planning, Adam Mickiewicz University, Fredry 10, 61-701 Poznań,

T. Caliński and R. Kala (eds.),
Proceedings of the International Conference on Linear Statistical Inference LINSTAT '93, 245–253.

often discussed in different aspects (see, e.g., Peres and Ting Hui-Ching, 1993; Yurramendi, 1993).

2. Results

Let us begin with some preliminary results in group majorization ($\mathcal{G}$–majorization). For a general concept of $\mathcal{G}$–majorization with applications to matrix orderings and suggestions how to verify it; see Giovagnoli and Wynn (1985).

Let $\mathcal{O}(ab)$ denote a group of $ab \times ab$ orthogonal matrices acting on $\mathbb{R}^{ab}$, and suppose that $\mathcal{G}$ is a closed subgroup of $\mathcal{O}(ab)$ (with respect to multiplication). For $\mathbf{x} \in \mathbb{R}^{ab}$, let $\mathcal{C}(\mathbf{x})$ denote the convex hull of the $\mathcal{G}$–orbit $\{g(\mathbf{x}): g \in \mathcal{G}\}$. The group $\mathcal{G}$ determines a preordering called $\mathcal{G}$–majorization, on the vector space of real $a \times b$ matrices. It is denoted by $\leq_G$ and, using the vec operator, can be defined as follows.

Definition 1. For any $a \times b$ real matrices $\mathbf{A}, \mathbf{B}$,

$$\mathbf{A} \leq_G \mathbf{B} \text{ if and only if } \operatorname{vec}(\mathbf{A}') \in \mathcal{C}[\operatorname{vec}(\mathbf{B}')].$$

Now, let us introduce two subgroups of the group $\mathcal{P}_{ab}$ of all permutation matrices of order ab. The first subgroup $\mathcal{G}_H$ is defined as

$$\mathcal{G}_H = \{\mathbf{\Pi} \in \mathcal{P}_{ab}: \mathbf{\Pi} = (\mathbf{\Pi}_0 \otimes \mathbf{I}_b)\operatorname{diag}(\mathbf{\Pi}_1, \ldots, \mathbf{\Pi}_a), \mathbf{\Pi}_0 \in \mathcal{P}_a, \mathbf{\Pi}_i \in \mathcal{P}_b,\ i = 1, \ldots, a\}, \tag{1}$$

where $\otimes$ stands for the Kronecker product. This group is related to a hierarchical classification and will be used in Section 3. The second subgroup of $\mathcal{P}_{ab}$ is related to a cross classification and is defined as

$$\mathcal{G}_C = \{\mathbf{\Pi} \in \mathcal{P}_{ab}: \mathbf{\Pi} = (\mathbf{\Pi}_1 \otimes \mathbf{\Pi}_2), \mathbf{\Pi}_1 \in \mathcal{P}_a, \mathbf{\Pi}_2 \in \mathcal{P}_b\}. \tag{2}$$

It will be used in Section 4.

Definition 2. A real valued function $f: \mathbb{R}^{ab} \to \mathbb{R}$ is called $\mathcal{G}$–symmetric if $f(\mathbf{A}) = f(\mathbf{B})$ for $\mathbf{B}$ such that $\operatorname{vec}(\mathbf{B}') = \mathbf{\Pi}\operatorname{vec}(\mathbf{A}')$ for every $a \times b$ matrix $\mathbf{A}$ and every permutation matrix $\mathbf{\Pi} \in \mathcal{G}$.

The following lemma (see Giovagnoli and Wynn, 1985) will be used in the next sections.

Lemma 1. *Let $\mathbf{A}, \mathbf{B}$ be $a \times b$ real matrices. If $\mathbf{A} \leq_G \mathbf{B}$ then $f(\mathbf{A}) \geq f(\mathbf{B})$ for all real–valued, concave, and $\mathcal{G}$–symmetric functions f.*

The notion of the Schur complement (see, e.g., Styan, 1983) plays an important role in the theory presented below. It possesses the following property (see, e.g., Marshall and Olkin, 1979, p. 469).

Lemma 2. *Let a positive definite matrix* $\mathbf{A}$ *be partitioned as follows*

$$\mathbf{A} = \begin{bmatrix} \mathbf{E} & \mathbf{F} \\ \mathbf{F}' & \mathbf{G} \end{bmatrix},$$

and $(\mathbf{A}/\mathbf{G})$ *be the Schur complement of* $\mathbf{G}$ *in* $\mathbf{A}$, *i.e.* $(\mathbf{A}/\mathbf{G}) = \mathbf{E} - \mathbf{F}\mathbf{G}^{-1}\mathbf{F}'$. *Then the function* $\psi(\mathbf{A}) = (\mathbf{A}/\mathbf{G})$ *is matrix-concave.*

Let us consider the general linear model

$$\mathbf{y} = \mathbf{X}\boldsymbol{\mu} + \mathbf{Z}\mathbf{u} + \mathbf{e}, \tag{3}$$

where $\mathbf{y} \in \mathbb{R}^n$ is a vector of normally distributed observations, $\mathbf{X}$ is a known $n \times p$ matrix of the rank p, $\boldsymbol{\mu} \in \mathbb{R}^p$ is a vector of unknown fixed parameters, $\mathbf{Z}$ is a known $n \times q$ matrix, $\mathbf{u}$ and $\mathbf{e}$ are uncorrelated and unobservable random vectors with the expectation zero and with dispersion matrices $\text{Cov}(\mathbf{e}) = \sigma_0^2\mathbf{I}_n, \sigma_0^2 > 0$, and $\text{Cov}(\mathbf{u}) = \text{diag}(\sigma_1^2\mathbf{I}_{q_1}, \ldots, \sigma_r^2\mathbf{I}_{q_r}), \sigma_i^2 \geq 0, i = 1, 2, \ldots, r$. The partition of $\text{Cov}(\mathbf{u})$ corresponds to partitioning of $\mathbf{u}$ into r subvectors, $\mathbf{u} = (\mathbf{u}_1' : \ldots : \mathbf{u}_r')'$, and $\mathbf{Z}$ into r submatrices, $\mathbf{Z} = (\mathbf{Z}_1 : \ldots : \mathbf{Z}_r)$. In result, the dispersion matrix of $\mathbf{y}$ takes the form $\mathbf{V} = \sum_{i=0}^r \sigma_i^2\mathbf{V}_i$, where $\mathbf{V}_i = \mathbf{Z}_i\mathbf{Z}_i'$, $i = 1, 2, \ldots, r$, and $\mathbf{V}_0 = \mathbf{I}_n$. In consequence, the moment matrix, playing a key role in the comparison of experiments, can be expressed as

$$\mathbf{M} = \mathbf{X}'\mathbf{V}^{-1}\mathbf{X} = \mathbf{X}'(\sum_{i=0}^r \sigma_i^2\mathbf{V}_i)^{-1}\mathbf{X}. \tag{4}$$

Moreover, let us define a matrix

$$\mathbf{S} = \begin{bmatrix} \mathbf{X}'\mathbf{X} & \mathbf{X}'\mathbf{Z} \\ \mathbf{Z}'\mathbf{X} & \mathbf{Z}'\mathbf{Z} \end{bmatrix}. \tag{5}$$

The main result, expressing properties of the matrix $\mathbf{M}$, is presented in the following

Theorem. *The matrix* $\mathbf{M} = \mathbf{M}(\mathbf{S})$ *is a matrix-concave function of* $\mathbf{S}$. *Moreover, if* $\mathbf{S}_* = \text{diag}(\mathbf{\Pi}_X, \mathbf{\Pi}_Z)\mathbf{S}\text{diag}(\mathbf{\Pi}_X', \mathbf{\Pi}_Z')$, *where* $\mathbf{\Pi}_X \in \mathcal{P}_p$ *and* $\mathbf{\Pi}_Z = \text{diag}(\mathbf{\Pi}_{Z_1}, \ldots, \mathbf{\Pi}_{Z_r})$ *with* $\mathbf{\Pi}_{Z_i} \in \mathcal{P}_{q_i}, i = 1, \ldots, r$, *then* $\mathbf{M}(\mathbf{S}_*) = \mathbf{\Pi}_X\mathbf{M}(\mathbf{S})\mathbf{\Pi}_X'$.

Proof. Notice that

$$\mathbf{M} = \mathbf{X}'(\sum_{i=0}^r \sigma_i^2\mathbf{V}_i)^{-1}\mathbf{X} = (1/\sigma_0^2)\mathbf{X}'[\mathbf{I}_n - \mathbf{U}(\mathbf{U}'\mathbf{U} + \mathbf{I}_q)^{-1}\mathbf{U}']\mathbf{X},$$

where $\mathbf{U} = (\mathbf{U}_1 : \ldots : \mathbf{U}_r)$ with $\mathbf{U}_i = (\sigma_i/\sigma_0)\mathbf{Z}_i, i = 1, 2, \ldots, r$ (see, e.g., Rao, 1973, p. 33). It means that

$$\mathbf{M} = (1/\sigma_0^2)[\mathbf{T}/(\mathbf{U}'\mathbf{U} + \mathbf{I}_q)], \tag{6}$$

where

$$\mathbf{T} = \begin{bmatrix} \mathbf{X}'\mathbf{X} & \mathbf{X}'\mathbf{U} \\ \mathbf{U}'\mathbf{X} & \mathbf{U}'\mathbf{U} + \mathbf{I}_q \end{bmatrix}, \tag{7}$$

and then, by Lemma 2, $\mathbf{M}$ is a matrix-concave function of $\mathbf{T}$. But $\mathbf{T}$ is related to $\mathbf{S}$,

$$\mathbf{T} = (1/\sigma_0^2)\mathbf{\Delta S \Delta} + \mathrm{diag}(\mathbf{O}_p, \mathbf{I}_q), \tag{8}$$

where $\mathbf{\Delta} = \mathrm{diag}(\sigma_0 \mathbf{I}_p, \sigma_1 \mathbf{I}_{q_1}, \ldots, \sigma_r \mathbf{I}_{q_r})$, which completes the proof of the first part. The second part follows immediately from (5), (6), (7), (8), and Lemma 2. □

Stępniak (1989) gave a necessary and sufficient condition for an experiment to be at least as good as another one (see also Torgersen, 1991, p. 416). Using this condition, Stępniak's definition can be rewritten in the following form.

Definition 3. An experiment $\mathcal{E}_2 = N_n(\mathbf{X}_2\boldsymbol{\mu}, \sum_{i=0}^r \sigma_i^2 \mathbf{V}_{2i})$ is said to be at least as good as an experiment $\mathcal{E}_1 = N_n(\mathbf{X}_1\boldsymbol{\mu}, \sum_{i=0}^r \sigma_i^2 \mathbf{V}_{1i})$ with respect to precision of linear estimation, in symbols $\mathcal{E}_1 \leq_{(pl)} \mathcal{E}_2$, if the matrix

$$\mathbf{M}_2 - \mathbf{M}_1 = \mathbf{X}_2'(\sum_{i=0}^r \sigma_i^2 \mathbf{V}_{21})^{-1}\mathbf{X}_2 - \mathbf{X}_1'(\sum_{i=0}^r \sigma_i^2 \mathbf{V}_{1i})^{-1}\mathbf{X}_1$$

is nonnegative definite for every $\sigma_0^2 > 0$ and $\sigma_i^2 \geq 0, i = 1, \ldots, r$.

In the case of experiments with known σ_i^2, the above condition coincides with a stronger criterion *to be more informative* (see, e.g., Torgersen, 1991, p. 436). From a statistical point of view it is more reasonable to compare moment matrices with the use of two–stage preordering, or to compare experiments using a class of functionals on $\mathbf{M}$, serving as optimality criteria (cf. Pukelsheim, 1987). In this paper we consider the class Φ consisting of all functionals φ satisfying the following conditions:

(i) $\varphi(\mathbf{M})$ is a concave function,
(ii) $\varphi(\mathbf{M})$ is symmetric, i.e. $\varphi(\mathbf{M}) = \varphi(\mathbf{\Pi M \Pi}')$ for all $\mathbf{\Pi} \in \mathcal{P}_p$,
(iii) $\varphi(\mathbf{M})$ is isotonic (increasing) with respect to the Löwner ordering.

Definition 4. Given $\varphi \in \Phi$ an experiment $\mathcal{E}_2 = N_n(\mathbf{X}_2\boldsymbol{\mu}, \sum_{i=0}^r \sigma_i^2 \mathbf{V}_{2i})$ is said to be at least as good as an experiment $\mathcal{E}_1 = N_n(\mathbf{X}_1\boldsymbol{\mu}, \sum_{i=0}^r \sigma_i^2 \mathbf{V}_{1i})$ with respect to φ-precision of linear estimation, in symbols $\mathcal{E}_1 \leq_\varphi \mathcal{E}_2$, if

$$\varphi(\mathbf{M}_2) - \varphi(\mathbf{M}_1) = \varphi[\mathbf{X}_2'(\sum_{i=0}^r \sigma_i^2 \mathbf{V}_{2i})^{-1}\mathbf{X}_2] - \varphi[\mathbf{X}_1'(\sum_{i=0}^r \sigma_i^2 \mathbf{V}_{1i})^{-1}\mathbf{X}_1] \geq 0$$

for every $\sigma_0^2 > 0$ and $\sigma_i^2 \geq 0, i = 1, \ldots, r$.

In the case of experiments with known σ_i^2, the above condition coincides with a comparison with respect to information functionals.

All results presented in next two sections are consequences of the Theorem proven above and the cited lemmas. To establish them, it suffices to specify appropriate matrix $\mathbf{S}$ being a function of an allocation matrix $\mathbf{N}$.

3. Comparison of Two-Stage Hierarchical Experiments

Let us consider the two-stage hierarchical classification model

$$\mathbf{y} = \mathbf{1}_n\theta + \mathbf{A}\boldsymbol{\alpha} + \mathbf{C}\boldsymbol{\gamma} + \mathbf{e},$$

where $\boldsymbol{\alpha} \in \mathbb{R}^a$ is a vector of effects of the primary factor, $\boldsymbol{\gamma} \in \mathbb{R}^{ab}$ is a vector of effects of the secondary factor, nested within levels of the primary factor, an $n \times a$ matrix $\mathbf{A} = \mathrm{diag}(\mathbf{1}_{n_1}, \ldots, \mathbf{1}_{n_a})$, $n_i = \sum_{j=1}^b n_{ij}, i = 1, \ldots, a$, $n = \sum_{i=1}^a n_i$, and an $n \times a$ matrix $\mathbf{C} = \mathrm{diag}(\mathbf{1}_{n_{11}}, \ldots, \mathbf{1}_{n_{1a}}, \mathbf{1}_{n_{21}}, \ldots, \mathbf{1}_{n_{ab}}), i = 1, \ldots, a,\ j = 1, \ldots, b$. Hence, matrices $\mathbf{A}$ and $\mathbf{C}$ are determined by the allocation matrix $\mathbf{N} = \{n_{ij}\}, i = 1, \ldots, a, j = 1, \ldots, b$, with n_{ij} representing the number of observations in cell i, j, i.e. the cell defined by the jth level of the secondary factor within the ith level of the primary factor.

The matrix $\mathbf{S}$, defined generally in (5), can now be written as

$$\mathbf{S} = (\mathbf{1}_n : \mathbf{A} : \mathbf{C})'(\mathbf{1}_n : \mathbf{A} : \mathbf{C}),$$

or as

$$\mathbf{S} = \begin{bmatrix} n & \mathbf{1}_b'\mathbf{N}' & [\mathrm{vec}(\mathbf{N}')]' \\ * & \mathrm{diag}(\mathbf{N}\mathbf{1}_b) & (\mathbf{I}_a \otimes \mathbf{1}_b')\mathrm{diag}[\mathrm{vec}(\mathbf{N}')] \\ * & * & \mathrm{diag}[\mathrm{vec}(\mathbf{N}')] \end{bmatrix},$$

where $*$ stand for such submatrices that $\mathbf{S}$ is symmetric.

Let us consider two models with allocation matrices $\mathbf{N}_1 = \{n_{ij}^{(1)}\}$ and $\mathbf{N}_2 = \{n_{ij}^{(2)}\} = \{n_{kl}^{(1)}\}$, where $i = \Pi_0(k)$ and $j = \Pi_k(l)$, i.e. $\mathbf{N}_2$ is a result of transformation $\boldsymbol{\Pi} \in \mathcal{G}_H$ applied to $\mathbf{N}_1$. Under this transformation at first elements of each row of $\mathbf{N}_1$ are permuted separately and then rows are permuted in whole. This can be written via the vec operation as $\mathrm{vec}(\mathbf{N}_2') = \boldsymbol{\Pi}\mathrm{vec}(\mathbf{N}_1')$, where $\boldsymbol{\Pi} \in \mathcal{G}_H$ is defined in (1). Then the matrix $\mathbf{S}$ corresponding to $\mathbf{N}_2, \mathbf{S}(\mathbf{N}_2)$, is related with $\mathbf{S}(\mathbf{N}_1)$ by the equation

$$\mathbf{S}(\mathbf{N}_2) = \boldsymbol{\Pi}_T\mathbf{S}(\mathbf{N}_1)\boldsymbol{\Pi}_T', \text{ where } \boldsymbol{\Pi}_T = \mathrm{diag}(1, \boldsymbol{\Pi}_0, \boldsymbol{\Pi}).$$

In the two-stage random hierarchical model vectors $\boldsymbol{\alpha}$ and $\boldsymbol{\gamma}$ are random while θ is a fixed unknown parameter and, in the notation of the general linear model (3), $\mathbf{X} = \mathbf{1}_n$ and $\mathbf{Z} = (\mathbf{A} : \mathbf{C})$. According to the equation (4), $\mathbf{M}$ is a real number and experiments are compared as in Definition 3.

Proposition 1. *In two–stage random hierarchical experiments* $\mathbf{M}$ *is a real–valued, concave and* $\mathcal{G}_H$*-symetric function of the allocation matrix* $\mathbf{N}$. *Moreover, if* $\mathbf{N}_1$ *and* $\mathbf{N}_2$ *are allocation matrices of experiments* $\mathcal{E}_1$ *and* $\mathcal{E}_2$, *respectively, then* $\mathbf{N}_2 \leq_{G_H} \mathbf{N}_1$ *implies* $\mathcal{E}_1 \leq_{(pl)} \mathcal{E}_2$.

Corollary 1. *Let the allocation matrices* $\mathbf{N}$, $\mathbf{N}_1$, *and* $\mathbf{N}_2$, *coresponding to two–stage random hierarchical experiments, be binary with elements* 0 *or* m *(a positive integer). Then* $\mathbf{M}$ *is a concave function of the vector* $\mathbf{N}\mathbf{1}_b$ *and symmetric with respect to permutations of its elements. Moreover, if* $\mathbf{N}_1$ *and* $\mathbf{N}_2$ *are allocation matrices of experiments* $\mathcal{E}_1$ *and* $\mathcal{E}_2$, *respectively, then* $\mathbf{N}_2\mathbf{1}_b \leq_m \mathbf{N}_1\mathbf{1}_b$ *implies* $\mathcal{E}_1 \leq_{(pl)} \mathcal{E}_2$.

In a special case, when the allocation matrix $\mathbf{N}$ consists of one column only, $\mathbf{N} = \mathbf{n}$, we obtain the one–way classification model. In consequence, the result of Shaked and Tong (1992) is derived as the following

Corollary 2. *Let $\mathbf{n}_1$ and $\mathbf{n}_2$ be allocation vectors of experiments $\mathcal{E}_1$ and $\mathcal{E}_2$, respectively, associated with the random one-way classification model. If $\mathbf{n}_2 \leq_m \mathbf{n}_1$ then $\mathcal{E}_1 \leq_{(pl)} \mathcal{E}_2$.*

In the two-stage mixed hierarchical model θ and $\boldsymbol{\alpha}$ are fixed while $\boldsymbol{\gamma}$ is a random vector. In the notation used in (3), we have $\mathbf{X} = (\mathbf{1} : \mathbf{A})$, with rank a, and $\mathbf{Z} = \mathbf{C}$. Now, the matrix $\mathbf{M}$ is of order $a+1$ and experiments will be compared according to Definition 4.

Proposition 2. *Let φ be a concave, symmetric and isotonic functional on moment matrices. Then $\varphi(\mathbf{M})$ is a concave and $\mathcal{G}_H$–symmetric function of the allocation matrix $\mathbf{N}$ associated with two-stage mixed hierarchical experiments. Moreover, if $\mathbf{N}_1$ and $\mathbf{N}_2$ are allocation matrices of experiments $\mathcal{E}_1$ and $\mathcal{E}_2$, respectively, then $\mathbf{N}_2 \leq_{G_H} \mathbf{N}_1$ implies $\mathcal{E}_1 \leq_\varphi \mathcal{E}_2$.*

Corollary 3. *Let allocation matrices $\mathbf{N}$, $\mathbf{N}_1$, and $\mathbf{N}_2$, corresponding to two–stage mixed hierarchical experiments, be binary with elements 0 or m (a positive integer). Then $\mathbf{M}$ is a concave function of the vector $\mathbf{N}\mathbf{1}_b$ and symmetric with respect to permutations of its elements. Moreover, if $\mathbf{N}_1$ and $\mathbf{N}_2$ are allocation matrices of experiments $\mathcal{E}_1$ and $\mathcal{E}_2$, respectively, then $\mathbf{N}_2\mathbf{1}_b \leq_m \mathbf{N}_1\mathbf{1}_b$ implies $\mathcal{E}_1 \leq_\varphi \mathcal{E}_2$.*

Proof. It suffices to note that $\mathbf{M} = (1/\sigma_0^2)(\mathbf{T}/\mathbf{T}_{22})$, where

$$\mathbf{T}_{22} = (\sigma_1^2/\sigma_0^2)\mathrm{diag}(\mathbf{N}\mathbf{1}_b) + \mathbf{I}_{ab}.$$

When $\mathbf{N}$ is binary, we obtain

$$\mathbf{M} = \xi \begin{bmatrix} n & \mathbf{1}_b'\mathbf{N}' \\ * & \mathrm{diag}(\mathbf{N}\mathbf{1}_b) \end{bmatrix},$$

where $\xi = [\sigma_1^2 m + \sigma_0^2]^{-1}$. Then $\mathbf{M}$ is a function of $\mathbf{N}\mathbf{1}_b$. □

In the two–stage fixed hierarchical model θ, $\boldsymbol{\alpha}$ and $\boldsymbol{\gamma}$ are fixed. Then $\mathbf{X} = (\mathbf{1}_n : \mathbf{A} : \mathbf{C})$, with rank ab, $\mathbf{M} = (1/\sigma_0^2)\mathbf{S}$, and experiments will be compared according to Definition 4.

Proposition 3. *Let φ be a concave, symmetric and isotonic functional on moment matrices. Then $\varphi(\mathbf{M})$ is a concave and $\mathcal{G}_H$–symmetric function of the allocation matrix $\mathbf{N}$ corresponding to two-stage fixed hierarchical experiments. Moreover, if $\mathbf{N}_1$ and $\mathbf{N}_2$ are allocation matrices of experiments $\mathcal{E}_1$ and $\mathcal{E}_2$, respectively, then $\mathbf{N}_2 \leq_{G_H} \mathbf{N}_1$ implies $\mathcal{E}_1 \leq_\varphi \mathcal{E}_2$.*

Corollary 4. *Let φ be a concave, symmetric and isotonic functional of moment matrices, and let $\mathbf{n}_1$ and $\mathbf{n}_2$ be allocation vectors of experiments $\mathcal{E}_1$ and $\mathcal{E}_2$,* *respectively, associated with fixed the one-way classification model. Then* $\mathbf{n}_2 \leq_m \mathbf{n}_1$ *implies* $\mathcal{E}_1 \leq_\varphi \mathcal{E}_2$.

4. Comparison of Two-Way Cross Experiments

Let us consider the two–way cross classification model

$$\mathbf{y} = \mathbf{1}_n\theta + \mathbf{A}\boldsymbol{\alpha} + \mathbf{B}\boldsymbol{\beta} + \mathbf{C}\boldsymbol{\gamma} + \mathbf{e},$$

where $\boldsymbol{\alpha} \in \mathbb{R}^a$ is a vector of effects due to the primary (row) factor, $\boldsymbol{\beta} \in \mathbb{R}^b$ is a vector of effects due to the secondary (column) factor, and $\boldsymbol{\gamma} \in \mathbb{R}^{ab}$ is a vector of effects for interactions of row with column. Moreover, matrices $\mathbf{A}$ and $\mathbf{C}$ are determined by the allocation matrix $\mathbf{N} = \{n_{ij}\}, i = 1, \ldots, a, j = 1, \ldots, b$, in a similar way as in the two-stage hierarchical model, while

$$\mathbf{B} = (\mathrm{diag}(\mathbf{1}'_{n_{11}}, \ldots, \mathbf{1}'_{n_{1b}}) : \ldots : \mathrm{diag}(\mathbf{1}'_{n_{a1}}, \ldots, \mathbf{1}'_{n_{ab}}))'.$$

The matrix (5), can now be written as

$$\mathbf{S} = (\mathbf{1}_n : \mathbf{A} : \mathbf{B} : \mathbf{C})'(\mathbf{1}_n : \mathbf{A} : \mathbf{B} : \mathbf{C}),$$

or as

$$\mathbf{S} = \begin{bmatrix} n & \mathbf{1}'_b\mathbf{N}' & \mathbf{1}'_a\mathbf{N} & [\mathrm{vec}(\mathbf{N}')] \\ * & \mathrm{diag}(\mathbf{N}\mathbf{1}_b) & \mathbf{N} & (\mathbf{I}_a \otimes \mathbf{1}'_b)\mathrm{diag}[\mathrm{vec}(\mathbf{N}')] \\ * & * & \mathrm{diag}(\mathbf{N}'\mathbf{1}_a) & (\mathbf{1}'_a \otimes \mathbf{I}_b)\mathrm{diag}[\mathrm{vec}(\mathbf{N}')] \\ * & * & * & \mathrm{diag}[\mathrm{vec}(\mathbf{N}')] \end{bmatrix}.$$

Let us consider two models with allocation matrices $\mathbf{N}_1$ and $\mathbf{N}_2 = \boldsymbol{\Pi}_1\mathbf{N}_1\boldsymbol{\Pi}'_2$, where $\boldsymbol{\Pi}_1 \in \mathcal{P}_a$ and $\boldsymbol{\Pi}_2 \in \mathcal{P}_b$, i.e. $\boldsymbol{\Pi}_1 \otimes \boldsymbol{\Pi}_2 \in \mathcal{G}_C$, defined in (2). Then

$$\mathbf{S}(\mathbf{N}_2) = \boldsymbol{\Pi}_T\mathbf{S}(\mathbf{N}_1)\boldsymbol{\Pi}'_T, \text{ where } \boldsymbol{\Pi}_T = \mathrm{diag}(1, \boldsymbol{\Pi}_1, \boldsymbol{\Pi}_2, \boldsymbol{\Pi}_1 \otimes \boldsymbol{\Pi}_2).$$

In the two–way random cross model, vectors $\boldsymbol{\alpha}$, $\boldsymbol{\beta}$ and $\boldsymbol{\gamma}$ are random, while θ is a fixed unknown parameter. In the notation used in (3), we have $\mathbf{X} = \mathbf{1}_n$ and $\mathbf{Z} = (\mathbf{A} : \mathbf{B} : \mathbf{C})$. According to the equation (4), $\mathbf{M}$ is a real number and experiments are compared as in Definition 3.

Proposition 4. *In two–way random cross experiments* $\mathbf{M}$ *is a real valued, concave and* $\mathcal{G}_C$*–symmetric function of the allocation matrix* $\mathbf{M}$. *Moreover, if* $\mathbf{N}_1, \mathbf{N}_2$ *are allocation matrices of experiments* $\mathcal{E}_1$ *and* $\mathcal{E}_2$, *respectively, then* $\mathbf{N}_2 \leq_{\mathcal{G}_C} \mathbf{N}_1$ *implies* $\mathcal{E}_1 \leq_{(pl)} \mathcal{E}$.

In the two–way mixed cross model the common parameter θ and one vector of effects, say $\boldsymbol{\alpha}$, are fixed, while the remaining effects are random. In the notation used in (3), we have $\mathbf{X} = (\mathbf{1}_n : \mathbf{A})$, with rank a, and $\mathbf{Z} = (\mathbf{B} : \mathbf{C})$. Now, $\mathbf{M}$ is of order $a + 1$ and experiments will be compared according to Definition 4.

Proposition 5. *Let φ be a concave, symmetric and isotonic functional of moment matrices. Then $\varphi(\mathbf{M})$ is a concave and $\mathcal{G}_C$-symmetric function of the allocation matrix $\mathbf{N}$ associated with two-way mixed cross experiments. Moreover, if $\mathbf{N}_1$ and $\mathbf{N}_2$ are allocation matrices of experiments $\mathcal{E}_1$ and $\mathcal{E}_2$, respectively, then $\mathbf{N}_2 \leq_{G_C} \mathbf{N}_1$ implies $\mathcal{E}_1 \leq_\varphi \mathcal{E}_2$.*

In the two-way fixed cross model θ, α, β and γ are fixed. In result we have $\mathbf{X} = (\mathbf{1}_n : \mathbf{A} : \mathbf{B} : \mathbf{C})$, with rank $ab, \mathbf{M} = (1/\sigma_0^2)\mathbf{S}$, and experiments will be compared according to Definition 4.

Proposition 6. *Let φ be a concave, symmetric and isotonic functional of moment matrices. Then $\varphi(\mathbf{M})$ is a concave and $\mathcal{G}_C$-symmetric function of the allocation matrix $\mathbf{N}$ corresponding to two-way fixed cross experiments. Moreover, if $\mathbf{N}_1$ and $\mathbf{N}_2$ are allocation matrices of experiments $\mathcal{E}_1$ and $\mathcal{E}_2$, respectively, then $\mathbf{N}_2 \leq_{G_C} \mathbf{N}_1$ implies $\mathcal{E}_1 \leq_\varphi \mathcal{E}_2$.*

5. General Comments

Let us notice that the matrix $t\mathbf{1}_a\mathbf{1}_b'$ is the minimal element with respect to $\mathcal{G}_H$-majorization and $\mathcal{G}_C$-majorization in the set of all $a \times b$ matrices with the sum of elements equal to abt. Hence, when $n = abm$ for some integer m, then the experiment with the allocation matrix $\mathbf{N}_* = m\mathbf{1}_a\mathbf{1}_b'$ is optimal among all experiments with n observations, having no more than a levels of the primary (row) factor, and no more than b levels of the secondary (column) factor (cf. Stępniak, 1983).

Properties presented in Propositions 1 and 4 were established by Hauke and Markiewicz (1993).

The main results of Sections 3 and 4 can be extended to multi-stage hierarchical and multi-way cross classifications using multi-dimensional allocation matrices and appropriately extended groups $\mathcal{G}_H$ and $\mathcal{G}_C$.

Acknowledgments

We are grateful to reviewers for their helpful suggestions which helped to improve the presentation of the results.

References

Bailey, R. A. (1981). Distributive block structures and their automorphismums. In: K. L. McAveney, Ed., *Combinatorial Mathematics VIII, Lecture Notes in Mathematics* **884**. Springer-Verlag, Berlin, 115–124.

Dawid, A. P. (1988). Symmetry models and hypotheses for structured data layouts. *Journal of the Royal Statistical Society B* **50**, 1–34.

Giovagnoli, A., Pukelsheim F. and Wynn H.P. (1987). Group invariant ordering and experimental designs. *Journal of Statistical Planning and Inference* **17**, 159–171.

Giovagnoli, A. and Wynn H.P. (1985). G-majorization with applications to matrix orderings. *Linear Algebra and its Applications* **67**, 111–135.

Hauke, J. and Markiewicz A. (1993). On comparison of allocations of units in experimental designs with hierarchical and cross classification. Raport No.11/93 of Department of Mathematical and Statistical Methods, Agricultural University of Poznań.

Marshall, A. W. and I. Olkin (1979). *Inequalities: Theory of Majorization and Its Applications.* Academic Press, New York.

Peres, C. A. and Ting Hui-Ching. (1993). Optimum allocation for point estimation of the intraclass correlation coefficient. *Bulletin of the International Statistical Institute, 49th Session. Contributed Papers*, 329–330.

Pukelsheim, F. (1987). Majorization ordering for linear regression designs. In: T. Pukkila and S. Puntanen, Eds., *Proceedings of the Second International Tampere Conference in Statistics.* Department of Mathematical Sciences, University of Tampere, 261–274.

Rao, C.R. (1973). *Linear Statistical Inference and Its Applications*, 2nd ed. Wiley, New York.

Shaked, M. and Tong Y.U. (1992). Comparison of experiments via dependence of normal variables with a common marginal distribution. *The Annals of Statistics* **20**, 614–618.

Stępniak, C. (1983). Optimal allocation of units in experimental designs with hierarchical and cross classification. *Annals of the Institute of Statistical Mathematics A* **35**, 461–473.

Stępniak, C. (1987). Reduction problems in comparison of linear models. *Metrika* **34**, 211–216.

Stępniak, C. (1989). Stochastic ordering and Schur-convex functions in comparison of linear experiments. *Metrika* **36**, 291–298.

Styan, G.P.H. (1983). Schur complements and linear statistical models. In: T. Pukkila and S. Puntanen, Eds., *Proceedings of the First Tampere Seminar in Linear Models.* Department of Mathematical Sciences, University of Tampere, 35–75.

Torgersen, E. (1991). *Comparison of Statistical Experiments.* Cambridge University Press, Cambridge.

Yurramendi, Y. (1993). Structural comparison of hierarchical classifications. *4th Conference of the International Federation of Classification Societies*, Paris. Collection of abstracts, 298–299.

CHARACTERIZATIONS OF OBLIQUE AND ORTHOGONAL PROJECTORS

GÖTZ TRENKLER
Department of Statistics
University of Dortmund
D-44221 Dortmund
Germany

Abstract. Projectors play an important role in statistics. The objective of this paper is to give an overview of results from the literature on oblique and orthogonal projectors. Furthermore some new characterizations are presented which are based on ranks, traces, eigenvalues and other properties.

Key words: Oblique projectors, Orthogonal projectors, Generalized inverses, Matrix orderings.

1. Introduction

In statistics the notion of a projector has become more and more important during the last years. It is central in regression analysis where the method of least squares is equivalent to the orthogonal projection of the observation vector of the dependent variable on the space generated by the observation matrix of the explanatory variables. Many monographs and papers have used this concept as an elegant way to present the theory of the linear model (e.g., Graybill, 1976; Seber, 1977; Christensen, 1987; Rao and Mitra, 1971; Mäkeläinen, 1970; Rao, 1974). It has also become extremely useful in the field of multivariate statistics, especially in canonical correlation analysis (Rao, 1973, Section 8f; Anderson, 1984, Section 12.2; Takeuchi, Yanai and Mukherjee, 1982), distribution of quadratic forms (Mardia, Kent and Bibby, 1979, Chapter 3; Rao, 1973, Section 3b; Styan, 1970) and tests of the linear hypothesis (Arnold, 1981; Baksalary, 1984). A comprehensive treatment of projectors in econometrics has been given by Pollock (1979). Further applications may be found in Theil (1971), Trenkler and Toutenburg (1992), Trenkler and Stahlecker (1993), Fisher (1991). Projectors also play a role in the coordinate free approach of statistics (Drygas, 1970; Eaton, 1983) or its geometrical visualization (Margolis, 1979; Bryant, 1984). There exists a number of books dealing with matrix theory with emphasis on projectors (Rao and Mitra, 1971; Ben-Israel and Greville, 1974; Campbell and Meyer, 1979; Graybill, 1969; Magnus and Neudecker, 1988; Searle, 1982; Basilevsky, 1983). A unified and abstract treatment of projectors is provided in some monographs on Hilbert or Banach space theory, e.g. Berberian (1961), Cater (1966), Riesz and Nagy (1990), Dunford and Schwarz (1967) and Conway (1990). Many of these concepts can be literally translated to the simpler language of matrix algebra, which is the main basis of this paper.

Subsequently we give an overview of the most important properties of oblique and orthogonal projectors. Furthermore some new characterizations are presented.

T. Caliński and R. Kala (eds.),
Proceedings of the International Conference on Linear Statistical Inference LINSTAT '93, 255–270.

For our considerations we content ourselves with the real vector space $\mathbb{R}^n$ of ordered n-tuples endowed with the ordinary inner product $(x, y) = x'y$ where $x, y \in \mathbb{R}^n$. A projector (oblique projector, idempotent matrix) is any $n \times n$ matrix satisfying $P^2 = P$. If in addition P is symmetric, i.e. $P = P'$, we call P an orthogonal projector.

If $\mathbb{R}^n$ is a direct sum of subspaces $\mathcal{U}$ and $\mathcal{V}$, every vector x can be expressed uniquely as $x = u + v$, $u \in \mathcal{U}$ and $v \in \mathcal{V}$, and we write $\mathbb{R}^n = \mathcal{U} \oplus \mathcal{V}$. Then the linear transformation $Px = u$ is a projector. Denoting by $\mathcal{R}(A)$ the column space and by $\mathcal{N}(A)$ the nullspace of a matrix A, we have $\mathcal{R}(P) = \mathcal{N}(I - P) = \mathcal{U}$ and $\mathcal{R}(I - P) = \mathcal{N}(P) = \mathcal{V}$. Since P is uniquely determined by $\mathcal{U}$ and $\mathcal{V}$, we also call P the projector on $\mathcal{U}$ along $\mathcal{V}$. Actually, we have a one–to–one correspondence between projectors and the projectors on $\mathcal{U}$ along $\mathcal{V}$ (cf. Ben–Israel and Greville, 1974, p.50). Sometimes we briefly call P the projector on $\mathcal{U}$. For example, the projector P on the column space $\mathcal{R}(A)$ satisfies $P^2 = P$ and $\mathcal{R}(P) = \mathcal{R}(A)$.

For any matrix A the symbols $A^+, A^-, \mathrm{r}(A)$ and $\mathrm{tr}(A)$ will stand for the Moore–Penrose inverse, a generalized inverse (g-inverse), the rank and the trace of A, respectively. In the latter case, of course, A is a square matrix. Then the orthogonal projector $P_{\mathcal{R}(A)}$ on the column space $\mathcal{R}(A)$ can be expressed in terms of $(A'A)^-$. We have $P_{\mathcal{R}(A)} = A(A'A)^-A'$ which is invariant with respect to the choice of $(A'A)^-$.

2. Oblique Projectors

In the following P is assumed to be an $n \times n$ matrix, where in some cases the column space of P is generated by a matrix A, i.e. $\mathcal{R}(P) = \mathcal{R}(A)$. In general proofs for the characterizations given below will be presented only if the latter are not known from the literature. Our first theorem gives some general characterizations.

Theorem 1. *P is a projector if and only if one of the following conditions is satisfied.*

(i) *P' is a projector.*

(ii) *$I - P$ is a projector.*

(iii) *P^k is a projector for all $k \in \mathbb{N}$.*

(iv) *APB is a projector for any matrices A and B of type $m \times n$ and $n \times m$, respectively, such that $BA = I$.*

(v) *$U = 2P - I$ is an involution, i.e. $U^2 = I$.*

(vi) $P^2(I - P) = P(I - P)^2 = 0$.

(vii) *P is a reflexive inverse of I_n, i.e. $PI_nP = P$.*

(viii) *$Px = x$ for all $x \in \mathcal{R}(P)$.*

(ix) $\mathcal{R}(P) \subset \mathcal{N}(I - P)$.

(x) $\mathcal{R}(I - P) \subset \mathcal{N}(P)$.

(xi) $\mathbb{R}^n = \mathcal{R}(P) \oplus \mathcal{R}(I - P)$.

(xii) $\mathcal{R}(P) \cap \mathcal{R}(I - P) = \{0\}$.

(xiii) *$P = F - E = H + G$, where E, F, G and H are projectors.*

(xiv) *There exists a matrix B such that $PB = B$, $\mathcal{R}(P) \subset \mathcal{R}(B)$.*

(xv) *There exists a matrix B such that $BP = B$, $\mathcal{R}(P') \subset \mathcal{R}(B')$.*

(xvi) *There exists a projector Q such that $\mathcal{R}(P) \subset \mathcal{R}(Q)$ and $PQP = P$.*

(xvii) *There exists a projector* Q *such that* PQ *is a projector and* $QPQ = P$.

(xviii) *There exist orthogonal projectors* R *and* S *such that* $I - RS$ *is nonsingular and* $P = (I - RS)^{-1}R(I - RS)$.

(xix) *There exist orthogonal projectors* R *and* S *such that* $(I - S)P = I - S$, $I - SR$ *is nonsingular and* $P = (I - SR)^{-1}(I - S)$.

(xx) *There exist orthogonal projectors* R *and* S *such that* $PR = R$, $R + S - SR$ *is nonsingular and* $P = R(R + S - SR)^{-1}$.

Proof. Conditions (i)–(vii) are straightforward, (vi) can be found in Halmos (1958, p. 78). For statements (viii)–(xii), which are also well known, we refer to Sibuya (1970). Characterization (xiii) was recently given by Hartwig and Putcha (1990). It shows that a projector is simultaneously a sum as well as a difference of two projectors. By taking $B = P$ necessity of (xiv) is obvious. To demonstrate sufficiency assume the existence of a matrix B such that $PB = B$ and $\mathcal{R}(P) \subset \mathcal{R}(B)$. Then we have $BB^+P = P$ and consequently $PP = BB^+PBB^+P = BB^+BB^+P = BB^+P = P$. Condition (xv) then is similarly shown. A weaker statement than (xiv) and (xv) can be found in Rao and Mitra (1971, p. 119). Observe that, without loss of generality, the matrix B can be chosen as a projector. In (xvi) necessity is immediate by taking $Q = P$. To prove sufficiency observe that $\mathcal{R}(P) \subset \mathcal{R}(Q)$ implies $QP = P$, which gives $PP = QPQP = QP = P$. Note that in Ben–Israel and Greville (1974, p. 54) a slightly weaker condition is stated. By taking $Q = P$ necessity of (xvii) is evident. For sufficiency observe that $PP = QPQQPQ = QPQPQ = QPQ = P$. Sufficiency of (xviii) is obvious. Necessity follows from Afriat (1957) by choosing $R = P_{\mathcal{R}(P)}$ and $S = I - P_{\mathcal{R}(P')}$. By the restrictions on R and S, respectively, conditions (xix) and (xx) are sufficient. The same choice of R and S as in (xviii) yields necessity of (xix) and (xx) by consulting Theorem 3 of Greville (1974). □

We note that conditions (xix) and (xx) may be generalized to the setting of Λ-orthogonal projectors (see Baksalary and Kala, 1979, where related results are achieved).

Our next theorem deals with rank, trace and eigenvalues of a projector.

Theorem 2. *P is a projector if and only if one of the following conditions is satisfied.*

(i) $\mathrm{r}(I_n - P) = n - \mathrm{r}(P)$.

(ii) $\mathrm{r}(P) = \mathrm{tr}(P)$ *and* $P^s = P^t$ *for some* $s, t \in \mathbb{N}, s \neq t$.

(iii) $\mathrm{r}(I_n - P) \leq \mathrm{tr}(I_n - P)$, $\mathrm{r}(P) = \mathrm{r}(P^2)$ *and all eigenvalues of* P *are nonnegative.*

(iv) $\mathrm{r}(I_n - P)^2 = \mathrm{r}(I_n - P)$, $\mathrm{r}(P) = \mathrm{r}(P^2)$ *and all eigenvalues of* P *are nonnegative.*

(v) *There exists a matrix* B *such that* $\mathrm{r}(P) \leq \mathrm{r}(B)$ *and* $PB = B$.

(vi) *There exists a matrix* B *such that* $\mathrm{r}(P) \leq \mathrm{r}(B)$ *and* $BP = B$.

Proof. Condition (i) is proved in Sibuya (1970), and the proof of (ii) may be taken from Khatri (1980) where, however, only sufficiency is shown. Necessity follows from the fact that for projectors trace and rank coincide (cf. Ben–Israel and Greville, 1974, p. 49). For a proof of (iii) and (iv) we refer to Hartwig and Putcha (1990, Corollary 7). Note that condition $\mathrm{r}(P) = \mathrm{r}(P^2)$ is equivalent to the existence of a group inverse (cf. Ben–Israel and Greville, 1974, p. 162). To prove (v) for necessity take $B = P$. Sufficiency can be seen as follows. Since $\mathcal{R}(B) = \mathcal{R}(PB) \subset \mathcal{R}(P)$ and $\mathrm{r}(P) \leq \mathrm{r}(B)$ we have $\mathcal{R}(B) = \mathcal{R}(P)$. Then the assertion follows by Theorem 1 (xiv). Condition (vi) may be shown along the same lines by taking transposes. For (v) and (vi) we also refer to Rao and Mitra (1971, p. 119) where a weaker condition is stated. □

Subsequently we give a description of projectors by means of g-inverses. Recall that a g-inverse A^- of a matrix A is given by the identity $AA^-A = A$. By $A\{1\}$ we denote the set of all g-inverses of A. A reflexive g-inverse A_r^- of A satisfies in addition $A_r^- AA_r^- = A_r^-$. Observe that for a projector P one choice of a reflexive g-inverse is $P_r^- = P$. We write $B \in A\{2\}$ if $BAB = B$.

Theorem 3. *P is a projector if and only if one of the following conditions is satisfied.*

(i) $P = C^-C$ *for some* $m \times n$ *matrix* C.

(ii) $P = C(DC)_r^- D$ *for some matrices* C *and* D *of type* $n \times p$ *and* $q \times n$*, respectively.*

(iii) $PP^- \in P\{1\}$.

(iv) $P^-P \in P\{1\}$.

(v) $P^-PPP^- \in P\{1\}$.

(vi) *There exist orthogonal projectors* G *and* H *such that* $P = GT(I-H)$*, where* $T \in (I-H)G\{2\}$.

Proof. (i) is obvious (see Mitra and Bhimasankaram, 1970). Condition (ii) is a special case of Theorem 3a in Mitra (1968). Statements (iii)–(v) are straightforward to prove. Sufficiency of (vi) is obvious. For necessity choose $G = P_{\mathcal{R}(P)}$, $H = I - P_{\mathcal{R}(P)}$ and $T = P$. Then by the identities $PG = G$, $GP = P$, $(I-H)P = P$ and $P(I-H) = P$ we get $GT(I-H) = P$ and $T(I-H)GT = T$. □

A result somewhat different from (vi) is shown by Greville (1974, Theorem 2).

Let us now discuss some properties of projectors in connection with the Moore–Penrose inverse.

Theorem 4. *P is a projector if and only if one of the following conditions is satisfied.*

(i) *There exist orthogonal projectors* R *and* S *such that* $P^+ = RS$.

(ii) $P^+P' = P^+$.

(iii) $P'P^+ = P^+$.

Proof. Condition (i) is originally stated in Penrose (1955). For a generalization we refer to Greville (1974, Theorem 1). We offer here a different proof (another proof is given by Farebrother; see Farebrother and Trenkler, 1993). Necessity follows by writing $P^+ = P^+PP^+ = P^+PPP^+ = RS$, where $R = P^+P$ and $S = PP^+$. Let, vice versa, $P^+ = RS$, where R and S are orthogonal projectors. Then by Campbell and Meyer (1979, p. 20) we have

$$P = (RS)^+ = (P_{\mathcal{R}(R')}S)^+(RP_{\mathcal{R}(S)})^+ ,$$

where $P_{\mathcal{R}(R')}$ and $P_{\mathcal{R}(S)}$ are the orthogonal projectors on the range of R' and S, respectively. However, since $P_{\mathcal{R}(R')} = R$ and $P_{\mathcal{R}(S)} = S$, we see that P is a projector. A quick proof is also offered by the referee: if $P^+ = RS$ then $\mathcal{R}(P') = \mathcal{R}(P^+) \subset \mathcal{R}(R)$ and $\mathcal{R}(P) = \mathcal{R}(P^{+'}) \subset \mathcal{R}(S)$, which implies $PR = P$ and $SP = P$, respectively, giving in the consequence $P^2 = PP = PRSP = PP^+P = P$. For (ii) and (iii) necessity is stated in Hartwig and Spindelböck (1984, proof of Corollary 2). It is evident, since P is the projector on $\mathcal{R}(P) = \mathcal{R}(P^{+'})$. To demonstrate sufficiency in (ii) consider the identity $P' = P'PP^+$. Then we get $P'P' = P'PP^+P'$ which equals $P'PP^+$ by assumption. Taking transposes we arrive at $PP = P$. Sufficiency of condition (iii) is derived from (ii) by replacing P by P'. □

Observe that a modification of (i) is the condition $P^+ = P_{\mathcal{R}(P')}P_{\mathcal{R}(P)}$ which is also equivalent for P to be a projector (cf. Greville, 1974, Formula 3.1). Our next characterization deals with representations of projectors by certain matrices.

Theorem 5. *Let $r = \mathrm{r}(P)$. P is a projector if and only if one of the following conditions is satisfied.*

(i) *$P = AB'$, where $B'A = I_r$ with A and B being $n \times r$ matrices.*

(ii)

$$P = V \begin{pmatrix} I_r & 0 \\ 0 & 0 \end{pmatrix} V^{-1} ,$$

where V is a nonsingular matrix.

(iii) *The Jordan canonical form of P can be written as*

$$\begin{pmatrix} I_r & 0 \\ 0 & 0 \end{pmatrix} .$$

(iv)

$$P = U \begin{pmatrix} I_r & K \\ 0 & 0 \end{pmatrix} U' ,$$

where U is an orthogonal matrix and K is of the type $r \times (n - r)$.

Proof. Condition (i) was originally shown by Langenhop (1967). Note that this representation of P is just the factorization of P into two matrices of full column and row rank, respectively (see also Ben–Israel and Greville, 1974, p. 49; Mäkeläinen, 1970, Theorem 2). It may alternatively be written as

$$P = \sum_{j=1}^{r} x_j y_j' ,$$

where $\{x_1, \ldots, x_r\}$, $\{y_1, \ldots, y_r\}$ are biorthogonal systems in $\mathbb{R}^n$ (cf. Lancaster and Tismenetsky, 1985, p. 195). Condition (ii) follows immediately from Ben–Israel and

Greville (1974, Theorem 9, p. 52). It shows that a projector is always diagonable, i.e. similar to the diagonal matrix of its eigenvalues (which are 0 and 1). For condition (iii) we refer to Lancaster and Tismenetsky (1985, Chapters 6 and 7). Only necessity of (iv) is nontrivial. Since P is of rank r there exist r linearly independent columns of P forming a basis of $\mathcal{R}(P)$. Apply the Gram–Schmidt orthonormalization procedure to obtain an $n \times r$ matrix U_1 that consists of orthonormal basis vectors. By construction we have $\mathcal{R}(P) = \mathcal{R}(U_1)$ and $PU_1 = U_1$. Choose now an orthonormal basis of $\mathcal{R}(P)^\perp$, whose columns are given in the $n \times (n-r)$ matrix U_2. Then $U = (U_1, U_2)$ is orthogonal, and $\mathcal{R}(PU_2) \subset \mathcal{R}(P) = \mathcal{R}(U_1)$. Hence, there exists an $r \times (n-r)$ matrix K such that $PU_2 = U_1 K$, which implies

$$PU = P(U_1, U_2) = (U_1, U_1 K) = U_1(I_r, K) = U \begin{pmatrix} I_r & K \\ 0 & 0 \end{pmatrix}$$

and, consequently,

$$P = U \begin{pmatrix} I_r & K \\ 0 & 0 \end{pmatrix} U' . \square$$

The representation in (iv) was given by Hartwig and Loewy (1992) in the context of complex matrices. It can be shown by Schur's unitary triangularization theorem (cf. Horn and Johnson, 1985, p. 79) and Hartwig (1993). The proof presented here, dealing with the real case, is due to Werner (1993).

It should be noted that this representation also offers an easy way to calculate the Moore–Penrose inverse and the orthogonal projector on the range of a projector P. It is readily seen that

$$P^+ = U \begin{pmatrix} C & 0 \\ D & 0 \end{pmatrix} U' ,$$

where $C = (I_r + KK')^{-1}$ and $D = K'C$,

$$P_{\mathcal{R}(P)} = U \begin{pmatrix} I_r & 0 \\ 0 & 0 \end{pmatrix} U' = U_1 U_1',$$

(cf. Theorem 15 (iv)).

Representation (iv) also has a nice statistical application. Consider the general linear model $(y, X\beta, \Omega)$ in which y is an $n \times 1$ observable random vector with expectation $E(y) = X\beta$ and with dispersion matrix $D(y) = \Omega$; X is a known $n \times p$ matrix, β is a $p \times 1$ vector of unknown parameters and Ω is an $n \times n$ nonnegative definite matrix, known or known except for a positive scalar factor. Further, let $\hat{\mu}$ stand for the best linear unbiased estimator (BLUE) of $\mu = X\beta$. Under the assumption that Ω is nonsingular we have $\hat{\mu} = Py$, where $P = X(X'\Omega^{-1}X)^+ X'\Omega^{-1}$ is an oblique projector. The least squares estimator (LSE) of μ has the representation $\mu^\star = P_{\mathcal{R}(X)} y = XX^+ y$. Baksalary and Kala (1978, 1980) investigated the problem how much BLUE $\hat{\mu}$ and LSE $\mu^\star$ differ. Following their approach we give a new upper bound for $||\mu^\star - \hat{\mu}||^2$, where $||.||$ denotes the Euclidean vector norm.

Since $r = \mathrm{r}(X) = \mathrm{r}(P)$, we have $\mathcal{R}(P) = \mathcal{R}(X)$, i.e. P and $P_{\mathcal{R}(X)}$ project on the same subspace. By the preceding construction we may write

$$P = U \begin{pmatrix} I_r & K \\ 0 & 0 \end{pmatrix} U' \quad \text{for some matrix } K \text{ and}$$

$$P_{\mathcal{R}(X)} = U \begin{pmatrix} I_r & 0 \\ 0 & 0 \end{pmatrix} U' ,$$

which gives $||\mu^\star - \hat{\mu}||^2 = y'U\Sigma U'y$, where

$$\Sigma = \begin{pmatrix} 0 & 0 \\ 0 & K'K \end{pmatrix} .$$

Hence $||\mu^\star - \hat{\mu}||^2 \leq \lambda_1 ||y||^2$, with λ_1 being the largest eigenvalue of $K'K$. We also see that $\mu^\star$ and $\hat{\mu}$ coincide if and only if $K = 0$. This means $PU_2 = 0$ or equivalently $\mathcal{R}(P)^\perp \subset \mathcal{N}(P)$, i.e. $\mathcal{R}(P') \subset \mathcal{R}(P) = \mathcal{R}(X)$ which may be rewritten as $\mathcal{R}(\Omega^{-1}X) = \mathcal{R}(X)$, one of the conditions for equality of $\mu^\star$ and $\hat{\mu}$ (see also Puntanen and Styan, 1989).

Our next theorem is concerned with the situation that the potential projector has some additional properties.

Theorem 6.
(i) *Let P be semisimple (i.e. P is similar to a diagonal matrix). Then P is a projector if and only if $P^k = P^{k+1}$ for some $k \in \mathbb{N}$.*
(ii) *Let $\mathrm{r}(P) = \mathrm{tr}(P)$. Then P is a projector if and only if $\mathrm{tr}(P^2P^+P') = \mathrm{r}(P)$.*
(iii) *Let $P^2 = P^3$. Then each of the following conditions is necessary and sufficient for P to be a projector:*
(a) $\mathrm{r}(P) = \mathrm{tr}(P)$,
(b) $\mathrm{r}(P) = \mathrm{r}(P^2)$.

Proof. The proof of (i) is easy and will be omitted. Note that the assertion 'for some $k \in \mathbb{N}$' can be replaced by 'for all $k \in \mathbb{N}$' (see also Rao and Mitra, 1971, p. 118). Condition (ii) is derived in Khatri (1985). Khatri also showed that if in addition P is semisimple with real eigenvalues then P is a projector if and only if $\mathrm{tr}(P^2) = \mathrm{r}(P)$. For the conditions in (iii) we refer to Banerjee and Nagase (1976) and Styan (1970). □

Sometimes the range $\mathcal{R}(P)$ of a matrix P is generated by another matrix A, or we have the inclusion $\mathcal{R}(P) \subset \mathcal{R}(A)$. Our problem now is to describe the cases in which P is a projector.

Theorem 7.
(i) *Assume that $P = AG$ for some matrices A and G (i.e. $\mathcal{R}(P) \subset \mathcal{R}(A)$). Then P is a projector on $\mathcal{R}(A)$ if and only if one of the following conditions is satisfied:*
(a) $P^2 = P$ *and* $\mathrm{r}(P) = \mathrm{r}(A)$,
(b) $PA = A$,
(c) $\mathcal{R}(A) \oplus \mathcal{R}(I - P) = \mathbb{R}^n$,
(d) $\mathrm{r}(I_n - P) = n - \mathrm{r}(A)$.
(ii) *P is a projector on $\mathcal{R}(A)$, where A is a given matrix, if and only if $PA = A$ and $\mathcal{R}(P) = \mathcal{R}(A)$.*
(iii) *P is the projector on $\mathcal{R}(A)$ along $\mathcal{R}(B)$, where A and B are given matrices such that $\mathbb{R}^n = \mathcal{R}(A) \oplus \mathcal{R}(B)$ if and only if one of the following conditions is satisfied:*

(a) $PA = A$ *and* $PB = 0$,

(b) *there exists a matrix* Q *with* $\mathrm{r}(A) = \mathrm{r}(QA)$, $\mathcal{R}(B) = \mathcal{N}(Q)$ *and* $P = A(QA)^{-}Q$,

(c) $P = A(A'Q_BA)^{-}A'Q_B$ *where* $Q_B = I - BB^{+}$.

Proof. For (i) we refer to Sibuya (1970, Proposition 3.2), and conditions (ii) and (iii) are shown in Pollock (1979, p. 53). □

Let us now proceed to characterize projectors by means of matrix orderings. We consider the most important ones, namely the Löwner ordering, the star ordering and the rank subtractive ordering. For given matrices A and B these orderings are given by the following conditions.

Löwner ordering (Löwner, 1934). $A \leq_L B$ if and only if $B - A = CC'$ for some matrix C.

Star ordering (Drazin, 1978). $A \leq_* B$ if and only if $A'A = A'B$ and $AA' = BA'$.

Rank subtractive ordering (Hartwig, 1980). $A \leq_{rs} B$ if and only if $\mathrm{r}(B - A) = \mathrm{r}(B) - \mathrm{r}(A)$.

Observe that only for the Löwner ordering the matrices A and B are required to be square. For projectors P and Q the rank subtractive ordering coincides with the notion of dominance (cf. Cater, 1966, p.70; Hartwig and Styan, 1987, Lemma 2.3), i.e. $P \leq_{rs} Q$ is equivalent to $P = PQ = QP$.

The subsequent characterization of projectors is due to Hartwig and Styan (1987, Theorem 3.3). Proofs can be found in their paper.

Theorem 8. *P is a projector if and only if one of the following conditions is satisfied.*

(i) $P \leq_{rs} I$.

(ii) $P \leq_* Q$ *for some projector* Q.

(iii) $P \leq_{rs} Q$ *for some projector* Q.

(iv) $P \leq_L Q$ *for some projector* Q *such that* $PQ = QP$ *and all eigenvalues of* P *are* 0 *or* 1.

Note that condition (i) of Theorem 8 just reexpresses assertion (i) of Theorem 2.

3. Orthogonal Projectors

Orthogonal projectors are projectors which are symmetric. Insofar all characterizations of the preceding paragraph may be regarded as necessary conditions. However for sufficiency we can also expect a variety of interesting descriptions.

Theorem 9. *P is an orthogonal projector if and only if one of the following conditions is satisfied.*

(i) $P'P = P$.

(ii) $I - P$ *is an orthogonal projector.*
(iii) $PP' = P'P$ *and* P *is a projector.*
(iv) P *and* $P'P$ *are projectors.*
(v) $PP'P = P$ *and* P *is a projector.*
(vi) $P'PP' = P'$ *and* P *is a projector.*
(vii) P *and* $(P + P')/2$ *are projectors.*
(viii) $PP' + P'P = P + P'$ *and* P *is a projector.*
(ix) VPV' *is an orthogonal projector for all orthogonal matrices* V.
(x) $I - 2P$ *is a symmetric orthogonal matrix.*
(xi) $P = I - A^2$ *for some symmetric tripotent matrix* A *(i.e.* $A = A'$ *and* $A^3 = A$*).*
(xii) $P^2 = P'$ *and* P *is tripotent.*
(xiii) $PP' = P'PP'$.
(xiv) $P = F - E = H + G$, *where* E, F, G *and* H *are orthogonal projectors.*

Proof. Condition (i) is well-known (see also Mitra and Rao, 1974). It follows from the fact that $P'P = P$ immediately implies $P = P'$. Condition (ii) is trivial. The first identity of (iii) states that P is a normal matrix, and together with the property of being a projector we get an orthogonal projector. This property is not frequently stated (cf. Conway, 1990, p.37). Necessity is trivial. To demonstrate sufficiency note that an equivalent condition for (i) is $P = PP'$. Characterization (iv) is a direct consequence of Theorem 3 in Baksalary, Kala and Kłaczyński (1983) by taking $M = I$. Conditions (v) and (vi) follow immediately from (iv) by using the well known cancellation law: $AZ'Z = BZ'Z$ if and only if $AZ' = BZ'$ (cf. Campbell and Meyer, 1979, p.3). Conditions (v) and (vi) can be rephrased as $P' \in P\{1\}$ and $P \in P'\{1\}$, respectively. To prove (vii) and (viii) we show: P is an orthogonal projector $\Rightarrow$ (vii) $\Rightarrow$ (viii) $\Rightarrow$ P is an orthogonal projector. The first implication is trivial. To derive the second observe that, by assumption, $(P + P')/2 = (P^2 + P'^2 + P'P + PP')/4$. Since P is a projector, we get $P + P' = (P + P' + P'P + PP')/2$ and, consequently, $P + P' = P'P + PP'$. Assume now that (viii) is valid. Taking traces we obtain $\mathrm{tr}(PP') = \mathrm{tr}(P)$ and then $\mathrm{tr}(PP') = \mathrm{tr}(P^2)$. However, in general we have for square matrices $\mathrm{tr}(A^2) \leq \mathrm{tr}(A'A)$ with equality if and only if A is symmetric (Schur's inequality, cf. Magnus and Neudecker, 1988, p.202). Thus P is symmetric. Condition (ix) is obvious, and (x), (xi) and (xii) are also straightforward. Condition (xiii) is a consequence of the cancellation rule. Necessity of (xiv) follows from the representation $P = I - (I - P) = P + 0$, whereas sufficiency is seen from Theorem 1 (xiii) and the symmetry of $H + G$. □

It should be remarked that the former identity $\mathrm{tr}(PP') = \mathrm{tr}(P^2)$ always may serve as an alternative condition for symmetry. Thus P is an orthogonal projector if and only if P is a projector and $\mathrm{tr}(PP') = \mathrm{tr}(P)$. If P is a projector the identity $\mathrm{tr}(PP') = \mathrm{tr}(P)$ can be replaced by $\mathrm{tr}(PP') = \mathrm{r}(P)$. Other conditions related to trace, rank and eigenvalues are given below.

Theorem 10. *P is an orthogonal projector if and only if one of the following conditions is satisfied.*
(i) P *is a projector and* $\mathrm{r}(I_n - P'P) = n - \mathrm{r}(P)$.

(ii) *P is a projector and $\lambda_1(PP') \leq 1$, where $\lambda_1(PP')$ is the maximal eigenvalue of PP'.*

(iii) *$P(P'P)^s = P'(PP')^t$ for some $s, t \in \mathbb{N}$, $s \neq t$ and* $\mathrm{tr}(P) = \mathrm{r}(P)$.

Proof. Condition (i) is a direct consequence of the fact that $\mathrm{r}(P) = \mathrm{r}(P'P)$, Theorem 2 (i) and Theorem 9 (iii). Condition (ii) follows from Theorem 3 in Baksalary, Schipp and Trenkler (1992) by taking $B = I$. The third condition originates from Khatri (1980, Theorem 3). □

Note that the second condition in (ii), namely $\lambda_1(PP') \leq 1$, can be replaced by $\lambda_1(PP') = 1$ provided $P \neq 0$ (cf. Conway, 1990, p.37).

Our next characterizations deal with vectors and in some cases with their Euclidean norm $||\cdot||$.

Theorem 11. *P is an orthogonal projector if and only if one of the following conditions is satisfied.*

(i) *P is a projector and $||Px|| \leq ||x||$ for all $x \in \mathbb{R}^n$.*

(ii) *$x'P'Px = x'Px$ for all $x \in \mathbb{R}^n$.*

(iii) *$Px = x$ for all $x \in \mathcal{N}(P)^\perp$, where $\mathcal{N}(P)^\perp$ is the orthogonal complement of $\mathcal{N}(P)$.*

(iv) *P is a projector and $x'Px \geq 0$ for all $x \in \mathbb{R}^n$.*

(v) *$||y - Py|| \leq ||y - x||$ for all $y \in \mathbb{R}^n$ and all $x \in \mathcal{R}(P)$.*

(vi) *$||x - Px|| \leq ||x - Qx||$ for all $x \in \mathbb{R}^n$ and all projectors Q such that $\mathcal{R}(P) = \mathcal{R}(Q)$.*

(vii) *$Px = x$ for all $x \in \mathcal{R}(P)$ and $Px = 0$ for all $x \in \mathcal{R}(P)^\perp$.*

(viii) *$||x - y||^2 = ||x - Px||^2 + ||Px - y||^2$ for all $x \in \mathbb{R}^n$ and some $y \in \mathcal{R}(P)$.*

(ix) *$(x - Px)'(Px - y) \geq 0$ for all $x \in \mathbb{R}^n$ and all $y \in \mathcal{R}(P)$.*

Proof. For condition (i) we refer to Ben–Israel and Greville (1974, p.71). Condition (ii) just corresponds with Theorem 9 (i) (see also Cater, 1966, p.137). It shows that P is a partial isometry. For (iii) we also refer to Cater (1966, p.138). Condition (iv) is given in Conway (1990, p.37). The inequality in (v) is one of the best known characterizations (cf. Mitra and Rao, 1974) for orthogonal projectors. It reveals that Py is that point in $\mathcal{R}(P)$ which is closest to y, where y is any vector from $\mathbb{R}^n$. Condition (vi) may be found in Sibuya (1970, Proposition 3.2). Condition (vii) is given by Christensen (1987, p.335) and condition (viii) is due to Mäkeläinen (1970, Lemma 2). For condition (ix) we refer to Zarantonello (1971), Lemma 1.1. □

Next we will give characterizations of orthogonal projectors in terms of the Moore–Penrose inverse and g-inverses.

Theorem 12. *P is an orthogonal projector if and only if one of the following conditions is satisfied.*

(i) *P is a projector and $P = P^+$.*

(ii) *P is a projector and $PP' = PP^+$.*

(iii) *$P^+ = P$ and $P^2 = P'$.*

(iv) $P^+ = P$ *and* $P^+ = P^2$.
(v) P *and* P^+ *are projectors.*
(vi) *There exists an* $n \times m$ *matrix* A *such that* $P = AA^+$.
(vii) $P^+ = P^2$ *and* P *is tripotent, i.e.* $P^3 = P$.
(viii) $P = PP^+$.
(ix) $P = P^+P$.
(x) P *is a projector and there exists a g-inverse* P^- *of* P *such that* $PP^- = PP'$.

Proof. For (i) necessity is clear and sufficiency is seen from $P = PP = PP^+$ implying that P is symmetric. Condition (ii) is necessary, and sufficiency is seen from Theorem 12 (i). Note that the second condition may alternatively be chosen as $P'P = PP = PP^+$ so that P is symmetric. Condition (iii) is necessary, and sufficiency follows from $P' = PP = PP^+$ so that P is symmetric. Condition (iv) is necessary, and conversely, if $P = P^+$ and $P^2 = P^+$, P is a projector. Symmetry follows from $P = PP = PP^+$. To show (v), necessity is trivial, but sufficiency is more involved. By Rao and Mitra (1971, p.68) the following two inclusions are necessary and sufficient for $P^+P^+ = P^+$: $\mathcal{R}(PP'P') \subset \mathcal{R}(P')$ and $\mathcal{R}(P'PP) \subset \mathcal{R}(P)$. Since P and P' are projectors, these inclusions may alternatively be written as $P'PP'P' = PP'P'$ and $PP'PP = P'PP$, which gives $P'PP' = PP'$ and $PP'P = P'P$. By the cancellation rule (cf. the proof of Theorem 9), from the first equation we get $P'P = P$ which coincides with condition (i) of Theorem 9. Different proofs for (v) are provided in Farebrother, Pordzik and Trenkler (1993). Another proof for sufficiency is given by the referee: since $\mathcal{R}(P^+) = \mathcal{R}(P')$ and $\mathcal{R}(P^{+'}) = \mathcal{R}(P)$ we obtain $P^+P' = P'$ and $PP^{+'} = P^{+'}$, which leads to Theorem 9 (v). Conditions (vi), (vii), (viii) and (ix) are evident, and their proof can be omitted. Note that instead of P^+ in (viii) we may choose a (1,3)-inverse, i.e. any g-inverse P^- such that PP^- is symmetric (least squares g-inverse). For (ix) a (1,4)-inverses would suffice, i.e. any g-inverse P^- such that P^-P is symmetric. Necessity of (x) is obvious and sufficiency follows from Theorem 9 (v). □

Let us now proceed by giving descriptions of projectors by means of certain factorizations.

Theorem 13. *P is an orthogonal projector if and only if one of the following conditions is satisfied.*
(i) $P = A(A'A)^{-1}A'$ *for some matrix* A *of type* $n \times m$, $\mathrm{r}(A) = m$.

(ii) $$P = V \begin{pmatrix} I_{\mathrm{r}(P)} & 0 \\ 0 & 0 \end{pmatrix} V',$$

for some orthogonal matrix V.
(iii) $P = BB'$ *for some* $n \times m$ *matrix* B *such that* $B'B = I_m$.

Proof. For condition (i) and (ii) we refer to Mäkeläinen (1970). Condition (iii) follows from (ii). It shows that every orthogonal projector P can be written as $P = \sum x_j x_j'$ where $x_i'x_j = \delta_{ij}$ (the Kronecker delta). □

Clearly, if P is symmetric, then $P^2 = P$ is necessary and sufficient for P to be an orthogonal projector. But there are also other conditions ensuring this property. Subsequently we also consider the case when P is nonnegative definite or is of the form $P = (A + A')/2$ for some matrix A.

Theorem 14.

(i) *Suppose that P is symmetric. Then P is an orthogonal projector if and only if one of the following conditions is satisfied:*

(a) *all eigenvalues of P are 0 or 1,*

(b) *there exists $k \in \mathbb{N}$ such that $P^k = P^{k+1}$,*

(c) *there exist $k, l \in \mathbb{N}$ such that $P^k = P^{k+2l-1}$,*

(d) *P^m is an orthogonal projector for all $m \in \mathbb{N}$.*

(ii) *Suppose that P is nonnegative definite, i.e. $P = AA'$ for some $n \times m$ matrix A. Then P is an orthogonal projector if and only if one of the following conditions is satisfied:*

(a) P is a projector,

(b) *A is a partial isometry,*

(c) *A' is a partial isometry,*

(d) *$A'A$ is a projector,*

(e) *$AA'A = A$, i.e. $A' \in A\{1\}$,*

(f) *$A'AA' = A'$, i.e. $A' \in A\{2\}$,*

(g) *$A' = A^+$,*

(h) *A^+ is a partial isometry.*

(iii) *Suppose that $P = (A + A')/2$ for some matrix A with $\mathrm{r}(A) = s$. Then P is an orthogonal projector of $\mathrm{r}(P) = r$ only if one of the following two conditions is satisfied:*

(a) *$AA' = A'A = 2A - A^2$,*

(b) *A is an orthogonal projector and $\mathrm{tr}_r(A) = 1$, where $\mathrm{tr}_r(A)$ is defined as the sum of the principal minors of order r in A.*

Proof. Conditions (a) – (d) from (i) follow directly from the spectral decomposition of P. Let us now proceed to conditions (a) – (h) of (ii). Condition (a) is mentioned for the sake of completeness. Conditions (b) – (h) are shown in Ben–Israel and Greville (1974, pp. 252–253). Recall that a partial isometry P is defined by the property $||Px|| = ||x||$ for all $x \in \mathcal{R}(P') = \mathcal{N}(P)^{\perp}$. Condition (iii) is proved in Khatri (1980, Theorem 1). □

As in the oblique projector case the range of a potential orthogonal projector can be generated by columns of another matrix. Our next theorem is concerned with the problem of characterizing these projectors in terms of the given matrix.

Recall the definition of a least squares g-inverse A_l^- of a matrix A: $AA_l^-A = A$ and AA_l^- is symmetric.

Theorem 15. *Let us be given an $n \times n$ matrix A. Then P is the orthogonal projector on $\mathcal{R}(A)$ (i.e. $P = P'P$ and $\mathcal{R}(P) = \mathcal{R}(A)$) if and only if one of the* *following conditions is satisfied.*

(i) $P = AB$ *for some matrix* B, $P = P'P$ *and* $\mathrm{r}(P) = \mathrm{r}(A)$.

(ii) $P = AB$ *for some matrix* B *and* $P'A = A$.

(iii) $||x - Px|| \leq ||x - Qx||$ *for any* $x \in \mathbb{R}^n$ *and any projector* Q *such that* $\mathcal{R}(Q) = \mathcal{R}(A)$.

(iv) $P = BB'$, *where the columns of* B *form an orthonormal basis for* $\mathcal{R}(P)$.

(v) $P = A(A'A)^- A'$, *where* $(A'A)^-$ *is any g-inverse of* $A'A$.

(vi) $P = AA_l^-$, *where* A_l^- *is a least squares g-inverse of* A.

Proof. For (i) – (iii) we refer to Sibuya (1970). Condition (iv) is shown in Seber (1977, p.394). Statement (v) is well-known (cf. Rao and Kleffe, 1988, p.6). Observe that $A(A'A)^- A'$ is invariant with respect to the choice of $(A'A)^-$, which implies $P = A(A'A)^+ A = AA^+$. Condition (vi) is stated in Rao and Mitra (1971, p.111).□

Another condition stated in Sibuya (1970), namely $PA = P'A = A$, as found out by one of the referees, is incorrect. For this consider $P = \left(\begin{smallmatrix}1 & 0\\ 0 & a\end{smallmatrix}\right)$ and $A = \left(\begin{smallmatrix}1 & 0\\ 0 & 0\end{smallmatrix}\right)$. Then $PA = P'A = A$, but $PP \neq P$ for all $a \neq 1$ or 0.

Finally again we pay some attention to matrix orderings and their application in connection with orthogonal projectors.

In addition to the orderings used in Section 2 we will now also consider the left-star and the right-star ordering introduced by Baksalary and Mitra (1991).

Left-star ordering. $A \leq_l B$ if and only if $A'A = A'B$ and $\mathcal{R}(A) \subset \mathcal{R}(B)$.

Right-star ordering. $A \leq_r B$ if and only if $AA' = BA'$ and $\mathcal{R}(A') \subset \mathcal{R}(B')$.

We may now state our last theorem.

Theorem 16. *P is an orthogonal projector if and only if one of the following conditions is satisfied.*

(i) $P \leq_l I$.

(ii) $P \leq_l Q$ *for some orthogonal projector* Q.

(iii) $P \leq_l P'P$.

(iv) $P \leq_r I$.

(v) $P \leq_r Q$ *for some orthogonal projector* Q.

(vi) $P \leq_\star I$.

(vii) $P \leq_{rs} P'P$.

(viii) $P \leq_\star Q$ *for some orthogonal projector* Q.

(ix) $P \leq_L Q$ *for some orthogonal projector* Q *and all eigenvalues of* P *are* 0 *or* 1.

(x) $0 \leq_L P \leq_L Q$ *and* $(Q - P)^2 = Q - P$ *for some orthogonal projector* Q.

Proof. Condition (i) just states $P' = P'P$. Necessity of (ii) is clear. Let now $P \leq_l Q$ for some orthogonal projector. Then $P'P = P'Q$ and $\mathcal{R}(P) \subset \mathcal{R}(Q)$. The latter inclusion implies $QP = P$ which gives $P = P'P$. If P is an orthogonal projector then $P \leq_l P = P'P$ which yields one direction in (iii). Conversely, let $P \leq_l P'P$. Then $P'P = P'P'P$ and $\mathcal{R}(P) \subset \mathcal{R}(P'P)$. By the cancellation rule we obtain $P' = P'P'$, and hence P' is a projector. Further, since $\mathcal{R}(P) \subset \mathcal{R}(P'P) = \mathcal{R}(P')$ it follows that $P'P = P$. Hence P is an orthogonal projector. Conditions (iv) and (vi) follow along the same lines. Statements (vii) – (x) can be found in Hartwig and Styan (1987, Theorem 3.4). □

4. Concluding Remarks

Many of the characterizations given above can be generalized to projectors which are defined with respect to a more general norm (cf. Mitra and Rao, 1974). Extensions should also be possible to the case of projectors associated with a subspace $\mathcal{R}(A) \oplus \mathcal{R}(B) \subset \mathbb{R}^n$, where A and B are disjoint matrices (cf. Rao, 1974; Rao and Yanai, 1979; Yanai, 1990). Since we are interested in projectors as a tool in statistics we have stated our results in the framework of the real vector space $\mathbb{R}^n$. Generalizations to the complex setting are straightforward. Finally we believe that the notion of matrix ordering deserves more attention in connection with the investigation of projectors.

Acknowledgements

Support by Deutsche Forschungsgemeinschaft, Grant No. TR 253/1-2, is gratefully acknowledged. The author would like to thank the referees for their helpful comments.

References

Afriat, S.N. (1957). Orthogonal and oblique projectors and the characteristics of pairs of vector spaces. *Proceedings of the Cambridge Philosophical Society* **53**, 800–816.

Anderson, T.W. (1984). *An Introduction to Multivariate Statistical Analysis*, 2nd ed. Wiley, New York.

Arnold, S.F. (1981). *The Theory of Linear Models and Multivariate Analysis*. Wiley, New York.

Baksalary, J.K. (1984). A study of the equivalence between a Gauss-Markoff model and its augmentation by nuisance parameters. *Mathematische Operationsforschung und Statistik, Series Statistics* **15**, 3–35.

Baksalary, J.K. and Kala, R. (1978). A bound for the Euclidean norm of the difference between the least squares and the best linear unbiased estimators. *The Annals of Statistics* **6**, 1390–1393.

Baksalary, J.K. and Kala, R. (1979). Two relations between oblique and Λ–orthogonal projectors. *Linear Algebra and its Applications* **24**, 99–103.

Baksalary, J.K. and Kala, R. (1980). A new bound for the Euclidean norm of the difference between the least squares and the best linear unbiased estimators. *The Annals of Statistics* **8**, 679–681.

Baksalary, J.K., Kala, R. and Kłaczyński, K. (1983). The matrix inequality $M \geq B^*MB$. *Linear Algebra and its Applications* **54** 77–86.

Baksalary, J.K. and Mitra, S.K. (1991). Left-star and right-star partial orderings. *Linear Algebra and its Applications* **149**, 73–98.

Baksalary, J.K., Schipp, B. and Trenkler, G. (1992). Some further results on Hermitian-matrix inequalities. *Linear Algebra and its Applications* **160** 119–129.

Banerjee, K.S. and Nagase, G. (1976). A note on the generalization of Cochran's theorem. *Communications in Statistics – Theory and Methods A* **5**, 837–842.

Basilevsky, A. (1983). *Applied Matrix Algebra in the Statistical Sciences*. North-Holland, New York.

Ben–Israel, A. and Greville, T.N.E. (1974). *Generalized Inverses: Theory and Applications*. Wiley, New York.

Berberian, S.K. (1961). *Introduction to Hilbert Space*. Oxford University Press, New York.

Bryant, P. (1984). Geometry, statistics, probability: Variations on a common theme. *The American Statistician* **38**, 38–48.

Campbell, S.K. and Meyer, C.D. (1979). *Generalized Inverses of Linear Transformations*. Pitman, London.

Cater, F.S. (1966). *Lectures on Real and Complex Vector Spaces*. W.B. Saunders, Philadelphia.

Christensen, R. (1987). *Plane Answers to Complex Questions: The Theory of Linear Models*. Springer-Verlag, New York.

Conway, J.B. (1990). *A Course in Functional Analysis*. Springer-Verlag, New York.

Drazin, M.P. (1978). Natural structures on semigroups with involution. *Bulletin of the American Mathematical Society* **84**, 139–141.

Drygas, H. (1970). *The Coordinate-Free Approach to Gauss–Markov Estimation*. Springer-Verlag, Berlin.

Dunford, N. and Schwarz, J.T. (1967). *Linear Operators. Part 1: General Theory*. Interscience Publishers, New York.

Eaton, M.L. (1983). *Multivariate Statistics. A Vector Space Approach*. Wiley, New York.

Farebrother, R.W., Pordzik, P. and Trenkler, G. (1993). Characterization of an orthogonal projection matrix. Submitted for publication.

Farebrother, R.W. and Trenkler, G. (1993). Characterization of an idempotent matrix. Submitted for publication.

Fisher, G. (1991). Teaching econometric theory from the coordinate-free viewpoint. In: D. Vere-Jones, Ed., *Teaching Statistics Beyond School Level*. ISI Publications, 303–312.

Graybill, F.A. (1969). *Introduction to Matrices with Applications in Statistics*. Wadsworth Publication, Belmont.

Graybill, F.A. (1976). *Theory and Application of the Linear Model*. Duxbury Press, North Scituate, Massachusetts.

Greville, T.N.E. (1974). Solutions of the matrix equation $XAX = X$, and relations between oblique and orthogonal projectors. *SIAM Journal of Applied Mathematics* **26**, 828–832.

Halmos, P.R. (1958). *Finite-Dimensional Vector Spaces*. D. van Nostrand, Princeton.

Hartwig, R.E. (1980). How to partially order regular elements. *Mathematica Japonica* **25**, 1–13.

Hartwig, R.E. (1993). Personal communication.

Hartwig, R.E. and Loewy, R. (1992). Maximal elements under the three partial orders. *Linear Algebra and its Applications* **175**, 39–61.

Hartwig, R.E. and Putcha, M.S. (1990). When is a matrix a difference of two idempotents? *Linear and Multilinear Algebra* **26**, 267–277.

Hartwig, R.E. and Spindelböck, K. (1984). Matrices for which A^* and A^+ commute. *Linear and Multilinear Algebra* **14**, 241–256.

Hartwig, R.E. and Styan, G.P.H. (1987). Partially ordered idempotent matrices. In: T. Pukkila and S. Puntanen, Eds., *Proceedings of the Second International Tampere Conference in Statistics*. Department of Mathematical Sciences, University of Tampere, 361–383.

Horn, R.A. and Johnson, C.R. (1985). *Matrix Analysis*. Cambridge University Press.

Khatri, C.G. (1980). Powers of matrices and idempotency. *Linear Algebra and its Applications* **33**, 57–65.

Khatri, C.G. (1985). A note on idempotent matrices. *Linear Algebra and its Applications* **70**, 185–195.

Lancaster, P. and Tismenetsky, M. (1985). *The Theory of Matrices*. Academic Press, Orlando.

Langenhop, C.E. (1967). On generalized inverses of matrices. *SIAM Journal of Applied Mathematics* **15**, 1239–1246.

Löwner, K. (1934). Über monotone Matrixfunktionen. *Mathematische Zeitschrift* **38**, 177–216.

Mäkeläinen, T. (1970). A specification analysis of the general linear model. *Commentationes Physico-Mathematicae* **38**, 55–100.

Magnus, J.R. and Neudecker, H. (1988). *Matrix Differential Calculus with Applications in Statistics and Econometrics*. Wiley, New York.

Mardia, K.V., Kent, J.T. and Bibby, J.M. (1979). *Multivariate Analysis*. Academic Press, London.

Margolis, M.S. (1979). Perpendicular projections and elementary statistics. *The American Statistician* **33**, 131–135.

Mitra, S.K. (1968). On a generalized inverse of a matrix and applications. *Sankhyā A* **30**, 107–114.

Mitra, S.K. and Bhimasankaram, P. (1970). Some results on idempotent matrices and a matrix equation connected with the distribution of quadratic forms. *Sankhyā A* **32**, 353–356.

Mitra, S.K. and Rao, C.R. (1974). Projections under seminorms and generalized Moore Penrose inverses. *Linear Algebra and its Applications* **9**, 155–167.

Penrose, R. (1955). A generalized inverse for matrices. *Proceedings of the Cambridge Philosophical Society* **51**, 406–419.

Pollock, D.S.G. (1979). *The Algebra of Econometrics.* Wiley, New York.

Puntanen, S. and Styan, G.P.H. (1989). The equality of the ordinary least squares estimator and the best linear unbiased estimator. *The American Statistician* **43**, 153–161.

Rao, C.R. (1973). *Linear Statistical Inference and Its Applications.* Wiley, New York.

Rao, C.R. (1974). Projectors, generalized inverses and the BLUE's. *Journal of the Royal Statistical Society B* **36**, 442–448.

Rao, C.R. and Kleffe, J. (1988). *Estimation of Variance Components and Applications.* North-Holland, Amsterdam.

Rao, C.R. and Mitra, S.K. (1971). *Generalized Inverse of Matrices and its Applications.* Wiley, New York.

Rao, C.R. and Yanai, H. (1979). General definition and decomposition of projectors and some applications to statistical problems. *Journal of Statistical Planning and Inference* **3**, 1–17.

Riesz, F. and Sz.–Nagy, B. (1990). *Functional Analysis.* Dover Publications, New York.

Searle, S.R. (1982). *Matrix Algebra Useful for Statistics.* Wiley, New York.

Seber, G.A.F. (1977). *Linear Regression Analysis.* Wiley, New York.

Sibuya, M. (1970). Subclasses of generalized inverses of matrices. *Annals of the Institute of Statistical Mathematics* **22**, 543–556.

Styan, G.P.H. (1970). Notes on the distribution of quadratic forms in singular normal variables. *Biometrika* **57**, 567–572.

Takeuchi, K., Yanai, H. and Mukherjee, B.N. (1982). *The Foundations of Multivariate Analysis.* Wiley Eastern, New Delhi.

Theil, H. (1971). *Principles of Econometrics.* Wiley, New York.

Trenkler, G. and Stahlecker, P. (1993). Dropping variables vs. use of proxy variables in linear regression. Submitted for publication.

Trenkler, G. and Toutenburg, H. (1992). Proxy variables and mean square error dominance in linear regression. *Journal of Quantitative Economics* **8**, 433–442.

Werner, H.J. (1993). Personal communication.

Yanai, H. (1990). Some generalized forms of least squares g-inverses, minimum norm g-inverse, and Moore-Penrose inverse matrices. *Computational Statistics & Data Analysis* **10**, 251–260.

Zarantonello, E.H. (1971). Projections on convex sets in Hilbert space and spectral theory. Part I: Projections on convex sets. Part II: Spectal theory. In: L.B. Rall, Ed., *Nonlinear Functional Analysis and Applications.* Academic Press, New York, 237–424.

ASYMPTOTIC PROPERTIES OF LEAST SQUARES PARAMETER ESTIMATORS IN A DYNAMIC ERRORS-IN-VARIABLES MODEL

JAN TEN VREGELAAR
Department of Mathematics
Agricultural University
Dreijenlaan 4
6703 HA Wageningen
The Netherlands

Abstract. Consider a parametrized dynamic model relating the output at time t to p previous outputs, q previous inputs and the input at time t, with p and q known. Both inputs and outputs are measured with error. A least squares approach for estimating the parameters is defined. Sufficient conditions for the strong consistency and asymptotic normality of the estimators (including proofs) are provided.

Key words: Consistency, Asymptotic normality, Errors-in-variables.

1. Introduction

In this paper we introduce the least squares parameter estimation method for a dynamic errors-in-variables model. In Ten Vregelaar (1990), an efficient algorithm for computing parameter estimates is provided. The main contributions of the present paper are the asymptotic properties of the estimation method which are proved under some fairly general conditions. Normality of errors, for instance, is not required.

The model and estimation method are presented in this section. In Section 2 we discuss the conditions which turn out to be sufficient for the strong consistency property of the least squares estimators. The asymptotic normality property is stated in Section 3. After a discussion in the fourth section, we give the proofs of both asymptotic properties in the Appendix.

Let us consider the dynamic model for scalar inputs ξ_t and outputs η_t (think of t representing time) described by

$$\eta_t = \sum_{i=1}^{p} \alpha_i \eta_{t-i} + \sum_{j=0}^{q} \beta_j \xi_{t-j}, \text{ for } t = m+1, m+2, \ldots, \tag{1}$$

where $m = \max(p, q)$ and p and q are known nonnegative integers. The model may be called ARMA-model (autoregressive-moving average), although this term usually refers to different models. The vector in $\mathbb{R}^{p+q+1}$ of unknown model parameters is

$$\theta = (\alpha_1, \alpha_2, \ldots, \alpha_p, \beta_0, \beta_1, \ldots, \beta_q)^T, \tag{2}$$

where T indicates transposed matrix.

T. Caliński and R. Kala (eds.),
Proceedings of the International Conference on Linear Statistical Inference LINSTAT '93, 271–283.

Next we suppose that both *inputs* and *outputs* are observed with errors denoted by δ and ε, respectively:

$$\left.\begin{array}{l} y_t = \eta_t + \varepsilon_t \\ x_t = \xi_t + \delta_t \end{array}\right\} \quad t = 1, 2, \ldots, m + N. \tag{3}$$

The positive integer N is the difference between the number of observation pairs (x_t, y_t) and m. The term *asymptotic* refers to letting N go to infinity.

The problem is to estimate the parameter vector θ from the data set $\{(x_1, y_1), (x_2, y_2), \ldots, (x_{m+N}, y_{m+N})\}$. We consider the model equations (1) for only those inputs and outputs that are observed, i.e. we take $t = m+1, m+2, \ldots, m+N$ in (1). We do not suppose that the inputs and outputs are drawn from some distribution, they are unknown constants. In other words, the model is *functional.*

By definition, the *least squares estimation* method for this problem chooses $\hat{\theta}, \hat{\xi}_1, \ldots, \hat{\xi}_{m+N}, \hat{\eta}_1, \ldots, \hat{\eta}_{m+N}$ to minimize $\sum_{t=1}^{m+N} \{(y_t - \eta_t)^2 + (x_t - \xi_t)^2\}$ subject to the model equations (1) for $t = m + 1, \ldots, m + N$.

For convenience, we introduce a compact notation for the model and observations. It is easy to verify that the equations (1) for $t = m + 1, \ldots, m + N$ can be written as the matrix vector equation

$$D\zeta = 0, \tag{4}$$

where $D \in \mathbb{R}^{N \times 2(m+N)}$ and $\zeta \in \mathbb{R}^{2(m+N)}$ are defined by

$$D = \begin{pmatrix} \gamma_0 & \gamma_1 & \cdots & \gamma_m & & & \\ & \gamma_0 & \gamma_1 & \cdots & \gamma_m & & \\ & & \cdot & \cdot & \cdots & \cdot & \\ & & & \gamma_0 & \gamma_1 & \cdots & \gamma_m \end{pmatrix}, \tag{5}$$

$$\zeta = (\eta_{m+N}, \xi_{m+N}, \ldots, \eta_2, \xi_2, \eta_1, \xi_1)^T. \tag{6}$$

The empty space in the right upper and left lower part of D represents zero entries. The γ's in (5) are defined by $\gamma_0 = (-1,\ \beta_0)$ and $\gamma_i = (\alpha_i,\ \beta_i)$, $i = 1, 2, \ldots, m$, where $\alpha_i = 0$ if $i > p$ and $\beta_j = 0$ if $j > q$. Corresponding to (6) we define a vector of observations

$$z = (y_{m+N}, x_{m+N}, \ldots, y_2, x_2, y_1, x_1)^T.$$

Now, the minimization problem can be written as

$$\min_{\theta, \zeta} (z - \zeta)^T (z - \zeta) \text{ subject to } D(\theta)\zeta = 0. \tag{7}$$

By application of Lagrange multipliers when keeping θ fixed, (7) reduces to

$$\min_{\theta} J_N(\theta), \text{ where } J_N(\theta) = z^T P(\theta) z / N, \text{ with } P = D^T (DD^T)^{-1} D. \tag{8}$$

The factor N in the denominator is introduced for convenience. Due to the matrix $(DD^T)^{-1}$, the object function $J_N(\theta)$ is highly nonlinear. The matrix $P \in \mathbb{R}^{2(m+N) \times 2(m+N)}$ represents the orthogonal projection on the column space of D^T. For later use, we give here an expression for $\dot{P}$ denoting $\partial P / \partial \theta_i$, where θ_i is an arbitrary component of the vector θ in (2). Using the Moore-Penrose inverse of D, $D^+ = D^T(DD^T)^{-1}$, and the orthogonal projection matrix $P^\perp = I - P$ (I is the identity matrix), one easily obtains

$$\dot{P} = D^+\dot{D}P^\perp + P^\perp\dot{D}^T(D^+)^T, \tag{9}$$

and

$$\dot{J}_N = z^T\dot{P}z/N = 2z^TD^+\dot{D}P^\perp z/N \tag{10}$$

as expression for any component of the gradient of J_N.

As pointed out by Aoki and Yue (1970a), solutions of (8) correspond to maximum likelihood estimates for θ when the noise process is white and Gaussian. However, they prove convergence results for *approximate* maximum likelihood methods. In the companion paper (Aoki and Yue, 1970b), they provide asymptotic results for the *true* maximum likelihood estimators indeed, but only in the special case of *zero input errors*. In the related papers (Söderström, 1981; Anderson, 1985), both authors choose the frequency domain approach and are mainly interested in the identifiability aspect.

For the special case $p = q = 0$ (the classical errors-in-variables model) the least squares estimator can be written explicitly (see Kendall and Stuart, 1977, p. 405; Linssen, 1980). The latter also provides an exhaustive exposition on the convergence results for this special case.

2. Consistency Property

Proving asymptotic properties is not straightforward, especially because of the matrix $(DD^T)^{-1}$ present in the object function $J_N(\theta)$ and its gradient (10). In this section we deal with 5 assumptions which turn out to be sufficient conditions for strong consistency. We introduce the vector of measurement errors

$$e = z - \zeta = (\varepsilon_{m+N}, \delta_{m+N}, \ldots, \varepsilon_2, \delta_2, \varepsilon_1, \delta_1)^T, \tag{11}$$

and assume from now that its components, which are for convenience also denoted as $e_1, e_2, \ldots, e_{2(m+N)}$, are random variables.

Assumption 1. The components of e are zero mean and independent, and have common (unknown) variance σ^2 and bounded 4th moment, as $N \to \infty$.

The variance-covariance matrix var e of e is thus $\sigma^2 I$. If var $\delta_i = \sigma_\delta^2$, var $\varepsilon_i = \sigma_\varepsilon^2$ ($i = 1, 2, \ldots, m+N$) and $\sigma_\delta^2/\sigma_\varepsilon^2$ is known, then one can easily transform the model into one satisfying var $e = \sigma^2 I$. The homogenity of the variance will be discussed further in Section 4.

The true value of the parameter vector, which is denoted by θ_0, is unknown. However, we make

Assumption 2. The parameter space Θ is a known convex and compact subset of $\mathbb{R}^{p+q+1}$, containing θ_0.

Now the (sequence of) least squares estimator(s) $\{\hat{\theta}_N\}$ is well-defined by

$$J_N(\hat{\theta}_N) = \min_{\theta\in\Theta} J_N(\theta). \tag{12}$$

Moreover, $\hat{\theta}_N$ is a random vector in the sense that it is measurable (see Bierens, 1981, p. 53) and notice that $J_N(\theta)$ defined in (8) is a random function now. By introducing the polynomials

$$A(\lambda) = -1 + \sum_{i=1}^{p} \alpha_i \lambda^i \text{ and } B(\lambda) = \sum_{j=0}^{q} \beta_j \lambda^j, \tag{13}$$

we rewrite (1) as $A(\lambda)\eta_t + B(\lambda)\xi_t = 0$, for λ representing the shift-back operator.

Assumption 3. (a) The sequence of inputs $\{\xi_t\}_{t=1}^{\infty}$ is bounded, (b) for all $\theta \in \Theta$, the zeros of $A(\lambda)$ lie outside the closed unit disk.

The second part is called the stability assumption of the model, due to

Corollary 1. *The sequence of outputs $\{\eta_t\}_{t=1}^{\infty}$ is bounded.*

This is the well-known BIBO-stability result (Bounded Input Bounded Output; see e.g. Zadeh and Desoer, 1963, p. 483). Assumption 3b has another useful consequence with respect to the matrix $(DD^T)^{-1}$ appearing in the object function $J_N(\theta)$ and its gradient.

Lemma 1. *Some constants ρ_1 and ρ_2, with $0 < \rho_1 < \rho_2 < \infty$, exist such that $\rho_1 I \leq (DD^T)^{-1} \leq \rho_2 I$ for all $\theta \in \Theta$ and $N \geq p+1$, where, for symmetric matrices M_1 and M_2, $M_1 \leq M_2$ denotes the Löwner ordering.*

Proof. Corresponding to (13), the matrices A and B in $\mathbb{R}^{N\times N}$ are defined by $A = -I + \sum_{k=1}^{p} \alpha_k S^k$, $B = \sum_{k=0}^{q} \beta_k S^k$, where

$$S = \begin{pmatrix} 0 & 1 & & \\ & \cdot & \cdot & \cdot \\ & & 0 & 1 \\ & & & 0 \end{pmatrix}$$

is the shiftmatrix in $\mathbb{R}^{N\times N}$.

It can be verified that $DD^T = AA^T + BB^T + EE^T$, where E is the matrix in $\mathbb{R}^{N\times 2m}$ defined by

$$E = \begin{pmatrix} \gamma_m & & \\ \cdot & \cdot & \\ \gamma_1 & \cdot & \gamma_m \end{pmatrix}. \tag{14}$$

For fixed θ, the result $k_1 I \leq AA^T \leq k_2 I$, where k_1 and k_2 are positive numbers not depending on N, is proved in Aoki and Yue (1970b). The stability Assumption 3b is used to obtain the *lower bound* $k_1 I$. It is easy to derive the *upper bounds* k_3I and k_4I for BB^T and EE^T, respectively. Then, for fixed θ, we have $k_1I \leq DD^T \leq (k_2 + k_3 + k_4)I$ provided $N \geq p+1$. When varying θ, the k_i become continuous functions on the compact set Θ. Hence their extreme values have arguments on Θ itself. This implies $c_1I \leq DD^T \leq c_2I$ for all $\theta \in \Theta$ and $N \geq p+1$, where c_1 and c_2 are positive constants (not depending on θ or N). The claim follows from choosing $\rho_1 = 1/c_2$ and $\rho_2 = 1/c_1$. □

Preparing the next assumption we rewrite the vector $D\zeta$ from (4) as

$$D\zeta = (H+K)\theta - \eta, \tag{15}$$

where H and K are matrices in $\mathbb{R}^{N\times(p+q+1)}$ given by

$$H = (S\eta, \ldots, S^p\eta, \xi, S\xi, \ldots, S^q\xi),$$

$$K = \begin{pmatrix} & & & & & & & 0 \\ & & \eta_m & & & & \cdot & \xi_m \\ \cdot & \cdot & \cdot & & & \cdot & \cdot\ \cdot\ \cdot & \cdot \\ \eta_m & \cdot & \cdot\ \eta_{m+1-p} & 0 & \xi_m & \cdot & \cdot\ \cdot & \xi_{m+1-q} \end{pmatrix},$$

and η and ξ are vectors in $\mathbb{R}^N$,

$$\eta = (\eta_{m+N}, \ldots, \eta_{m+2}, \eta_{m+1})^T, \quad \xi = (\xi_{m+N}, \ldots, \xi_{m+2}, \xi_{m+1})^T.$$

We have $D\zeta = 0$ for $\theta = \theta_0$, since the equations (1) for $t = m+1, \ldots, m+N$ hold at the true value θ_0. Then (15) implies that

$$D\zeta = (H+K)(\theta - \theta_0). \tag{16}$$

Assumption 4. For all $\theta \in \Theta$, the matrix $H^T(DD^T)^{-1}H/N$ in $\mathbb{R}^{(p+q+1)\times(p+q+1)}$ converges as $N \to \infty$. The limiting matrix, say G, is positive definite on Θ.

This is a sort of identifiability assumption. The corollary below is one of the tools for proving consistency.

Corollary 2. $\mathbf{E}J_N(\theta)$ *converges as* $N \to \infty$ *for all* $\theta \in \Theta$ *with limiting function*

$$J(\theta) = \sigma^2 + (\theta - \theta_0)^T G(\theta)(\theta - \theta_0). \tag{17}$$

Proof. Let var z denote the variance-covariance matrix of the vector z. Then

$$\begin{aligned} \mathbf{E}J_N &= \mathbf{E}(z^TPz/N) = \left[\operatorname{tr}(\text{var } zP) + (\mathbf{E}z)^T P\mathbf{E}z\right]/N \\ &= \sigma^2 + \zeta^T D^T(DD^T)^{-1}D\zeta/N \\ &= \sigma^2 + (\theta-\theta_0)^T(H+K)^T(DD^T)^{-1}(H+K)(\theta-\theta_0)/N, \end{aligned}$$

cf. Assumption 1 and (16). The rest follows from Assumption 4, because of the vanishing effect of K as $N \to \infty$. □

Remark 1. Generalizing Aoki and Yue (1970a), it can be shown that Assumption 4 is equivalent to the input sequence $\{\xi_t\}_{t=1}^{\infty}$ being *persistently exciting* of order $p+q$ and the polynomials A and B of (13) being relatively prime. See Ten Vregelaar (1988) for details.

We give the consistency property, and defer the proof to the Appendix.

Theorem 1. *Under Assumptions* 1-4, *any sequence of estimators* $\{\hat{\theta}_N\}$, *defined by* (12), *is strongly consistent for the true value* θ_0 *of the parameter vector, notation* $\hat{\theta}_N \overset{\text{a.s.}}{\rightarrow} \theta_0$ *(almost surely), i.e.* $\mathbf{P}(\hat{\theta}_N \rightarrow \theta_0) = 1$.

Remark 2. It can be shown (see the Appendix) that σ^2 is estimated strongly consistently by $J_N(\hat{\theta}_N)$,

$$J_N(\hat{\theta}_N) \overset{\text{a.s.}}{\rightarrow} J(\theta_0) = \sigma^2. \tag{18}$$

3. Asymptotic Normality

A common way of proving $\{\hat{\theta}_N\}$ to be asymptotically normal is to start from a first order Taylor expansion for the gradient, which is denoted by J'_N from now on (the gradient is chosen to be a column vector),

$$\sqrt{N} J'_N(\hat{\theta}_N) = \sqrt{N} J'_N(\theta_0) + H_N \sqrt{N}(\hat{\theta}_N - \theta_0), \tag{19}$$

where H_N is a matrix of second derivatives with its (i, j) entry evaluated at some mean value point $\tilde{\theta}_N^i$,

$$(H_N)_{ij} = \frac{\partial^2 J_N}{\partial\theta_j \partial\theta_i}(\tilde{\theta}_N^i), \ i, j = 1, 2, \ldots, p+q+1 \ .$$

In addition to the assumptions in Section 2, we impose the following conditions.

Assumption 1a. The components of e have, for some $\delta > 0$, bounded $(4+\delta)$th moment, as $N \rightarrow \infty$ and have zero 3th moment.

The bounded $(4+\delta)$th moment condition replaces the bounded 4th moment condition from Assumption 1.

Assumption 2a. The true value θ_0 of the parameter vector is an interior point of the parameter space Θ.

By consequence $\sqrt{N} J'_N(\theta_0) = 0$ a.s. for sufficiently large N. Therefore, cf. (19),

$$\sqrt{N} J'_N(\theta_0) + H_N \sqrt{N}(\hat{\theta}_N - \theta_0) = 0 \text{ a.s. for sufficiently large } N. \tag{20}$$

Let J''_N denote the $(p+q+1) \times (p+q+1)$ matrix of second derivatives of J_N. It can be verified (see the Appendix) that the sequence $\{\mathbf{E}J''_N\}$ is bounded, uniformly on Θ. To obtain the convergence of H_N the pointwise convergence of $\mathbf{E}J''_N$ is assumed.

Assumption 5. For all $\theta \in \Theta$, $\mathbf{E}J''_N$ converges as $N \to \infty$.

The asymptotic normality of $\sqrt{N}(\hat{\theta}_N - \theta_0)$ will follow from that of $\sqrt{N}J'_N(\theta_0)$. The result is stated in the theorem below, its proof can be found in the Appendix. If A is some positive definite matrix, $A^{\frac{1}{2}}$ (square root of A) denotes any square matrix satisfying $(A^{\frac{1}{2}})^T A^{\frac{1}{2}} = A$. We introduce the notation H_0 for the matrix $J''(\theta_0)$, which is the matrix of second derivatives of the function J of (17), evaluated in θ_0, $H_0 = J''(\theta_0) = 2G(\theta_0)$.

Theorem 2. *Under Assumptions* 1, 1a, 2, 2a, 3, 4 *and* 5, *any sequence of estimators* $\{\hat{\theta}_N\}$ *defined by* (12) *is asymptotically normally distributed, i.e.*

$$\{[\text{var}\ (\sqrt{N}J'_N(\theta_0))]^{-\frac{1}{2}}\}^T H_0\sqrt{N}(\hat{\theta}_N - \theta_0) \xrightarrow{d} N(0, I)\ .$$

Remark. If in addition to the assumptions in Theorem 2, var $(\sqrt{N}J'_N(\theta_0))$ converges with (positive definite) limit $V(\theta_0)$, then we obtain

$$\sqrt{N}(\hat{\theta}_N - \theta_0) \xrightarrow{d} N(0, H_0^{-1}V(\theta_0)H_0^{-1}). \tag{21}$$

4. Discussion

4.1. Generalizations

We speak of the so-called MIMO-model (Multiple Input - Multiple Output) if η_t and ξ_t in (1) are vectors in $\mathbb{R}^s$ and $\mathbb{R}^r$, respectively. The parameters α_i and β_j are matrices in $\mathbb{R}^{s\times s}$ and $\mathbb{R}^{s\times r}$. It is straightforward to define the least squares estimation method for this problem, analogously to that of Section 1. The strong consistency and asymptotic normality are obtained under only slight modifications of the assumptions done in this paper. For details we refer to Ten Vregelaar (1988).

The latter also deals with equality constraints on the parameters (components of θ) on the one hand and partial input noise on the other hand. Partial input noise refers to the case that only part of the components of the input vectors ξ_t are measured with errors. This makes sense if $r > 1$. Along the lines in this paper, asymptotic results can be obtained for those problems as well, even for the combined case of equality constaints and partial input noise.

In Section 2, it was noticed that a model with unequal variances for input observations and output observations is easily transformed to one with common variance if the quotient of the variances is known. If this or other additional information is missing, the likelihood for the very special nondynamical model ($p = q = 0$) with normal errors has no maximum (cf. Solari, 1969). Now, assuming var $e = \sigma^2 V$ with a known positive definite matrix V, we apply the transformation $\tilde{e} = W^{-T}e$, where W satisfies $V = W^T W$. Then, var $\tilde{e} = \sigma^2 I$ holds. The so-called *weighted least squares* method is obtained by application of least squares to the transformations $\tilde{z} = W^{-T}z$ and $\tilde{\zeta} = W^{-T}\zeta$, and then $\min\|\tilde{z} - \tilde{\zeta}\|^2$ under $\tilde{D}\tilde{\zeta} = 0$ with $\tilde{D} = DW^T$. This gives rise to the object function $\tilde{z}^T\tilde{P}\tilde{z}/N$, where $\tilde{P} = \tilde{D}^T(\tilde{D}\tilde{D}^T)^{-1}\tilde{D}$. By somewhat modifying the assumptions, we can find asymptotic results again, especially if the distribution of the errors is normal, in which case the weighted least squares and maximum likelihood methods coincide. However, it should be noticed that the matrix $\tilde{D}$ does not possess the nice structure of the matrix D in (5), which is unattractive from a computational point of view.

4.2. Special case

We consider the so-called *normal* case, referring to the vector of measurement errors e having the multivariate normal distribution with mean vector 0 and covariance matrix $\sigma^2 I$. Obviously, the Assumptions 1 and 1a are automatically satisfied. The least squares and the *maximum likelihood estimators* for the components of θ coincide. Furthermore, it is straightforward to give a consistent estimator for the asymptotic covariance matrix of $\sqrt{N}(\hat{\theta}_N - \theta_0)$ (see Ten Vregelaar, 1988). The proof of the asymptotic normality property can be given without the great effort of showing Lemma 6 (see the Appendix). As in the proof of Lemma 6, we introduce the quadratic form $S_N = z^T Lz$, with $L = \sum_{i=1}^{p+q+1} \lambda_{N,i} \frac{\partial P(\theta_0)}{\partial \theta_i} / \sqrt{N}$. The n-th cumulant κ_n of $S_N/\sqrt{\text{var } S_N}$ is $\kappa_n = (2\sigma^2)^{n-1}(n-1)!(\sigma^2 \text{tr} L^n + n\zeta^T L^n \zeta)/(\text{var } S_N)^{n/2}, n = 1, 2, \ldots$, (see Searle, 1971, p. 56). The asymptotic normality of S_N results from $\kappa_1 = 0$, $\kappa_2 = 1$ and $\kappa_n \to 0$ as $N \to \infty$ for $n = 3, 4, \ldots$. Bounds for $\text{tr} L^n$ and $\zeta^T L^n \zeta$ are obtained from (i) in the proof of Lemma 4 (see the Appendix).

For the combined special case $p = q = 0$ and e normally distributed as $N(0, \sigma^2 I)$ we obtain the expression $V_N(\beta_0) = \text{var } (\sqrt{N} J_N'(\beta_0)) = 4\sigma^4/(1+\beta_0^2)^2 + 4\sigma^2 \xi^T \xi/(1+\beta_0^2)N$, where $\xi = (\xi_1, \ldots, \xi_N)^T$, which converges to $V(\beta_0) = 4\sigma^4/(1+\beta_0^2)^2 + 4\sigma^2 G(\beta_0)$, where $G(\beta) = \lim_{N\to\infty} (\xi^T \xi/N)/(1+\beta^2)$, (see Assumption 4). From $\hat{\beta}_N \overset{\text{a.s.}}{\to} \beta_0$, $\hat{J}_N \overset{\text{a.s.}}{\to} J(\beta_0) = \sigma^2$, and $\hat{J}_N'' \overset{\text{a.s.}}{\to} J''(\beta_0) = 2G(\beta_0) = H_0$, we obtain as strongly consistent estimator for the asymptotic variance $H_0^{-1} V(\beta_0) H_0^{-1}$ of $\sqrt{N}(\hat{\beta}_N - \beta_0)$ (see (21)), $(\hat{J}_N'')^{-1}[2\hat{J}_N(\hat{J}_N'' + 2\hat{J}_N/(1+\hat{\beta}_N^2)^2)](\hat{J}_N'')^{-1} = v^2 + (1+\hat{\beta}_N^2)v$, where $v = N\hat{J}_N/(x^T x - N\hat{J}_N)$ and $x = (x_1, \ldots, x_N)^T$.

4.3. Remarks

The object function resulting from the least squares method introduced for the dynamic errors-in-variables model defined in this paper, does not allow straightforward methods for proving consistency and asymptotic normality. However, the assumptions for obtaining strong consistency are not stronger than those made by Aoki and Yue (1970a), who have considered two simplified object functions; their assumption of normal errors can be weakened. It seems interesting to take into account other approaches (e.g. instrumental variable methods) for more or less the same problem setting.

It is also interesting to know how to estimate consistently the asymptotic covariance matrix of $\sqrt{N}(\hat{\theta}_N - \theta_0)$ in the general case. So far, only for the special cases of normal errors and the model without dynamics we were able to do so.

Appendix

The convergence (in some sense) of J_N, J_N' and J_N'' will be part of the arguments leading to Theorems 1 and 2. Due to the matrix $(DD^T)^{-1}$, it seems not possible, in general, to write J_N, J'_N and J''_N as sums of independent variables, convenient for applying some laws of large numbers or some central limit theorem. Therefore, we use some concepts from mathematical analysis in order to obtain the so-called *uniform a.s. convergence* of the stochastic variables mentioned above. For any matrix X, we use the notation $||X||^2 = \text{tr } X^T X$.

Lemma 2. *Let Θ be a compact and convex subset of $\mathbb{R}^k$ and let $\{f_n\}$ be a sequence of differentiable real functions with uniformly bounded gradient on Θ, i.e. for some $c > 0$, $||f'_n(\theta)|| \leq c$ holds for all $n \in \mathbf{N}$ and all $\theta \in \Theta$. If $f_n(\theta) \to f(\theta)$ for all $\theta \in \Theta$ then also $f_n \to f$ uniformly on Θ.*

Proof. See Dieudonné (1969), Theorem 7.5.6. □

The analogue of Lemma 2 for stochastic functions is stated in

Lemma 3. *Suppose $\{f_n(\theta)\}$ is a sequence of real, a.s. differentiable (with respect to θ) stochastic functions on a compact and convex set $\Theta \subset \mathbb{R}^k$ and let $\{\Omega, \mathcal{F}, \mathbf{P}\}$ be the probability space involved. If $\{f'_n\}$ is a.s. bounded, uniformly on Θ, i.e. there exist a positive number c and a null set $E \subseteq \Omega$ ($\mathbf{P}(E) = 0$) such that for every $\omega \in \Omega \setminus E$ an integer n_0 exists with $||f'_n(\theta, \omega)|| \leq c$ holds for all $n > n_0$ and all $\theta \in \Theta$, then $f_n(\theta) \stackrel{\text{a.s.}}{\to} 0$ for all $\theta \in \Theta$ implies $f_n \stackrel{\text{a.s.}}{\to} 0$, uniformly on Θ, i.e. $\sup_{\theta\in\Theta} |f_n(\theta)| \stackrel{\text{a.s.}}{\to} 0$.*

For the proof see Theorem 3.10b in Ten Vregelaar (1988). Lemmas 2 and 3 will now be applied to obtain the uniform a.s. convergence of $J_N(\theta)$, which leads to the proof of Theorem 1. We introduce the stochastic function

$$L_N(\theta) = J_N(\theta) - \mathbf{E}J_N(\theta). \tag{22}$$

Lemma 4. *Under Assumptions* 1, 2, 3 *and* 4, *we have $J_N \stackrel{\text{a.s.}}{\to} J$, uniformly on Θ, i.e. $\sup_{\theta\in\Theta} |J_N(\theta) - J(\theta)| \stackrel{\text{a.s.}}{\to} 0$.*

Proof. We distinguish three steps.

(i) *There exists a $k_1 > 0$ such that $-k_1 I \leq \dot{P} \leq k_1 I$ for all $\theta \in \Theta$ and $N \geq p+q+1$.* We start from

$$(\dot{P})^2 = D^+ \dot{D} P^\perp \dot{D}^T (D^+)^T \perp \dot{D}^T (DD^T)^{-1} \dot{D} P^\perp$$

which is easily seen from (9). From $P^\perp \leq I$, $\dot{D}\dot{D}^T \leq I$, $\dot{D}^T\dot{D} \leq I$ and $(DD^T)^{-1} \leq \rho_2 I$ (see Lemma 1) it follows that $(\dot{P})^2 \leq \rho_2^2 D^T D + \rho_2 I$. In view of (5), for any x, we have $||Dx|| \leq (1 + \sum_{i=1}^p |\alpha_i| + \sum_{i=0}^q |\beta_i|) ||x||$. Therefore, defining $k_1^2 = \rho_2^2 \max_{\theta\in\Theta} (1 + \sum_{i=1}^p |\alpha_i| + \sum_{i=0}^q |\beta_i|)^2 + \rho_2$, we obtain $(\dot{P})^2 \leq k_1^2 I$, for all $\theta \in \Theta$ and $N \geq p+q+1$, which is equivalent to (i).

(ii) $\mathbf{E}J_N \to J$, *uniformly on* Θ. The pointwise convergence of $\mathbf{E}J_N = \sigma^2 + \zeta^T P \zeta / N$ to J (see Corollary 2) implies the uniform convergence if the gradient of

$\mathbf{E}J_N$ is uniformly bounded on Θ by application of Lemma 2. The latter results from

$$|\frac{\partial}{\partial\theta_i}\mathbf{E}J_N| = |\zeta^T\dot{P}\zeta/N| \leq k_1\zeta^T\zeta/N \leq k_1(m+N)(\text{const})/N \leq \text{const},$$

where const denotes an appropriate constant upper bound, which inequality holds for any component θ_i of Θ (cf. (i), Assumption 3 and Corollary 1).

(iii) $L_N \overset{\text{a.s.}}{\rightarrow} 0$, *uniformly on* Θ. Applying Lemma 3, we need to show

$$L_N \overset{\text{a.s.}}{\rightarrow} 0, \text{ for all } \theta \in \Theta, \text{ and} \tag{23}$$

$$\{L'_N\} \text{ is a.s. bounded, uniformly on } \Theta. \tag{24}$$

From (22), it follows via $z = \zeta + e$, that

$$L_N(\theta) = 2e^TP(\theta)\zeta/N + e^TP(\theta)e/N - \sigma^2. \tag{25}$$

For every $\varepsilon > 0$,

$$\begin{aligned}\mathbf{P}(|e^TP\zeta/N| \geq \varepsilon) &\leq \mathbf{E}(e^TP\zeta)^4/(\varepsilon^4N^4) \leq \text{const}\max_i \mathbf{E}e_i^4\|P\zeta\|^4/(\varepsilon^4N^4)\\ &\leq \text{const}\|\zeta\|^4/(\varepsilon^4N^4) \leq \text{const}(m+N)^2/(\varepsilon^4N^4)\end{aligned}$$

holds, due to Theorem 2 in Whittle (1960) and Assumptions 1 and 3. Therefore, $\sum_{N=1}^{\infty}\mathbf{P}(|e^TP\zeta/N| \geq \varepsilon) < \infty$ for every $\varepsilon > 0$, which implies $e^TP(\theta)\zeta/N \overset{\text{a.s.}}{\rightarrow} 0$ for all $\theta \in \Theta$. With respect to the remaining terms of the right-hand-side in (25), we have for every $\varepsilon > 0$

$$\begin{aligned}\mathbf{P}(|e^TPe/N - \sigma^2| \geq \varepsilon) &\leq \text{var}\,(e^TPe)/(\varepsilon^2N^2)\\ &\leq \text{const}\max_i \mathbf{E}e_i^4\|P\|^2/(\varepsilon^2N^2) \leq \text{const}/(\varepsilon^2N),\end{aligned}$$

using Theorem 2 in Whittle (1960), Assumption 1 again, and $\|P\|^2 = \text{tr}\ P = N$. By definition $e^TPe/N \overset{\text{P}}{\rightarrow} \sigma^2$ (convergence in probability) holds. Since $P/N \geq 0$ and $\text{tr}\ P/N = 1$, $e^TP(\theta)e/N \overset{\text{a.s.}}{\rightarrow} \sigma^2$ for all $\theta \in \Theta$, which follows from Corollary 3 in Varberg (1968). This proves (23). To show (24), let us consider an arbitrary component of L'_N, $\dot{L}_N = (2e^T\dot{P}\zeta + e^T\dot{P}e)/N$. In view of (i), $|e^T\dot{P}e/N| \leq k_1e^Te/N$ holds, with $e^Te/N \overset{\text{a.s.}}{\rightarrow} 2\sigma^2$, by the Kolmogorov strong law of large numbers. Furthermore, we have

$$|e^T\dot{P}\zeta/N| \leq (\zeta^T(\dot{P})^2\zeta/N)^{\frac{1}{2}}(e^Te/N)^{\frac{1}{2}} \leq \text{const}(e^Te/N)^{\frac{1}{2}} \overset{\text{a.s.}}{\rightarrow} \text{const}\ \sigma,$$

using the Cauchy-Schwarz inequality, (i) and Assumption 3. This yields (24), since $\dot{L}_N$ (hence L'_N) is bounded by a stochastic variable not depending on θ and being a.s. convergent to a positive number.

The proof of Lemma 4 is complete now, since $|J_N - J| \leq |L_N| + |\mathbf{E}J_N - J|$.□

Proof of Theorem 1. By application of Lemma 3.1.3 in Bierens (1981), the claim results from Lemma 4 and the facts that the limit function J, on the one hand is continuous on Θ (as a uniform limit of continuous functions, see (ii)), and, on the other hand, has a unique minimum point θ_0 on Θ due to Assumption 4. □

The statement (18) is an immediate consequence of Lemma 4 and Theorem 1. This method of proving consistency has been applied before by Jennrich (1969).

The proof of Theorem 2 is based on some other auxiliary results.

Lemma 5. *Under Assumptions* 1, 2, 3, 4 *and* 5, *we have* $J'_N \xrightarrow{\text{a.s.}} J'$, *uniformly on* Θ, *and* $J''_N \xrightarrow{\text{a.s.}} J''$, *uniformly on* Θ, *where the convergences of the vector* J'_N *and the matrix* J''_N *hold componentwise.*

Proof. Let i, j, k denote arbitrary elements of $\{1, 2, \ldots, p+q+1\}$, and define

$$P^{ij} = \frac{\partial^2 P}{\partial\theta_j \partial\theta_i} \text{ and } P^{ijk} = \frac{\partial^3 P}{\partial\theta_k \partial\theta_j \partial\theta_i}.$$

Again, three steps are distinguished.

(i) *There exist positive numbers* k_2 *and* k_3 *such that* $(P^{ij})^2 \leq k_2^2 I$ *and* $(P^{ijk})^2 \leq k_3^2 I$ *hold for all* $\theta \in \Theta$ *and* $N \geq p+q+1$. This claim follows in a similar way as (i) in the proof of Lemma 4.

(ii) $\mathbf{E}J'_N \to J'$, *uniformly on* Θ *and* $\mathbf{E}J''_N \to J''$, *uniformly on* Θ. From tr $P = N$ we obtain tr $P^{ij} = 0$, hence $|(\mathbf{E}J''_N)_{i,j}| = |\zeta^T P^{ij}\zeta/N| \leq k_2\zeta^T\zeta/N \leq$ const. Furthermore $|\frac{\partial}{\partial\theta_k}(\mathbf{E}J''_N)_{i,j}| = |\zeta^T P^{ijk}\zeta/N| \leq$ const. Thus both $\mathbf{E}J''_N$ and its derivative with respect to θ are uniformly bounded on Θ. In view of Lemma 2 and Assumption 5, this implies the uniform convergence of $\mathbf{E}J''_N$. According to Theorem 8.6.3 in Dieudonne (1969), the uniform limit J of $\mathbf{E}J_N$ is a C^2-function and J' and J'' must be the uniform limits of $\mathbf{E}J'_N$ and $\mathbf{E}J''_N$ respectively (notice that $\mathbf{E}J'_N(\theta_0) = 0$).

(iii) $L'_N \xrightarrow{\text{a.s.}} 0$, *uniformly on* Θ *and* $L''_N \xrightarrow{\text{a.s.}} 0$, *uniformly on* Θ. Applying Lemma 3 again, we prove the a.s. pointwise convergence to zero of L'_N and L''_N and notice that the uniform a.s. boundedness of L''_N and L'''_N (the third derivative) is obtained from (i) in the same way as (24). Differentiating (25) yields $\frac{\partial L_N}{\partial\theta_i} = 2e^T P^i\zeta/N + e^T P^i e/N$ and $\frac{\partial^2 L_N}{\partial\theta_j\partial\theta_i} = 2e^T P^{ij}\zeta/N + e^T P^{ij}e/N$.

From the first step of the proof of Lemma 4 and (i), it follows $e^T P^i\zeta/N \xrightarrow{\text{a.s.}} 0$ and $e^T P^{ij}\zeta/N \xrightarrow{\text{a.s.}} 0$ for all $\theta \in \Theta$, in the same way as the pointwise convergence to zero of $e^T P\zeta/N$, in the last step of the proof of Lemma 4. Analogously to the derivation of $e^T Pe/N \xrightarrow{P} \sigma^2$, we obtain $e^T P^i e/N \xrightarrow{P} 0$ and $e^T P^{ij}e/N \xrightarrow{P} 0$, since tr P^i = tr $P^{ij} = 0$, $||P^i||^2 = \text{tr } (P^i)^2 \leq 2(m+N)k_1^2$ and $||P^{ij}||^2 = \text{tr } (P^{ij})^2 \leq 2(m+N)k_2^2$, respectively. Due to Theorem 3 in Varberg (1968), the convergences hold in a.s. sense as well.

To complete the proof note that $||J'_N - J'|| \leq ||L'_N|| + ||\mathbf{E}J'_N - J'||$ and $||J''_N - J''|| \leq ||L''_N|| + ||\mathbf{E}J''_N - J''||$, since $L'_N = J'_N - (\mathbf{E}J_N)' = J'_N - \mathbf{E}J'_N$ and $L''_N = J''_N - (\mathbf{E}J'_N)' = J''_N - \mathbf{E}J''_N$. □

Lemma 6. *Provided the Assumptions* 1, 1a, 2, 3 *and* 4, *we have*

$$\{[(\text{var } (\sqrt{N}J'_N(\theta_0))]^{-\frac{1}{2}}\}^T\sqrt{N}J'_N(\theta_0) \xrightarrow{d} N(0, I).$$

Proof. Let us introduce $S_N = \sqrt{N}\lambda_N^T J'_N(\theta_0)$, where $\{\lambda_N\}$ is an arbitrary sequence in $\mathbf{R}^{p+q+1}$ with $||\lambda_N|| = 1$. The statement to be proven is equivalent to the asymptotic normality of S_N. Using (10), we write S_N as a quadratic form in z, $S_N = z^T L z$, with $L = \sum_{i=1}^{p+q+1} \lambda_{N,i} \frac{\partial P(\theta_0)}{\partial \theta_i} / \sqrt{N}$. Now, let us consider the subsequence $\{S_{N^2}\}$ of $\{S_N\}$ and denote $S_{N^2} = \tilde{z}^T \tilde{L} \tilde{z}$, where $\tilde{z} \in \mathbb{R}^{2(m+N^2)}$, and $\tilde{L} = \sum_{i=1}^{p+q+1} \lambda_{N^2,i} \frac{\partial \tilde{P}(\theta_0)}{\partial \theta_i} / N$. Here $\tilde{P} = \tilde{D}^+ \tilde{D}$ (cf. (8)) and $\tilde{D} \in \mathbb{R}^{N^2 \times 2(m+N^2)}$ is the ***blown up*** version of (5). Below, matrices $\tilde{M}$ and $\tilde{R}$ will be defined such that: (i) $\tilde{L} = \tilde{M} + \tilde{R}$, (ii) var $\tilde{z}^T \tilde{R} \tilde{z}$/var $S_{N^2} \to 0$ as $N \to \infty$ and (iii) $\tilde{z}^T \tilde{M} \tilde{z}$ is asymptotically normal.

Applying the result of Thomas (1986 pp. 190–191), we obtain that then S_{N^2} is asymptotically normal as well. In order to define $\tilde{M}$ and $\tilde{R}$, we split up the matrix D as $D = [\Gamma, E]$, with Γ being the submatrix of D consisting of its first $2N$ columns, and $E \in \mathbb{R}^{N \times 2m}$ being as in (14). Next, we introduce $\tilde{\Gamma} \in \mathbb{R}^{N^2 \times 2(m+N^2)}$, $\tilde{\Gamma} = (\text{diag}(\Gamma, \Gamma, \ldots, \Gamma) : 0)$, where the number of Γ blocks is N.

The matrices $\tilde{M}$ and $\tilde{R}$ in (i) are defined as $\tilde{M} = \sum_{i=1}^{p+q+1} \lambda_{N^2,i} \frac{\partial \tilde{Q}(\theta_0)}{\partial \theta_i} / N$, with $\tilde{Q} = \tilde{\Gamma}^+ \tilde{\Gamma}$, and $\tilde{R} = \tilde{L} - \tilde{M}$.

Due to Assumption 1a, we obtain (via $\tilde{z} = \tilde{\zeta} + \tilde{e}$) that

$$\text{var } \tilde{z}^T \tilde{R} \tilde{z} / \text{var } S_{N^2} = \frac{\text{var } \tilde{e}^T \tilde{R} \tilde{e} + 4\sigma^2 \|\tilde{R} \tilde{\zeta}\|^2}{\text{var } \tilde{e}^T \tilde{L} \tilde{e} + 4\sigma^2 \|\tilde{L} \tilde{\zeta}\|^2} .$$

Applying of Theorem 2 in Whittle (1960) again, yields

$$\text{var } \tilde{z}^T \tilde{R} \tilde{z} / \text{var } S_{N^2} \leq \text{const}(\|\tilde{R}\|^2 + \|\tilde{R} \tilde{\zeta}\|^2) / \|\tilde{L} \tilde{\zeta}\|^2 .$$

After some cumbersome calculations, in which Lemma 1 plays a key role again, the property (ii) can be shown. In particular, $\|\tilde{R}\|^2$ and $\|\tilde{R}\tilde{\zeta}\|^2$ are $O(1/N)$ and $\|\tilde{L}\tilde{\zeta}\|^2 \geq a$ for some $a > 0$ for sufficiently large N.

Now consider the property (iii). One can easily verify that $\tilde{M} = \text{diag}(M, M, \ldots, M, 0)$, where the number of M blocks is N, the zero block is of order $2m$, and $M = \sum_{i=1}^{p+q+1} \lambda_{N^2,i} \frac{\partial Q(\theta_0)}{\partial \theta_i} / N \in \mathbb{R}^{2N \times 2N}$ with $Q = \Gamma^+ \Gamma$. Hence $\tilde{z}^T \tilde{M} \tilde{z} = \sum_{i=1}^{N} X_i$, where $X_i = \tilde{z}_i^T M \tilde{z}_i$ and $\tilde{z}_1, \ldots, \tilde{z}_N$ are vectors in $\mathbb{R}^{2N}$ defined by $\tilde{z} = (\tilde{z}_1^T, \ldots, \tilde{z}_N^T, (z^*)^T)^T$ ($z^* \in \mathbb{R}^{2m}$).

Since $\tilde{z}^T \tilde{M} \tilde{z}$ is a sum of independent stochastic variables, the Liapounov central limit theorem (cf. Serfling, 1980, p.30) can be applied; from $\sum_{i=1}^{N} |X_i - \mathbf{E}X_i|^{2+\delta} / (\text{var } \tilde{z}^T \tilde{M} \tilde{z})^{1+\frac{1}{2}\delta} \sim O(N^{-\frac{1}{2}\delta})$, it follows the asymptotic normality of $\tilde{z}^T \tilde{M} \tilde{z}$. An upper bound for the numerator can be found by application of an easy extension of Theorem 2 in Whittle (1960); the positive δ is the one of Assumption 1a. (cf. Ten Vregelaar, 1988 for details).

So far, the asymptotic normality of S_{N^2}, i.e. $S_{N^2}/\sqrt{\text{var } S_{N^2}} \xrightarrow{d} N(0,1)$ has been proved (notice that $\mathbf{E}S_{N^2} = 0$). The result can be extended to the sequence $\{S_N\}$ by choosing a modified decomposition of D. □

Proof of Theorem 2. We start from the equation (20). Premultiplying by $\{[(\text{var } \sqrt{N} J_N'(\theta_0))]^{-\frac{1}{2}}\}^T$, gives, via Lemma 6, that

$$\{[(\text{var } \sqrt{N} J_N'(\theta_0))]^{-\frac{1}{2}}\}^T H_N \sqrt{N}(\hat{\theta}_N - \theta_0) \xrightarrow{d} N(0, I). \tag{26}$$

From $J_N'' \xrightarrow{\text{a.s.}} J''$, uniformly on Θ (see Lemma 5) and Theorem 1, we obtain $H_N \xrightarrow{\text{a.s.}} H_0$. This implies $H_N^{-1} \xrightarrow{\text{a.s.}} H_0^{-1}$ and $H_0 H_N^{-1} \xrightarrow{\text{a.s.}} I$.

Therefore, $\{[(\text{var}\,\sqrt{N}J_N'(\theta_0))]^{-\frac{1}{2}}\}^T H_0 H_N^{-1}\{[(\text{var}\,\sqrt{N}J_N'(\theta_0))]^{\frac{1}{2}}\}^T \xrightarrow{\text{a.s.}} I$ holds, which in view of (26) proves the theorem. □

References

Anderson, B.D.O. (1985). Identification of scalar errors-in-variables models with dynamics. *Automatica* **21**, 709–716.

Aoki, M. and Yue, P.C. (1970a). On a priori error estimates of some identification methods. *IEEE Transactions* **AC-15**, 541–548.

Aoki, M. and Yue, P.C. (1970b). On certain convergence questions in system identification. *SIAM Journal of Control* **8**, 239–256.

Bierens, H.J. (1981). Robust Methods and Asymptotic Theory in Nonlinear Econometrics. *Lecture Notes in Economics and Mathematical Systems* **192**, Springer-Verlag, Berlin.

Dieudonné, J. (1969). *Foundations of Modern Analysis.* Academic Press, New York.

Jennrich, R.I. (1969). Asymptotic properties of nonlinear least squares estimators. *The Annals of Mathematical Statistics* **40**, 633–643.

Kendall, M.G. and Stuart, A. (1977). *The Advanced Theory of Statistics. Inference and Relationship*, 4th ed. Griffin, London.

Linssen, H.N. (1980). Functional Relationships and Minimum Sum Estimation. Ph.D. thesis, Eindhoven University of Technology, Eindhoven.

Searle, S.R. (1971). *Linear Models.* Wiley, New York.

Serfling, R.J. (1980). *Approximation Theorems of Mathematical Statistics.* Wiley, New York.

Söderström, T. (1981). Identification of stochastic linear systems in presence of input noise. *Automatica* **17**, 713–725.

Solari, M.E. (1969). The maximum likelihood solution of the problem of estimating a linear functional relationship. *Journal of the Royal Statistical Society B* **31**, 372–375.

Ten Vregelaar, J.M. (1988). Least squares parameter estimation in a dynamic model from noisy observations. Ph.D. thesis, Eindhoven University of Technology, Eindhoven.

Ten Vregelaar, J.M. (1990). On computing objective function and gradient in the context of estimating the parameters in a dynamic errors-in-variables model. *SIAM Journal on Scientific and Statistical Computation.* To appear.

Thomas, J.B. (1986). *Introduction to Probability.* Springer-Verlag, New York.

Varberg, D.E. (1968). Almost sure convergence of quadratic forms in independent random variables. *The Annals of Mathematical Statistics* **39**, 1502–1506.

Whittle, P. (1960). Bounds for the moments of linear and quadratic forms in independent variables. *Theory of Probability and its Applications* **5**, 302–305.

Zadeh, L.A. and Desoer, C.A. (1963). *Linear Systems Theory, the State-Space Approach.* McGraw-Hill, New York.

A GENERIC LOOK AT FACTOR ANALYSIS

MICHEL LEJEUNE
Ecole Supérieure de Commerce de Paris
Ecole Nationale de la Statistique et de l'Administration Economique
3 avenue Pierre Larousse
F-92241 Malakoff Cedex
France

Abstract. In contrast with the traditional Principal Components framework we look at Factor Analysis as a modelization technique to explain the variability within a data matrix by means of simple structures, i.e. by matrices of rank one. This viewpoint makes the following essential elements of this methodology apparent: (i) a reference matrix $\mathbf{X}_0$ with respect to which deviations are to be measured, and (ii) a matrix norm $\|\cdot\|_{\mathbf{\Lambda},\mathbf{\Gamma}}$ defining the measure of the total variation.

This formalization is simple and, at the same time, it allows for extensions of the technique. Finally this presentation makes explicit the choices that one has to face among possible variations, and clarifies some specific properties of Correspondence Analysis in relation with these choices.

Key words: Factor Analysis, Principal Components, Correspondence Analysis.

1. Introduction

Factor Analysis is an essential set of techniques for the analysis of multidimensional data. Among its large variety of uses let us mention the reduction of dimensionality to enhance further analysis (for instance clustering, regression), the revelation of fundamental structures in the data, and the solving of scaling problems (e.g. for time seriation in archeological data or site ordering in ecology). For uses of Factor Analysis see Fisher (1936), Frane and Hill (1976), and Kim and Mueller (1978).

The basic concept of principal axes defined by Pearson (1901) and, further, the technique of Principal Component Analysis (PCA), originated by Hotelling (1933), are generally used as a reference for the various methods designated by the generic term of Factor Analysis. For example Correspondence Analysis (CA), as developped by Benzécri (1973), can be described as a weighted version of PCA (see Section 2). Although the efforts of the French school of 'Analyse des Données' towards a general framework for Factor Analysis contributed to the extension of these techniques, the choice of PCA suffers two drawbacks. The first one relates to the cases by variables structure of the data matrix which implies a non symmetric status for rows and for columns. The second one, actually linked to the previous one, follows from the particular centering process of columns by substracting means of corresponding variables. These drawbacks restrict the possibilities of generalization and, at the same time, they somewhat complicate the setup in instances like CA, not closely connected to PCA.

In our presentation we introduce a new formalization of Factor Analysis that comprises a wider scope of techniques in the domain of multivariate analysis and also simplifies the numerical understanding of it. This occurs by defining a so-

T. Caliński and R. Kala (eds.),
Proceedings of the International Conference on Linear Statistical Inference LINSTAT '93, 285–291.

called reference matrix as a substitute for the centering process and by using a matrix norm instead of the concept of sum of squares. In particular, the notion of a reference matrix allows to exhibit deviational structures in the data with respect to any given model. Then we will show how the mathematical solution of PCA still holds with some appropriate adjustments. Finally we will see that a special choice of the reference matrix, with respect to the special norm, explains peculiar properties of CA.

It should be noted here that the classical method of Spearman (1904), extended by Thurstone (1947), is not fitting into PCA framework because of the strong structural assumption of common and specific factors.

2. The Generalized PCA Framework

The development of Correspondence Analysis brought with itself the need for a general formalization, as soon as it was recognized as a variation of PCA. This trend appeared early among various authors and the later presentation by Escoufier (1985), where the geometric interpretation is predominant, is the most favored nowadays.

In PCA rows are statistical units, the variability of which is measured by their sum of squares in $\mathbb{R}^p$. Let $\mathbf{X}$ be an $n \times p$ data matrix with rows centered, and possibly standardized columnwise, and let $\mathbf{x}_i$ denote the i-th row of $\mathbf{X}$. Then the total variability of the data is measured by the overall sum of squares

$$\sum_{i=1}^{n} ||\mathbf{x}_i||^2 = \text{trace}(\mathbf{X}'\mathbf{X})$$

and its geometric representation is a *cloud* of points in the Cartesian p-dimensional set of axes, whose variability is described by the sum of distances to the origin.

This setting allows for variations from ordinary PCA while (i) putting weights on points, i.e. rows, and (ii) choosing among the variety of Euclidean distances. Thus Factorial Analysis, as a generalization of PCA, will be a triple $(\mathbf{X}, \mathbf{Q}, \mathbf{D})$, where $\mathbf{X}$ is the $n \times p$ raw data matrix, $\mathbf{Q}$ is a $p \times p$ matrix standing for the Euclidean norm in $\mathbb{R}^p$, and $\mathbf{D}$ is an $n \times n$ diagonal matrix putting weights on rows. Let $\mathbf{X}_0$ be the data matrix centered for rows with respect to the weighted means. The p-vector of weighted means characterizes the *gravity center* of the cloud of points, which is chosen as the origin, and the overall sum of squares above is replaced by the *total inertia*, $\text{trace}(\mathbf{X}_0'\mathbf{D}\mathbf{X}_0\mathbf{Q})$, of this cloud, taking into account the particular distance and the weighing of points. Obviously, splitting $\mathbf{Q}$ into $\mathbf{T}\mathbf{T}'$ and transforming the centered data matrix $\mathbf{X}_0$ into $\mathbf{W} = \mathbf{D}^{\frac{1}{2}}\mathbf{X}_0\mathbf{T}$, yields the usual norm, $\text{trace}(\mathbf{W}'\mathbf{W})$, for total inertia, with the data matrix $\mathbf{W}$ thus containing the transformed coordinates of row points in order to recover the usual geometry.

For non standardized *Principal Component Analysis*, $\mathbf{Q}$ and $\mathbf{D}$ are identity matrices and for standardized PCA, $\mathbf{Q}$ is a diagonal matrix containing the inverses of the variances of the columns. As for *Correspondence Analysis*, it is customary to present it as PCA performed not on the raw data x_{ij} but on *profiles* defined, say for row i, by components $x_{ij}/x_{i.}, j = 1, \ldots, p$, using the usual dot notation for index summation. For rows the choice of $\mathbf{Q}$ and $\mathbf{D}$ is

$$\mathbf{Q} = \text{diag}(1/x_{.j}) \text{ and } \mathbf{D} = \text{diag}(x_{i.}/x_{..}). \qquad (1)$$

With this choice the center of gravity is defined by the row marginal profile $x_{.j}/x_{..}$, $j = 1, \ldots, p$, and the total inertia is equal to the chi-square coefficient of the data matrix divided by its grand total $x_{..}$. The distance induced in $\mathbb{R}^p$ between two row entities is the so-called chi-square distance

$$d^2(i,i') = \sum_{j=1}^{p} \frac{1}{x_{.j}} (\frac{x_{ij}}{x_{i.}} - \frac{x_{i'j}}{x_{i'.}})^2. \tag{2}$$

From this formula it can be seen that, even if the weight matrix $\mathbf{D}$ remains the same, this formal representation is not unique. As a matter of fact some authors use $x_{ij}/x_{i.}x_{.j}$ as an initial transformation instead of profiles, so that $\mathbf{Q}$ becomes equal to diag$(x_{.j})$. Now the reason for which the weight matrix $\mathbf{D}$ should be kept as in (1) relates to the interpretability of the distance (2) between rows; with this choice two rows are confounded in $\mathbb{R}^p$ when their profiles are identical. This weighing artifact is, indeed, the main originality in the setup of CA. Yet, in order to view distances properly in factorial mappings one must recover the usual Euclidean norm by using, say for row i, coordinates

$$\frac{x_{ij}}{x_{.i}\sqrt{x_{.j}}}, \; j = 1, \ldots, p.$$

Finally note that the definition of CA via the $(\mathbf{X}, \mathbf{Q}, \mathbf{D})$ triple is not symmetric with respect to rows and columns, although the analysis is. Thus PCA formulation is not quite well suited here.

In the new formalization one will see that it is not necessary to introduce the weight matrix $\mathbf{D}$ because the factorial decomposition is not subordinated to it. The choice of $\mathbf{D}$ only relates to the way one wants to make spatial representations in factorial planes. In addition, the implicit centering of the data in PCA can be removed and this will provide a wider generalization.

3. Extended Formulation of Factor Analysis

Suppose we have an $n \times p$ matrix $\mathbf{X}$, whose rows and columns are some physical entities, not necessarily cases and variables, and our aim is to analyze the variations among these entities. To be able to do this the *uniformity* reference, i.e. zero variation, must be given by a matrix $\mathbf{X}_0$ that we will call the *reference matrix*. Deviations will be measured with respect to this matrix, i.e. we will analyze the difference $\mathbf{X} - \mathbf{X}_0$.

In practical situations the analyst must define the matrix $\mathbf{X}_0$ himself, which is profitable because it requires to give some ideas about the kind of variation one is looking for. As a matter of fact, in many occasions, we do not face a situation where rows and columns can be clearly identified as cases and variables. Consider, for instance, a data table containing rates of change across years (rows) of GNP for a set of countries (columns). Then $\mathbf{X}_0$ can consist of identical rows, each one being equal to the average rates of the various countries over the years, so that one will be looking at variations across years. Countries are then compared in terms of yearly variations with respect to their mean level over the whole period. On the other hand $\mathbf{X}_0$ can be chosen with identical columns in order to consider deviations among countries with respect to the average rates of change on the whole set of countries.

Furthermore, $\mathbf{X}_0$ can be chosen with proportional rows and columns - and this is the choice of CA - insofar that we are interested in the variational pattern of both rows and columns.

More generally, it is possible to use $\mathbf{X}_0$ as any reference matrix with respect to which deviations will be studied. One may also use an external source as a baseline. In the previous example one could wish to use yearly rates of the world economy as a reference column to be substracted from each column of the data matrix. Thus countries would be analyzed in terms of how different they have performed relatively to the world growth. In other situations $\mathbf{X}_0$ can be derived from some well established theory and one is interested in investigating how the various entities behave in relationship to this theory.

Once deviations are defined by $\mathbf{X} - \mathbf{X}_0$, a measure of the *total variation* must be introduced to aggregate all cell deviations, i.e. one has to choose a *matrix norm* $\|\,.\,\|_*$. The total variation is then equal to $\|\mathbf{X} - \mathbf{X}_0\|_*^2$. For the usual sum of squares the choice is $\|\mathbf{X} - \mathbf{X}_0\|^2 = \text{trace}[(\mathbf{X} - \mathbf{X}_0)'(\mathbf{X} - \mathbf{X}_0)]$, which can be extended by putting different weights on either rows or columns, $\|\mathbf{X} - \mathbf{X}_0\|^2_{\mathbf{\Lambda},\mathbf{\Gamma}} = \text{trace}[(\mathbf{X} - \mathbf{X}_0)'\mathbf{\Lambda}(\mathbf{X} - \mathbf{X}_0)\mathbf{\Gamma}]$, i.e. by plugging, into the norm, diagonal matrices $\mathbf{\Lambda}$ or $\mathbf{\Gamma}$, respectively. Note that it is not desirable to use any Euclidean norm, except for diagonal matrices, because the appealing additivity of contributions of rows or columns to total variation would then be lost.

For standardized PCA $\mathbf{\Lambda}$ is equal to $\mathbf{I}_n$, or, for convenience, to $(1/n)\mathbf{I}_n$, and $\mathbf{\Gamma} = \text{diag}(1/s_j^2)$, whereas for CA, $\mathbf{\Lambda} = \text{diag}(1/x_{i.})$ and $\mathbf{\Gamma} = \text{diag}(1/x_{.j})$, which does reflect the symmetry between rows and columns.

Once the reference matrix and the norm are chosen, the Factor Analysis of the data is defined as the additive hierarchical decomposition of $\mathbf{X} - \mathbf{X}_0$ in successive models consisting of matrices of rank one, as is stated in the following definition.

Definition 1. Given the reference matrix $\mathbf{X}_0$ and the matrix norm $\|\cdot\|_{\mathbf{\Lambda},\mathbf{\Gamma}}$, the Factor Analysis of the matrix $\mathbf{X}$ of rank r is the decomposition of $\mathbf{X} - \mathbf{X}_0$ in matrices $\mathbf{T}_1, \mathbf{T}_2, \ldots, \mathbf{T}_r$, of rank one such that:

$$\|(\mathbf{X} - \mathbf{X}_0) - \mathbf{T}_1\|_{\mathbf{\Lambda},\mathbf{\Gamma}} = \min!,$$
$$\|(\mathbf{X} - \mathbf{X}_0 - \mathbf{T}_1) - \mathbf{T}_2\|_{\mathbf{\Lambda},\mathbf{\Gamma}} = \min!,$$
$$\ldots$$
$$\|(\mathbf{X} - \mathbf{X}_0 - \mathbf{T}_1 - \ldots - \mathbf{T}_{r-2}) - \mathbf{T}_{r-1})\|_{\mathbf{\Lambda},\mathbf{\Gamma}} = \min!,$$

and $\mathbf{X} = \mathbf{X}_0 + \mathbf{T}_1 + \mathbf{T}_2 + \ldots + \mathbf{T}_{r-1} + \mathbf{T}_r$.

The fact that there will be up to r matrices of rank one will appear in the general solution below. We will denote the Factor Analysis by a triple $(\mathbf{X}, \mathbf{X}_0, \|\cdot\|_{\mathbf{\Lambda},\mathbf{\Gamma}})$ assuming implicitly the decomposition of $\mathbf{X} - \mathbf{X}_0$.

4. The General Solution

Let us begin with the usual Euclidean case dealing with the ordinary sum of squares of deviations across all cells in $\mathbf{X} - \mathbf{X}_0$. The $\mathbf{T}_k$'s are given by the *singular value decomposition* of $\mathbf{X} - \mathbf{X}_o$ (see Eckart and Young, 1936),

$$\mathbf{X} - \mathbf{X}_o = \sum_{k=1}^{r} \sqrt{\lambda_k} \mathbf{v}_k \mathbf{u}_k' , \tag{3}$$

where $\mathbf{v}_k$ is the unit eigenvector of $(\mathbf{X} - \mathbf{X}_o)(\mathbf{X} - \mathbf{X}_o)'$ with respect to the k-th largest eigenvalue λ_k, and $\mathbf{u}_k$ is the corresponding unit eigenvector of $(\mathbf{X} - \mathbf{X}_o)'(\mathbf{X} - \mathbf{X}_o)$. Because of the orthogonality of eigenvectors it is easily checked that the total variation of $\mathbf{X} - \mathbf{X}_o$ is equal to the sum of the λ_k's, while λ_k is the total variation within $\mathbf{T}_k$. Another nice property of the Euclidean norm is that the best approximation of $\mathbf{X} - \mathbf{X}_o$ by a matrix of rank q ($q < r$) is given by $\mathbf{T}_1 + \mathbf{T}_2 + \ldots + \mathbf{T}_q$ and the variation explained in that way is equal to $\lambda_1 + \lambda_2 + \ldots + \lambda_q$.

Proposition 1. *For the Factor Analysis* $(\mathbf{X}, \mathbf{X}_o, \|\cdot\|_{\mathbf{\Lambda},\mathbf{\Gamma}})$ *the matrices of the decomposition are given by*

$$\mathbf{T}_k = \sqrt{\lambda_k} \mathbf{\Lambda}^{-\frac{1}{2}} \mathbf{v}_k \mathbf{u}_k' \mathbf{\Gamma}^{-\frac{1}{2}}, \quad k = 1, \ldots, r,$$

where $\mathbf{v}_k$ *and* $\mathbf{u}_k$ *are the unit eigenvectors, for the k-th largest eigenvalue* λ_k, *of* $\mathbf{\Lambda}^{\frac{1}{2}}(\mathbf{X} - \mathbf{X}_o)\mathbf{\Gamma}(\mathbf{X} - \mathbf{X}_o)'\mathbf{\Lambda}^{\frac{1}{2}}$ *and* $\mathbf{\Gamma}^{\frac{1}{2}}(\mathbf{X} - \mathbf{X}_o)'\mathbf{\Lambda}(\mathbf{X} - \mathbf{X}_o)\mathbf{\Gamma}^{\frac{1}{2}}$, *respectively.*

Since $\|\mathbf{X} - \mathbf{X}_o\|_{\mathbf{\Lambda},\mathbf{\Gamma}} = \|\mathbf{\Lambda}^{\frac{1}{2}}(\mathbf{X} - \mathbf{X}_o)\mathbf{\Gamma}^{\frac{1}{2}}\|$, this proposition follows from pre and postmultiplication of (3) by $\mathbf{\Lambda}^{\frac{1}{2}}$ and $\mathbf{\Gamma}^{\frac{1}{2}}$, respectively. The part of the total variation, in the sense of the norm $\|\cdot\|_{\mathbf{\Lambda},\mathbf{\Gamma}}$, explained by $\mathbf{T}_1, \mathbf{T}_2, \ldots, \mathbf{T}_q$ is equal to $\lambda_1 + \lambda_2 + \ldots + \lambda_q$, since

$$\|\mathbf{T}_k\|^2_{\mathbf{\Lambda},\mathbf{\Gamma}} = \text{trace}[(\mathbf{\Lambda}^{\frac{1}{2}}\mathbf{T}_k\mathbf{\Gamma}^{\frac{1}{2}})'(\mathbf{\Lambda}^{\frac{1}{2}}\mathbf{T}_k\mathbf{\Gamma}^{\frac{1}{2}})] = \lambda_k .$$

Of course there are many ways of looking at Factor Analysis but this one fits into the usual framework of linear modeling: we have some data to explain and we split it into the model part and the residual part; accordingly we have a measure for the degree of explanation achieved, with respect to the total variation in the data, which guides us for the choice of the number of terms to retain.

Consider now the roles played by rows (or columns) which are the entities of main interest. Because of the additivity of the contributions of rows (or columns) to the total variation $\|\mathbf{X} - \mathbf{X}_o\|_{\mathbf{\Lambda},\mathbf{\Gamma}}$, the elements that are necessary for further interpretation of the decomposition are well defined. The *relative contribution* of the row i to the factor k is defined as the part of the total variation λ_k in the k-th factor accounted for by the i-th row. It is equal to v_{ki}^2, where v_{ki} is the i-th component of $\mathbf{v}_k$. The *squared cosine* of the row i with the factor s is defined as the part of the i-th row variation explained by the s-th factor. It is equal to $\lambda_s v_{si}^2 / \sum_{k=1}^{r} \lambda_k v_{ki}^2$.

As for the *coordinates* of rows or columns on factors, they can be defined in several ways. Their choice is purely a matter of convenience for the interpretation of mappings and should be considered as a subsequent step without interferring with the formal definition of the Factor Analysis itself. One can allow for weights on rows and/or columns as in CA to make the distances between the points representing rows (or columns) relevant. One can also simply use, say for rows, the *scores* $\sqrt{\lambda_s}\mathbf{v}_s$ on the s-th factor, reflecting contributions in terms of squared distances to the origin.

5. Coherence of Reference and Norm

Finally we exhibit a special feature of $\mathbf{X}_o$ which will highlight some special properties of CA.

Definition 2. In the Factor Analysis $(\mathbf{X}, \mathbf{X}_o, \|\cdot\|_{\mathbf{\Lambda},\mathbf{\Gamma}})$, $\mathbf{X}_o$ is said to be coherent with the norm $\|\cdot\|_{\mathbf{\Lambda},\mathbf{\Gamma}}$ if it is such that $\|\mathbf{X} - \mathbf{X}_o\|_{\mathbf{\Lambda},\mathbf{\Gamma}} = \min!$

Proposition 2. *For Correspondence Analysis the reference matrix is coherent with the matrix norm.*

This means that the independence model is the closest rank one fit to the matrix $\mathbf{X}$ in the sense of the chi-square norm. This can be verified by differentiating the expression

$$\sum_{i,j} \frac{(x_{ij} - u_i v_j)^2}{x_{i.} x_{.j}}$$

with respect to v_i and u_j and checking that $v_i = x_{i.}$ and $u_j = x_{.j}/x_{..}$ is a solution. This special feature implies that we can start the analysis from the decomposition of the original matrix $\mathbf{X}$ itself, considering $\mathbf{X}_o$ as the first term. This will explain some specificities of CA. For convenience we will denote this factor by (λ_o, u_o, v_o).

This coherence shows why all eigenvalues in CA are less or equal to one, since one can easily check that the norm of $\mathbf{X}_o$ is equal to 1. Because the norm of $\mathbf{X} - \mathbf{X}_o$ is $\chi^2/x_{..}$, the total variation in $\mathbf{X}$ is equal to $1 + \chi^2/x_{..}$, where χ^2 is the chi-square coefficient of $\mathbf{X}$. Thus the residuals, after fitting the independence model, account for $\chi^2/(\chi^2 + x_{..})$, which is Pearson's (squared) coefficient. Coherence is also simplifying the so-called transition equations relating eigenvectors (and further, relating coordinates) in $\mathbb{R}^n$ and $\mathbb{R}^p$. For instance from $\mathbb{R}^p$ to $\mathbb{R}^n$ we have

$$\sqrt{\lambda_k}\mathbf{v}_k = \mathbf{\Lambda}^{\frac{1}{2}}(\mathbf{X} - \mathbf{X}_o)\mathbf{\Gamma}^{\frac{1}{2}}\mathbf{u}_k = \mathbf{\Lambda}^{\frac{1}{2}}\mathbf{X}\mathbf{\Gamma}^{\frac{1}{2}}\mathbf{u}_k, \quad k = 1, \ldots, r.$$

In contrast one should note that coherence does not hold for PCA, in general, and the status of the matrix $\mathbf{X}_o$ is of a different kind. In PCA one has to start the decomposition from the centered matrix.

References

Benzécri, J.P. (1973). *L'Analyse des Données; Tome II: L'Analyse des Correspondances.* Dunod, Paris.

Eckart, C. and Young, G. (1936). The approximation of one matrix by another of lower rank. *Psychometrika* **1**, 211–218.

Escoufier, Y. (1985). L'analyse des correspondances: ses propriétés et ses extensions. *Bulletin of the International Statistical Institute, 45th Session.*

Fisher, R.A. (1936). The utilization of multiple measurements in taxonomic problems. *Annals of Eugenics* **7**, 179–88.

Frane, W.J. and Hill, M. (1976). Factor analysis as a tool for data analysis. *Communications in Statistics - Theory and Methods* *A* **5**, 487–506.

Hotelling, H. (1933). Analysis of a complex of statistical variables into principal components. *Journal of Educational Psychology* **24**, 417–441, 498–520.

Kim, J. and Mueller, C.W. (1978). *Factor Analysis.* Sage University Paper Series on Quantitative Applications in the Social Sciences 07-013,014. Sage Publication, Beverley Hills.

Pearson, K. (1901). On lines and planes of closest fit to a system of points in space. *Philosophical Magazine* **6**, 559–72.

Spearman, C. (1904). General intelligence, objectively determined and measured. *American Journal of Psychology* **15**, 201–93.

Thurstone, L. L. (1947). *Multiple Factor Analysis.* University Press, Chicago.

ON Q–COVARIANCE AND ITS APPLICATIONS

ANDRZEJ KRAJKA and DOMINIK SZYNAL
Institute of Mathematics
Maria Curie-Skłodowska University
Pl. Marii Skłodowskiej-Curie 1
20-031 Lublin
Poland

Abstract. We introduce a new concept of covariance (Q–covariance) which contains, as a particular case, the classical one. At the same time we give alternative formulae for covariance of square integrable random variables. Some applications of that notion is also considered.

Key words: Dependence, Covariance, Autocovariance, Stationary process, Mixture of distributions, Correlation, Uncorrelatedness, Sample, Range, Randomized block design.

1. Introduction

The concept of the covariance $\mathrm{Cov}(X,Y)$ between random variables X and Y is defined by the well known formula

$$\mathrm{Cov}(X,Y) = EXY - EXEY, \tag{1.1}$$

for $X \in L^1, Y \in L^1, XY \in L^1$ (L^1 – the space of all integrable random variables). Hoeffding (1940), extended that concept to a larger class of random variables by the formula

$$\mathrm{Cov}^{(F)}(X,Y) = \int_{-\infty}^{\infty}\int_{-\infty}^{\infty}[F_{X,Y}(x,y) - F_X(x)F_Y(y)]dxdy \tag{1.2}$$

valid whenever the above integral is finite. A more general formula was given by Mardia and Thompson (1972):

$$\mathrm{Cov}^{(F)}(X^r,Y^s) = rs\int_{-\infty}^{\infty}\int_{-\infty}^{\infty}x^{r-1}y^{s-1}[F_{X,Y}(x,y) - F_X(x)F_Y(y)]dxdy, \tag{1.3}$$

$r,s \geq 1$, provided that the latest integral is finite. Moreover, they proved that $\mathrm{Cov}(X,Y) = \mathrm{Cov}^{(F)}(X,Y)$, whenever $X \in L^1$, $Y \in L^1$ and $XY \in L^1$, and they showed that the F–covariance (1.2) (or (1.3)) characterizes the dependence between X and Y when the usual covariance (1.1) is undefined. The notion of the F–covariance appeared to be very useful in establishing the properties of dependent random variables (for instance associated random variables) and in analysing of statistical data (Esary, Proschan and Walkup, 1967; Joag-Dev and Proschan, 1983; Mardia, 1967). Nevertheless, the F–covariance loses a part of information contained in random variables. For example, if we know that $X \in L^1$, but $E|Y| = \infty$, then

T. Caliński and R. Kala (eds.),
Proceedings of the International Conference on Linear Statistical Inference LINSTAT '93, 293–300.

the fact of integrability of X is not explicite underline in the concept of F–covariance. We deal with a concept of the covariance, which we call the Q–covariance, allowing to use information, provided by random variables, in a higher degree than the classical and the F–covariance do. Furthermore, we note in Krajka and Szynal (1993), (using multivariate Pareto type distributions, cf. Mardia, 1962) that the introduced Q–covariance characterizes the dependence when neither the usual covariance nor the F–covariance can do it. Some statistical applications are also mentioned.

2. Q–covariance

Let $(\Omega, \mathcal{A}, P)$ be a probability space, and let L^0 be the space of all random variables X. L^r stands for the space of random variables such that $E \mid X \mid^r < \infty$, $r > 0$. For any $p \in (0,1)$, $x(p)$ and $y(p)$ stand for the quantile functions of random variables X and Y, respectively (i.e. $P[X < x(p)] \leq p \leq P[X \leq x(p)]$, $P[Y < y(p)] \leq p \leq P[Y \leq y(p)]$).

For $X \in L^1$ and $Y \in L^0$ with continuous distribution functions we write

$$\begin{aligned} L_{X,Y}(p) &= E(X - EX)I[Y \geq y(p)], \\ \overline{L}_{X,Y}(p) &= E(X - EX)I[Y < y(p)], \end{aligned} \tag{2.1}$$

where $I[\cdot]$ denotes the indicator function. Similarly we define $L_{Y,X}(\cdot)$ and $\overline{L}_{Y,X}(\cdot)$ for $X \in L^0$ and $Y \in L^1$ (cf. Kowalczyk, 1977; Kowalczyk and Pleszczyńska, 1977).

For discrete random variables $X \in L^1$ and $Y \in L^0$ we use

$$\begin{aligned} L_{X,Y}(p) &= EXI[Y > y(p)] + \\ &+ \{1 - p - P[Y > y(p)]\}E[X \mid Y = y(p)] - (1-p)EX, \\ \overline{L}_{X,Y}(p) &= pEX - EXI[Y > y(p)] - \\ &- \{p - P[Y > y(p)]\}E[X \mid Y = y(p)]. \end{aligned} \tag{2.2}$$

Among properties of the functions $L_{X,Y}(\cdot)$ and $L_{Y,X}(\cdot)$ we mention their continuity and boundness of the total variation on (0,1).

Now we introduce the notion of Q–covariance.

Definition 1. Let (X, Y) be a pair of random variables with $X \in L^1$ and $Y \in L^0$. The Q–covariance between random variables X and Y, $\mathrm{Cov}^{(Q)}(X,Y)$, is defined by the formula:

$$\mathrm{Cov}^{(Q)}(X,Y) = -\int_0^1 y(p)dL_{X,Y}(p) = \int_0^1 y(p)d\overline{L}_{X,Y}(p), \tag{2.3}$$

provided that one of the above integrals is finite.

The Q–covariance between $Y \in L^1$ and $X \in L^0$, $\mathrm{Cov}^{(Q)}(Y,X)$, is defined by

$$\mathrm{Cov}^{(Q)}(Y,X) = -\int_0^1 x(p)dL_{Y,X}(p) = \int_0^1 x(p)d\overline{L}_{Y,X}(p), \tag{2.4}$$

provided that one of the above integrals is finite.

We give here two instructive examples, when the covariance (1.1) does not exists but the Q-covariance exists.

Example 1 (Krajka and Szynal, 1993). Let (X, Y) be a pair of random variables with $X = YI[|\, Y \,| \leq 1]$, and Y having the Cauchy density function $f(y) = 1/[\pi(1 + y^2)]$, $-\infty < y < \infty$.

We see that $E \mid X \mid < \infty$, $EX = 0$, $E \mid Y \mid = \infty$, and $E \mid XY \mid < \infty$, so the covariance (1.1), as a characteristic of dependence, can not be used here. But using (2.1), we see that

$$L_{X,Y}(p) = \begin{cases} EYI[\tan \pi(p - 1/2) < Y < 1], & p \in (1/4, 3/4), \\ 0, & p \in (0,1)\backslash(1/4, 3/4), \end{cases}$$

and

$$-\int_0^1 y(p) dL_{X,Y}(p) = (4 - \pi)/(2\pi).$$

Example 2. Let (X, Y) be a pair of discrete random variables with $X = YI[|\, Y \,| \leq 1]$, and Y having the probability function $p_Y(n) = P[Y = \pm n] = 3/(\pi^2 n^2), n \in N$.

We see that $P[X = \pm 1] = 3/\pi^2$, $P[X = 0] = 1 - 6/\pi^2$, $EX = 0$. Using (2.2), we get

$$L_{X,Y}(p) = \begin{cases} 3/\pi^2 - 1/2 + p, & 1/2 - 3/\pi^2 \leq p < 1/2, \\ 3/\pi^2 + 1/2 - p, & 1/2 \leq p < 1/2 + 3/\pi^2, \\ 0, & \text{otherwise}, \end{cases}$$

which gives

$$-\int_0^1 y(p) dL_{X,Y}(p) = 6/\pi^2.$$

There are examples of pairs (X,Y) of random variables for which $\text{Cov}(X, Y)$ and $\text{Cov}^{(F)}(X, Y)$ are undefined but $\text{Cov}^{(Q)}(X, Y)$ and $\text{Cov}^{(Q)}(Y, X)$ can be calculated (cf. Krajka and Szynal, 1993).

A concept of the Q-uncorrelation is contained in the following definition.

Definition 2. A random variable X is Q-uncorrelated of a random variable Y if $\text{Cov}^{(Q)}(X, Y) = 0$. If $\text{Cov}^{(Q)}(Y, X) = 0$, then we say that Y is Q-uncorrelated of X. Random variables X and Y are called Q-uncorrelated if $\text{Cov}^{(Q)}(X, Y) = \text{Cov}^{(Q)}(Y, X) = 0$.

Examples Q-uncorrelated and dependent random variables are as follows.

Example 3. Let X_1 and X_2 have the probability functions:

$$P[X_1 = k] = (1 - p_1)p_1^{k-1}, \quad P[X_2 = k] = (1 - p_2)p_2^{k-1}, \quad k = 1, 2, \ldots,$$

and Y_1 and Y_2 have the probability functions

$$P[Y_1 = k] = c_2/2k^2 + c_4/2k^4, \quad P[Y_2 = k] = c_2/2k^2 + c_6/2k^6, \quad k = 1, 2, \ldots,$$

where $c_r = (\sum_1^\infty 1/k^r)^{-1}, r \geq 2$, i.e. $c_2 = 6/\pi^2$, $c_4 = 90/\pi^4$, $c_6 = 945/\pi^6$.

Suppose that for $i = 1, 2$, (X_i, Y_i) is an independent pair of random variables (cf. Behboodian, 1990). Define (X, Y) as a mixture of (X_1, Y_1) and (X_2, Y_2) with the probability function

$$P[X = k, Y = l] = \alpha P[X_1 = k]P[Y_1 = l] + (1 - \alpha)P[X_2 = k]P[Y_2 = l], \quad \alpha \in [0, 1].$$

We see that

$$P[X=k]=\alpha P[X_1=k]+(1-\alpha)P[X_2=k], \quad P[Y=l]=\alpha P[Y_1=l]+(1-\alpha)P[Y_2=l],$$

and $EX = \alpha/(1 - p_1) + (1 - \alpha)/(1 - p_2)$, $EY = \infty$, $EXY = \infty$. Therefore, $\mathrm{Cov}(X, Y)$ is undefined, but from (2.2), we have

$$\begin{aligned} L_{X,Y}(p) &= \alpha(1-\alpha)[1/(1-p_1) - 1/(1-p_2)]\{P[Y_1 > y(p)]P[Y_2 = y(p)] \\ &- P[Y_2 > y(p)]P[Y_1 = y(p)] + (1-p)(P[Y_1 = y(p)] \\ &- P[Y_2 = y(p)])\}/\{\alpha P[Y_1 = y(p)] + (1-\alpha)P[Y_2 = y(p)]\}, \end{aligned}$$

where $y(p) = \sup\{y : P[Y < y] \leq p\}$, which implies that

$$\mathrm{Cov}^{(Q)}(X, Y) = \frac{\alpha(1-\alpha)}{2}[1/(1-p_1) - 1/(1-p_2)](c_4 c_3^{-1} - c_6 c_5^{-1}).$$

Hence we conclude that X is Q–uncorrelated of Y, if $p_1 = p_2$.

Some properties of the Q–covariance are collected in the following

Lemma 1.

(i) *For X and Y being independent* $\mathrm{Cov}^{(Q)}(X, Y) = 0$, $\mathrm{Cov}^{(Q)}(Y, X) = 0$ *when these Q–covariances are defined.*

(ii) *For any strictly increasing (decreasing) function h we have*

$$\begin{aligned} \mathrm{Cov}^{(Q)}(X, h(Y)) &= \int_0^1 h(y(p))d\overline{L}_{X,Y}(p) \\ (\mathrm{Cov}^{(Q)}(X, h(Y)) &= \int_0^1 h(y(p))d\overline{L}_{X,Y}(1-p)). \end{aligned}$$

(iii) *Let X, Y and Z be random variables such that* $\mathrm{Cov}^{(Q)}(X, Z)$ *and* $\mathrm{Cov}^{(Q)}(Y, Z)$ *are defined. Then for $a, b, c, d, e \in R$,*

$$\mathrm{Cov}^{(Q)}(aX + bY + c, dZ + e) = ad\mathrm{Cov}^{(Q)}(X, Z) + bd\mathrm{Cov}^{(Q)}(Y, Z).$$

The following theorem gives a representation of the Q–covariance.

Theorem 1 (Krajka and Szynal, 1993). *Let (X,Y) be a pair of random variables with continuous and strictly monotone marginal distribution functions. If $X \in L^1$ and $L_{X,Y}(p),\ p \in (0,1)$, is differentiable, then*

$$\mathrm{Cov}^{(Q)}(X,Y) = E[Xy(P)(\frac{f(X,y(P))}{f_X(X)f_Y(y(P))} - 1)], \tag{2.5}$$

whenever

$$Xy(P)(\frac{f(X,y(P))}{f_X(X)f_Y(y(P))} - 1) \in L^1.$$

P denotes here a random variable uniformly distributed on $(0,1)$, independent of X and $f(\cdot,\cdot)$ denotes the density function of (X,Y).

Proof. By the differentiablity of $L_{X,Y}(\cdot)$ and (2.1) we get

$$\begin{aligned}
\mathrm{Cov}^{(Q)}(X,Y) &= \int_0^1 y(p)d\overline{L}_{X,Y}(p) \\
&= \int_0^1 y(p)\overline{L}'_{X,Y}(p)dp \\
&= \int_0^1 \{y(p)[\int_{-\infty}^{\infty} f(x)\int_{-\infty}^{y(p)} (x-EX)f(y \mid x)dy\,dx]'\}dp \\
&= \int_0^1 \{y(p)\int_{-\infty}^{\infty} (x-EX)f(y(p) \mid x)y'(p)f(x)\,dx\,\}dp \\
&= \int_0^1 \{y(p)\int_{-\infty}^{\infty} x[f(x \mid y(p)) - f(x)]dx\,\}dp \\
&= E\{Xy(P)[\frac{f(X,y(P))}{f_X(X)f_Y(y(P))} - 1]\}. \square
\end{aligned}$$

Theorem 1 suggests introducing the following equivalent definition of Q-covariance $\mathrm{Cov}^{(Q)}(X,Y)$ between discrete random variables X and Y with $X \in L^1$ and $Y \in L^0$ as the quantity

$$\mathrm{Cov}^{(Q)}(X,Y) = E\{Xy(P)[\frac{p_{X,Y}(X,y(P))}{p_X(X)p_Y(y(P))} - 1]\}, \tag{2.6}$$

where $p_{X,Y}(\cdot,\cdot)$ is the joint probability function of (X,Y), $p_X(.)$, $p_Y(.)$ are the marginal probability functions, and P is a random variable uniformly distributed on $(0,1)$, independent of X.

The relation among $\mathrm{Cov}(X,Y)$, $\mathrm{Cov}^{(F)}(X,Y)$, $\mathrm{Cov}^{(Q)}(X,Y)$ and $\mathrm{Cov}^{(Q)}(Y,X)$, given by (1.1), (1.2), (2.3) and (2.4), respectively, is formalized as the following theorem.

Theorem 2. *Let (X,Y) be a pair of random variables with continuous distribution functions such that $X \in L^1, Y \in L^1$ and $XY \in L^1$. Then*

$$\mathrm{Cov}(X,Y) = \mathrm{Cov}^{(F)}(X,Y) = \mathrm{Cov}^{(Q)}(X,Y) = \mathrm{Cov}^{(Q)}(Y,X).$$

3. Applications

3.1. An Estimate of Hölder's (Schwarz's) Type

Note that under the assumption of Theorem 1 we have

$$| \mathrm{Cov}^{(Q)}(X,Y) | \leq (\int_0^1 | y(p) |^r \, dp)^{1/r} (\int_0^1 | \overline{L}'_{X,Y}(p) |^s \, dp)^{1/s},$$

whenever $Y \in L^r, \overline{L}'_{X,Y}(P) \in L^s$, where $1/r + 1/s = 1, \ \ r,s \geq 1$.

Suppose that $X \in L^r, \ \ Y \in L^r$ and $g(X, y(P)) = [\frac{f(X,y(P))}{f_X(X)f_Y(y(P))} - 1] \in L^s$, where $1/r + 1/s = 1, \ \ r,s \geq 1$.

Taking into account that $L'_{X,Y}(p) = EXg(X,P)$, and using Theorem 2, we get

$$\begin{aligned}
| \mathrm{Cov}(X,Y) | &= | \mathrm{Cov}^{(Q)}(X,Y) | \\
&\leq (E| y(P) |^r)^{1/r} [\int_0^1 | EXg(X,P) |^s \, dp]^{1/s} \\
&\leq E^{1/r} | y(P) |^r \, [\int_0^1 (E^{1/r} | X |^r \, E^{1/s} | g(X,P) |^s)^s dp]^{1/s} \\
&\leq E^{1/r} | y(P) |^r \, E^{1/r} | X |^r \, E^{1/s} | g(X,P) |^s .
\end{aligned}$$

Thus we have proved

Lemma 2. *Under the assumption of Theorem 1 with* $X \in L^r, \ \ Y \in L^r, \ \ g(X, y(P)) \in L^s, \ 1/r + 1/s = 1, \ \ r,s \geq 1$, *we have*

$$| \mathrm{Cov}(X,Y) | \leq E^{1/r} | y(P) |^r \, E^{1/r} | X |^r | \, E^{1/s} | g(X,P) |^s .$$

Letting $r = s = 2$ *we get the following Schwarz's type inequality*

$$| \mathrm{Cov}(X,Y) | \leq E^{1/2} | y(P) |^2 \, E^{1/2} | X |^2 \, E^{1/2} | g(X,P) |^2 .$$

3.2. Q–Covariance Stationary Process

Let $X(t), t \in T$ be a stochastic process with finite moments of the first order. By the Q–autocovariance function of that process we mean the quantity:

$$K^{(Q)}(s,t) = \mathrm{Cov}^{(Q)}(X(s), X(t)), \ \ s,t \in T.$$

With that function we can extend the commonly used L^2 – theory of stochastic processes to a larger class of L^1 – stochastic processes. In particular, we can consider so-called Q–covariance stationary processes, that is L^1 – stochastic processes for which there exists a function $R^{(Q)}(.)$ such that for all s and t in T we have $K^{(Q)}(s,t) = R^{(Q)}(t-s)$ or, more precisely, $R^{(Q)}(.)$ has the property that for every t and u in T

$$\mathrm{Cov}^{(Q)}(X(t), X(t+u)) = R^{(Q)}(u).$$

3.3. Q–Covariance in Statistics

The Q–covariance can be useful in analysing samples from a two-dimensional population when one of marginal distributions is not integrable. Moreover, the formulae for the Q–covariance can be alternatively used in classical cases. We discuss that problem considering the covariance for the ranges of correlated samples. The correlation between ranges of correlated normal deviates was attacked by many authors, particularly in the analysis of experimental designs such as randomized block design with n treatments and k blocks, by short-cut techniques. Knowledge of the correlation coefficient $\rho_w(n,\rho)$ between two ranges in samples of size n drawn from a bivariate population with correlation ρ, allows to determine the so-called 'scale factor' c and 'equivalent degrees of freedom' ν appearing in the approximate distribution used by a short-cut method. Values $\rho_w(n,\rho)$, c and ν were many times tabulated. Among papers discussing these problems we mention David (1951); Harter (1960); Hartley (1950); Kurtz, Link, Tukey and Wallace (1965, 1966); Mardia (1967).

Now we show in what way the correlation of ranges can be determined by the Q–covariance techniques.

Let $((X_1,Y_1),...,(X_n,Y_n))$ be a sample of size n from a population with joint distribution function $F_{X,Y}$. Write

$$\begin{aligned} X_{(1)} &= \min(X_1,...,X_n), & X_{(n)} &= \max(X_1,...,X_n), & V_{(n)} &= X_{(n)} - X_{(1)},\\ Y_{(1)} &= \min(Y_1,...,Y_n), & Y_{(n)} &= \max(Y_1,...,Y_n), & W_{(n)} &= Y_{(n)} - Y_{(1)}. \end{aligned}$$

We are interested in $\mathrm{Cov}^{(Q)}(V_{(n)},W_{(n)})$ and $\mathrm{Cov}(V_{(n)},W_{(n)})$. For $\mathrm{Cov}^{(Q)}(V_{(n)},W_{(n)})$ we can only give the equality

$$\mathrm{Cov}^{(Q)}(V_{(n)},W_{(n)}) = \mathrm{Cov}^{(Q)}(X_{(n)},W_{(n)}) - \mathrm{Cov}^{(Q)}(X_{(1)},W_{(n)}).$$

Now suppose that $F_{X,Y}(=F)$ has a density function $f_{X,Y}(=f)$ and that the marginal distribution functions are strictly increasing. If $X \in L^1, Y \in L^1$, and $XY \in L^1$, then we have

$$\begin{aligned} \mathrm{Cov}(V_{(n)},W_{(n)}) &= \mathrm{Cov}(X_{(n)},Y_{(n)}) - \mathrm{Cov}(X_{(1)},Y_{(n)})\\ &- \mathrm{Cov}(X_{(n)},Y_{(1)}) + \mathrm{Cov}(X_{(1)},W_{(1)}). \end{aligned}$$

Using the formula (2.5), taking into account that

$$\begin{aligned} F_{X_{(n)},Y_{(n)}}(x,y) &= F^n_{X,Y}(x,y),\\ F_{X_{(n)},Y_{(1)}}(x,y) &= F^n_X(x) - [F_X(x) - F_{X,Y}(x,y)]^n,\\ F_{X_{(1)},Y_{(n)}}(x,y) &= F^n_Y(y) - [F_Y(y) - F_{X,Y}(x,y)]^n,\\ F_{X_{(1)},Y_{(1)}}(x,y) &= 1 - [1-F_X(x)]^n - [1-F_Y(y)]^n\\ &- [1 - F_X(x) - F_Y(y) + F_{X,Y}(x,y)]^n, \end{aligned}$$

and putting $P_n = P^{1/n}$, $\overline{P}_n = 1-(1-P)^{1/n}$, and

$$\begin{aligned} g_n(X,P) &= \frac{f(X,y(P_n))}{f_X(X)f_Y(y(P_n))}, & \overline{g}_n(X,P) &= \frac{f(X,y(\overline{P}_n))}{f_X(X)f_Y(y(\overline{P}_n))},\\ G_n(X,P) &= \frac{F(X,y(P_n))}{F_Y(y(P_n))}, & \overline{G}_n(X,P) &= \frac{F_X(X)-F(X,y(\overline{P}_n))}{1-F_Y(y(\overline{P}_n))}, \end{aligned}$$

$$h_n(X,P) = \frac{\int_{-\infty}^{y(P_n)} f(X,v)dv}{f_X(X)F_Y(y(P_n))}, \quad \overline{h}_n(X,P) = \frac{f_X(X) - \int_{-\infty}^{y(\overline{P}_n)} f(X,v)dv}{f_X(X)(1-F_Y(y(\overline{P}_n)))},$$

$$H_n(X,P) = \frac{\int_{-\infty}^{x} f(u,y(P_n))du}{f_Y(y(P_n))}, \quad \overline{H}_n(X,P) = \frac{\int_{-\infty}^{x} f(u,y(\overline{P}_n))du}{f_Y(y(\overline{P}_n))},$$

we get

$$\begin{aligned}\mathrm{Cov}(V_{(n)}, W_{(n)}) &= EXy(P_n)g_n(X,P)\{G_n^{n-1}(X,P) - [1-G_n(X,P)]^{n-1}\} \\ &+(n-1)EXy(P_n)h_n(X,P)\{H_n(X,P)G_n^{n-2}(X,P) - [1-H_n(X,P)][1-G_n(X,P)]^{n-} \\ &+ EXy(\overline{P}_n)\overline{g}_n(X,P)\{[1-\overline{G}_n(X,P)]^{n-1} - \overline{G}_n^{n-1}(X,P)\} \\ &+(n-1)EXy(\overline{P}_n)\overline{h}_n(X,P)\{[1-\overline{H}_n(X,P)][1-\overline{G}_n(X,P)]^{n-2} - \overline{H}_n(X,P)\overline{G}_n^{n-2}(X, \\ &+EX\{F_X^{n-1}(X) - [1-F_X(X)]^{n-1}\}[Ey(\overline{P}_n) - Ey(P_n)].\end{aligned}$$

References

Behboodian, J. (1990). Examples of uncorrelated dependent random variables using a bivariate mixture. *The American Statistician* **44**, 218.

David, H.A. (1951). Further applications of range to the analysis of variance. *Biometrika* **38**, 393–409.

Esary, J.D., Proschan, F. and Walkup, D.W. (1967). Association of random variables, with applications. *The Annals of Mathematical Statistics* **38**, 1466–1474.

Harter, H.L. (1960). Tables of range and studentized range. *The Annals of Mathematical Statistics* **31**, 1122–1147.

Hartley, H.O. (1950). The use of the range in analysis of variance. *Biometrika* **37**, 271–280.

Hoeffding, W. (1940). Mastabinvariante Korrelations – Theorie. *Schriften Math. Inst. Univ. Berlin* **5**, 187–233.

Joag-Dev, K. and Proschan, F. (1983). Negative association of random variables with applications. *The Annals of Statistics* **11**, 286–295.

Kowalczyk, T. (1977). General definition and sample counterparts of monotonic dependence functions of bivariate distributions. *Mathematische Operationsforschung und Statistik, Series Statistics* **8**, 351–369.

Kowalczyk, T. and Pleszczyńska, E. (1977). Monotonic dependence functions of bivariate distributions. *The Annals of Statistics* **5**, 1221–1227.

Krajka A. and Szynal, D. (1993). On a notion of Q–covariance. Manuscript.

Kurtz, T.E., Link, R.F., Tukey, J.W. and Wallace, D.L. (1965). Short-cut multiple comparisons for balanced single and double classifications: Part 2, derivations and approximations. *Biometrika* **52**, 485–498.

Kurtz, T.E., Link, R.F., Tukey, J.W. and Wallace, D.L. (1966). Correlation of ranges of correlated deviates. *Biometrika* **53**, 191–196.

Mardia, K.V. (1962). Multivariate Pareto distributions. *The Annals of Mathematical Statistics* **33**, 1008–1015.

Mardia, K.V. (1967). Correlation of the ranges of correlated samples. *Biometrika* **54**, 529–539.

Mardia, K.V. (1970). *Families of Bivariate Distributions*, Griffin, London.

Mardia, K.V. and Thompson, J.W. (1972). Unified treatment of moment formulae. *Sankhyā A* **34**, 121–132.

LIST OF REFEREES

Jean-Marc Azaïs (Toulouse, France)
Rosemary A. Bailey (London, England)
Tadeusz Bednarski (Wrocław, Poland)
Norbert Benda (Berlin, Germany)
Barbara Bogacka (Poznań, Poland)
Tadeusz Caliński (Poznań, Poland)
Leo C.A. Corsten (Wageningen, The Netherlands)
Carles M. Cuadras (Barcelona, Spain)
Anita Dobek (Poznań, Poland)
Hilmar Drygas (Kassel, Germany)
Charles W. Dunnett (Hamilton, Canada)
Yves Escoufier (Montpellier, France)
Subir Ghosh (Riverside, U.S.A)
Alessandra Giovagnoli (Perugia, Italy)
Stanisław Gnot (Wrocław, Poland)
Jan Hauke (Poznań, Poland)
Berthold Heiligers (Augsburg, Germany)
Zygmunt Kaczmarek (Poznań, Poland)
Sanpei Kageyama (Hiroshima, Japan)
Radosław Kala (Poznań, Poland)
Krystyna Katulska (Poznań, Poland)
Christos P. Kitsos (Athens, Greece)
Krzysztof Kłaczyński (Poznań, Poland)
Henning Knautz (Hamburg, Germany)
Teresa Kowalczyk (Warsaw, Poland)
Mirosław Krzyśko (Poznań, Poland)
Lynn R. LaMotte (Baton-Rouge, U.S.A)
Augustyn Markiewicz (Poznań, Poland)
Thomas Mathew (Baltimore, U.S.A)
Stanisław Mejza (Poznań, Poland)
Joãgo Tiago Mexia (Almada, Portugal)
Anna Molińska (Poznań, Poland)
Hervé Monod (Versailles, France)
Christine Müller (Berlin, Germany)
Hans Nyquist (Umeå, Sweden)
Andrej Pázman (Bratislava, Slovakia)
Paweł Pordzik (Poznań, Poland)
Friedrich Pukelsheim (Augsburg, Germany)
Dieter Rasch (Wageningen, The Netherlands)
Bikas K. Sinha (Calcuta, India)

Bimal K. Sinha (Baltimore, U.S.A)
Czesław Stępniak (Lublin, Poland)
Richard Tomassone (Paris, France)
Erik Torgersen † (Oslo, Norway)
Götz Trenkler (Dortmund, Germany)
Armin Tuchscherer (Dummerstorf, Germany)
Ditrich von Rosen (Uppsala, Sweden)
Jan ten Vregelaar (Wageningen, The Neherlands)
Peter Westfall (Lubbock, U.S.A)
Ryszard Zieliński (Warsaw, Poland)
Wojciech Zieliński (Warsaw, Poland)
Roman Zmyślony (Wrocław, Poland)
Stefan Zontek (Wrocław, Poland)

CONTRIBUTOR INDEX

This index contains names of authors of the individual papers collected in this volume; the numbers refer to the first pages of the articles.

SUBJECT INDEX

This index contains key-words of the individual papers collected in this volume; key-words are quoted only once per article.

Zeitfracht Medien GmbH
Ferdinand-Jühlke-Straße 7
99095 Erfurt, Deutschland
produktsicherheit@kolibri360.de